CHEMICAL APPROACHES to the SYNTHESIS of PEPTIDES and PROTEINS

NEW DIRECTIONS in ORGANIC and BIOLOGICAL CHEMISTRY

Series Editor: C.W. Rees, CBE, FRS
Imperial College of Science, Technology and Medicine, London, UK

Advisory Editor: Alan R. Katritzky, FRS
University of Florida, Gainesville

Published and Forthcoming Titles

Chirality and the Biological Activity of Drugs
Roger J. Crossley

Enzyme-Assisted Organic Synthesis
Manfred Schneider and Stefano Servi

C-Glycoside Synthesis
Maarten Postema

Organozinc Reagents in Organic Synthesis
Ender Erdik

Activated Metals in Organic Synthesis
Pedro Cintas

Capillary Electrophoresis: Theory and Practice
Patrick Camilleri

Cyclization Reactions
C. Thebtaranonth and Y. Thebtaranonth

Mannich Bases: Chemistry and Uses
Maurilio Tramontini and Luigi Angiolini

Vicarious Nucleophilic Substitution and Related Processes in Organic Synthesis
Mieczyslaw Makosza

Aromatic Fluorination
James H. Clark, David Wails, and Tony W. Bastock

Lewis Acids and Selectivity in Organic Synthesis
M. Santelli and J.-M. Pons

Dianion Chemistry in Organic Synthesis
Charles M. Thompson

Asymmetric Synthetic Methodology
David J. Ager and Michael B. East

Synthesis Using Vilsmeier Reagents
C. M. Marson and P. R. Giles

The Anomeric Effect
Eusebio Juaristi and Gabriel Cuevas

Chiral Sulfur Reagents
M. Mikołajczyk, J. Drabowicz, and P. Kiełbasiński

Chemical Approaches to the Synthesis of Peptides and Proteins
Paul Lloyd-Williams, Fernando Albericio, and Ernest Giralt

Concerted Organic Mechanisms
Andrew Williams

CHEMICAL APPROACHES to the SYNTHESIS of PEPTIDES and PROTEINS

Paul Lloyd-Williams
Fernando Albericio
Ernest Giralt
University of Barcelona
Barcelona, Spain

CRC Press
Boca Raton   New York

Library of Congress Cataloging-in-Publication Data

Lloyd-Williams, Paul.
    Chemical approaches to the synthesis of peptides and proteins / by Paul Lloyd-Williams,
Fernando Albericio, Ernest Giralt.
        p. cm.— (New directions in organic and biological chemistry)
    Includes bibliographical references and index.
    ISBN 0-8493-9142-3 (alk. paper)
    1. Peptides—Synthesis. 2. Proteins—Synthesis. I. Albericio, Fernando.
    II. Giralt, Ernest. III. Title. IV. Series.
    QP551.L56 1997
    547'.75—dc21   97-7644
                                                                                         CIP

No claim to original U.S. Government works
International Standard Book Number 0-8493-9142-3
Library of Congress Card Number 97-7644
Printed in the United States of America 1 2 3 4 5 6 7 8 9 0
Printed on acid-free paper

About The Authors

Paul Lloyd-Williams studied chemistry at the University of Manchester, U.K., obtaining his B.Sc. in 1978 and his Ph.D. in synthetic organic chemistry in 1985. After postdoctoral work at the University of Manchester and the University of Arizona, he moved to Barcelona where he joined Professor Giralt's group in 1988. He is currently Assistant Professor of Organic Chemistry at the University of Barcelona.

Dr. Lloyd-Williams is interested in chemical methods for amino acid and peptide synthesis and in the total synthesis of peptide-based natural products with biological activity. He is coauthor of 25 scientific papers and four review articles.

Fernando Albericio studied at the University of Barcelona, Spain, where he was awarded his first degree in chemistry in 1975 and his Ph.D. in organic chemistry in 1981. After postdoctoral work at the Université d'Aix–Marseille, France, and then at the University of Minnesota, he returned to the University of Barcelona as an associate professor. In 1991 he was appointed director of peptide research at Milligen/Biosearch in the U.S., a position that he maintained until his return to the University of Barcelona in 1993. He was promoted to full professor in 1994.

Professor Albericio's research involves peptide synthesis and synthetic methodology, particularly in the design of new handles, protecting groups, and coupling reagents for solid-phase peptide synthesis. He has published over 200 papers and several review articles and is currently editor of *Letters in Peptide Science*. He received the Leonidas Zervas Award in 1994.

Ernest Giralt studied at the University of Barcelona, Spain, where he was awarded his first degree in chemistry in 1970 and his Ph.D. in organic chemistry in 1974. After postdoctoral work at the Université de Montpellier, France, he returned to the University of Barcelona as an assistant professor. He was subsequently promoted to associate professor in 1977 and to full professor in 1986. He was a visiting professor at the University of California, San Diego, and a research associate at the Scripps Research Institute, U.S., in 1991.

Professor Giralt's major research interests lie in the fields of peptide synthesis and structure determination, in particular using nuclear magnetic resonance spectroscopy. He has published some 200 papers, several review articles, and two books. He was one of the founding members of the European Peptide Society and is currently on the editorial boards of the *International Journal of Peptide and Protein Science*, the *Journal of Peptide Science*, *Biomedical Peptides, Proteins and Nucleic Acids,* and *Current Topics in Peptide and Protein Research*. He received the Narcís Monturiol Prize in 1992 and the Leonidas Zervas Award in 1994.

Preface

The synthesis of natural products has long been one of the organic chemist's main objectives. Within this field of endeavor, peptides and proteins pose special problems, and it is only over the last few decades that reliable methods for their preparation have been developed. Chemical synthesis confirms the covalent structure and provides material both for three-dimensional structure determination by nuclear magnetic resonance or X-ray crystallography and for biological evaluation. The synthesis of analogues allows the relation between molecular structure and pharmacological activity to be determined, and, in favorable cases, compounds with activity, selectivity, or biostability superior to the original natural product can be produced at the industrial scale.

Today, the chemist's imagination is not restricted by the structure of natural molecules or their closely related analogues. In the field of protein *de novo* design it might be said that "anything goes." L-Amino acids can be combined with D-amino acids; natural amino acids can be combined with nonnatural ones; and peptide sequences that are unknown in nature can be investigated. Although it is too early to say, the results of such endeavors will, in the long run, very likely be of fundamental importance. We can hope that, among other things, they will provide us with a better knowledge of the rules that control protein folding and allow us to produce compounds with tailor-made properties that may serve as ion channels, switches, biocompatible materials, new catalysts, and so on.

All this, however, hinges upon having efficient synthetic methods for the preparation of peptides and proteins. Despite the advances of the last 30 years, methods for the general and efficient synthesis of these molecules regardless of their size or sequence are still elusive. In this book we try to give an overview of the state of the art in the mid 1990s. While we have tried to make our coverage comprehensive, it is, of course, subjective. Given the huge amount of information on peptide synthesis available today, it is impossible not to be selective. We concentrate on those methods or approximations that we consider to be most relevant now or that we feel will become so in the future.

There are six chapters, the first of which is an introduction that presents essential basic concepts such as protecting groups, polymeric supports, coupling reagents, and so forth. Chapter 2 is one of the longest and is dedicated to the linear solid-phase synthesis of peptides. It is structured so that the solid support, protection schemes, methods for incorporating the first amino acid, chain elongation (including side reactions), and some selected examples are treated sequentially. Peptide synthesis in solution, an approach that should not be considered to be completely superseded by solid-phase methods but, rather, to be complementary to them, is covered in Chapter 3.

The synthesis of large peptides and proteins is especially challenging and remains one of the main goals of contemporary peptide science. In Chapter 4, three of the most promising approaches to such macromolecules are reviewed, namely, convergent solid-phase synthesis, the coupling of peptide segments in solution, and peptide chemical ligation. In the case of chemical ligation, we differentiate between those methods that give rise to backbone-engineered peptides and those that give rise to native structures. This chapter also includes methods for the synthesis of branched peptides, including dendrimers, which are acquiring an increasing importance in fields as diverse as immunology, new materials, and protein folding.

The disulfide bridge is a key structural feature both in natural proteins and in artificial designed peptides, since it ensures that the molecule adopts the correct three-dimensional structure. From a synthetic point of view, the formation of disulfide bridges, especially in peptides with several cysteine residues, can be difficult. Chapter 5 reviews the different possible approaches, starting with the formation of a single intramolecular bridge and proceeding via the formation of several intramolecular bridges to intermolecular disulfides.

Up to now the emphasis has been on those methods that allow the unequivocal synthesis of one single peptide structure with a high degree of purity. However, Chapter 6 is concerned with the world of chemical diversity, and the different methods available for the solid-phase synthesis of peptide libraries are reviewed.

This, in a nutshell, is the book you have in your hands. We know that there are omissions. Some of these, such as treatments of peptide mimetics, of purification and characterization procedures, of enzymatic synthesis, semisynthesis, or *in vivo* expression, are deliberate. Others are not, and we realize that we may have overlooked methods or examples that are at least as relevant as those we have chosen. In spite of all its limitations we hope this book will be useful to you.

Barcelona
January 1997

Paul Lloyd-Williams
Fernando Albericio
Ernest Giralt

Contents

Chapter 1

Introduction

Proteins play a crucial role in almost all fundamental processes in the living cell. Although the genetic information is encoded in nucleic acids, the mechanisms of DNA replication and of gene expression are controlled by proteins. The enzymes that mediate the chemical reactions necessary for life are proteins, as are several of the hormones that control the biochemical balance in complex organisms. The storage and transport of a variety of materials ranging from macromolecules to electrons are also controlled by proteins. In animals, the immune system uses different types of proteins to identify and reject foreign invaders, and muscles, tendons, skin, bones, nails, and hair are all comprised of several kinds of structural proteins. Although they carry out an almost bewildering range of functions in living things, all proteins are made up of the same basic building blocks, being biopolymers of the 20 DNA-encoded amino acids. Beyond this, however, they are not just unstructured chains of their constituent monomers but, rather, adopt characteristic, highly organized three-dimensional arrangements in solution that are intimately related to their biological function.

Peptides are simply smaller versions of proteins. Their three-dimensional structures tend to be less well defined, but many, such as the peptide–hormones vasopressin, oxytocin, and calcitonin, the neuroactive peptides found in the brain, and the toxins of certain animals and bacteria, are biologically important. There are currently several peptide or peptide-based drugs in widespread clinical use, and peptide molecules show much promise as potential therapeutic agents against infectious disorders such as malaria or foot-and-mouth disease. While there is no clear dividing line between peptides and proteins, an acceptable working distinction is that proteins are large peptides, where *large* is a relative term and may mean anything from perhaps 30 to several hundred amino acid residues.

Since the difference between peptides and proteins is essentially one of size or of length of the amino acid backbone, the problems involved in the chemical synthesis of proteins are basically those of the synthesis of peptides or of *peptide chemistry*. Having said that, the difficulties one may encounter in the synthesis of a protein of, say, 150 residues are not the same as those that the synthesis of a peptide of, say, ten residues might present. The synthesis of proteins has challenged chemists for over a century,[1] since the first simple peptides were synthesized by Theodor Curtius[2] and Emil Fischer.[3] Although the total syntheses of oxytocin,[4] of insulin,[5-8] and of ribonuclease A[9,10] were milestones in synthetic organic chemistry, they did not lead to general methods for protein synthesis but, rather, were isolated examples of success in what remained a dauntingly difficult field. However, in the last 30 years the picture has changed dramatically and peptide chemistry has undergone a revolution, brought about, in the main, by two fundamental developments.

Bruce Merrifield's[11] invention of solid-phase peptide synthesis (SPPS) together with the enormous improvements in liquid chromatographic techniques, particularly the advent of high-performance liquid chromatography,[12] have changed peptide chemistry from being a specialist area in which a few research groups were active, into a field where virtually any scientist whose research leads to the need for synthetic peptides may attempt to synthesize them. The refinement of Merrifield's original method has led to the development of automatic peptide synthesizers. These remove much of the drudgery from modern peptide synthesis by carrying out the tedious, repetitive operations required to couple each of the amino acids of any given peptide. Although the chemist must decide the overall synthetic strategy at the outset, many modern synthesizers simply require that the desired sequence be programmed into a dedicated computer. The machine then carries out all of the various chemical steps needed to elaborate the peptide chain automatically, often without the operator having to intervene.

Modern synthetic methods, whether manual or machine assisted, using solid supports or in solution, allow many peptides to be synthesized without undue difficulty. However, large peptides and proteins, or those peptides having a high incidence of the more sensitive amino acids, are a different matter. Side reactions can always occur even in quite simple sequences, but for larger molecules much more serious complications can manifest themselves. In solid-phase synthesis, incomplete deprotection and coupling reactions tend to become more pronounced as the length of the peptide chain increases. The failure of these key steps can present insurmountable obstacles. Synthesis in solution is only rarely a practical alternative for large peptides, since it is slow, labor-intensive, and dogged by the problem of poor solubility of the synthetic intermediates. However, although the chemical synthesis of complex peptides is neither a routine nor a trivial matter, and in many cases may require the best efforts of the most-seasoned practitioners, today it can be attempted with at least a reasonable expectation of success.

In addition to chemical synthesis, biotechnological techniques now provide an alternative and often very efficient means of producing proteins. In favorable cases they are currently the best way of producing useful amounts of material for research and industrial purposes. It is even possible to adapt the methods so that modified protein structures can be produced. However, despite their power and potential, they also have their drawbacks and disadvantages. The isolation of the desired protein from the fermentation medium can be difficult, and biotechnology is not really appropriate for the generation of the large number of analogues that are routinely needed for structure–activity relationship studies. Such variation of structural characteristics is fundamentally important if therapeutic agents are to be produced and is best done by chemical synthesis. One of the most exciting contemporary areas of research is "*de novo*" protein design. This involves the design and synthesis of nonnatural protein analogues, either in order to mimic the natural molecules or to investigate specifically some structural or mechanistic aspect of protein function.

For this, chemical synthesis is obviously the method of choice. Although there will certainly be further improvements in biotechnology, chemical approaches to the synthesis of peptides and proteins are unlikely to be superseded or made obsolete in the foreseeable future.

1.1 AMINO ACIDS AND PEPTIDE AND PROTEIN STRUCTURE

The 20 DNA-encoded or proteinogenic α-amino acids are shown in Table 1.1. All except proline have the same basic structure that incorporates a primary amino group and differs only in the nature of the side chain. Proline, on the other hand, is unique in having a cyclic structure with a secondary amine. With the exception of glycine, all are chiral, due to the presence of at least one stereogenic carbon atom, and belong to the L-stereochemical series. The chiral α-amino acids all have the (S) configuration at the α-carbon atom, except cysteine in which it is (R) as a consequence of the manner in which the Cahn–Ingold–Prelog convention functions.[13] Two amino acids, threonine and isoleucine, have a second stereogenic center at the β-carbon atom, giving rise to four possible diastereomers for each. In the case of L-Thr this second atom has the (R) configuration, so that L-Thr has the stereochemistry 2S, 3R, and in L-Ile the β-carbon atom has the (S) configuration so that L-Ile has the stereochemistry 2S, 3S.

The biosynthesis of peptides and proteins with the 20 proteinogenic amino acids is carried out ribosomally under nucleic acid control. However, certain nonproteinogenic amino acid residues are sometimes found in peptides and proteins, and this is a consequence of posttranslational enzymatic modification of DNA-encoded residues. In collagen, for example, proline can be hydroxylated giving 4(R)-hydroxy-L-proline. The proteins involved in blood coagulation often contain γ-carboxyglutamic acid, formed by the carboxylation of glutamic acid. The hydroxyl groups of Ser, The, and Tyr can be sulfated or phosphorylated in a variety of biologically active peptides.

In several types of lower organisms, such as algae, sponges, yeasts, and fungi, peptides are often biosynthesized enzymatically[14] rather than on the ribosomes. Apart from the proteinogenic amino acids, these peptides may contain other, modified amino acids of which many hundreds are known to exist. They may be divided into several broad classes of which some may simply be enantiomers of the proteinogenic amino acids, such as D-alanine 1. Others may be derivatives methylated either at the amino group or at the α-carbon atom, such as, for example, N-methylleucine 2 or aminoisobutyric acid 3, respectively. Some are α-amino acids with modified side chains, which may be quite simple, as in the case of norvaline 4, or more complex as in 4-(E)-butenyl-4(R)-methyl-N-methyl-L-threonine 5, a component of cyclosporin.[15] Another group of these amino acids is constituted by those that have the amino group at some position other than the α-carbon, for example, isostatine 6, a component of the didemnins.[16] Often, several of these structural features are combined in one amino acid.

Irrespective of the type of amino acid present, the essential structural feature of peptides and proteins is that they consist of chains of the monomers linked together by amide bonds, also known as peptide bonds. There are several characteristics of this linkage that influence the overall three-dimensional structure of the molecule. The carbon–nitrogen bond of amides has appreciable double-bond character, as a consequence of the resonance stabilization brought about by conjugation between the lone pair of the nitrogen with the carbonyl group. The amide bond is flat, therefore, with the carbonyl carbon, oxygen, nitrogen, and amide hydrogen all lying in the same plane. The partial double-bond character of the carbon–nitrogen bond in amides gives rise to an energy barrier to free rotation of approximately 25 kcal mol^{-1}. The peptide bond, consequently, exists as two rotational isomers, the most stable of which is usually the lower-energy *trans* amide. The two isomers are shown in Figure 1.1. In peptide bonds involving proline the difference in energy between *trans* and *cis* amides is lowered, as is the barrier to rotation, and proline-containing peptides often exhibit *cis–trans* isomerism. This can normally be detected by nuclear magnetic resonance (NMR) spectroscopy.

The hydrogen bonds involving the polyamide backbone are an important stabilizing factor in the secondary structures of peptides and proteins. The amide linkage has a significant dipole moment; the carbonyl oxygen is a good hydrogen bond acceptor and the NH a good hydrogen bond donor. Secondary structures such as α-helices and β-sheets are stabilized by the intra- or intermolecular hydrogen bonds of the peptide backbone. Furthermore, the polarity of the amide linkage can impart a net dipole to regular structures such as helices, with the amino terminus having a partial positive charge and the carboxyl terminus a partial negative one. Such dipoles can be important in protein tertiary structure.

1.2 AMINO ACID ACTIVATION AND COUPLING

Generally speaking, peptides are formed by the linking of α-amino acids by amide bonds. The mixing of two α-amino acids in solution leads only to salt formation at ambient temperatures, and the transformation of these salts into amides would require conditions far too harsh for the formation of peptides. It is necessary therefore

Table 1.1 The Proteinogenic Amino Acids

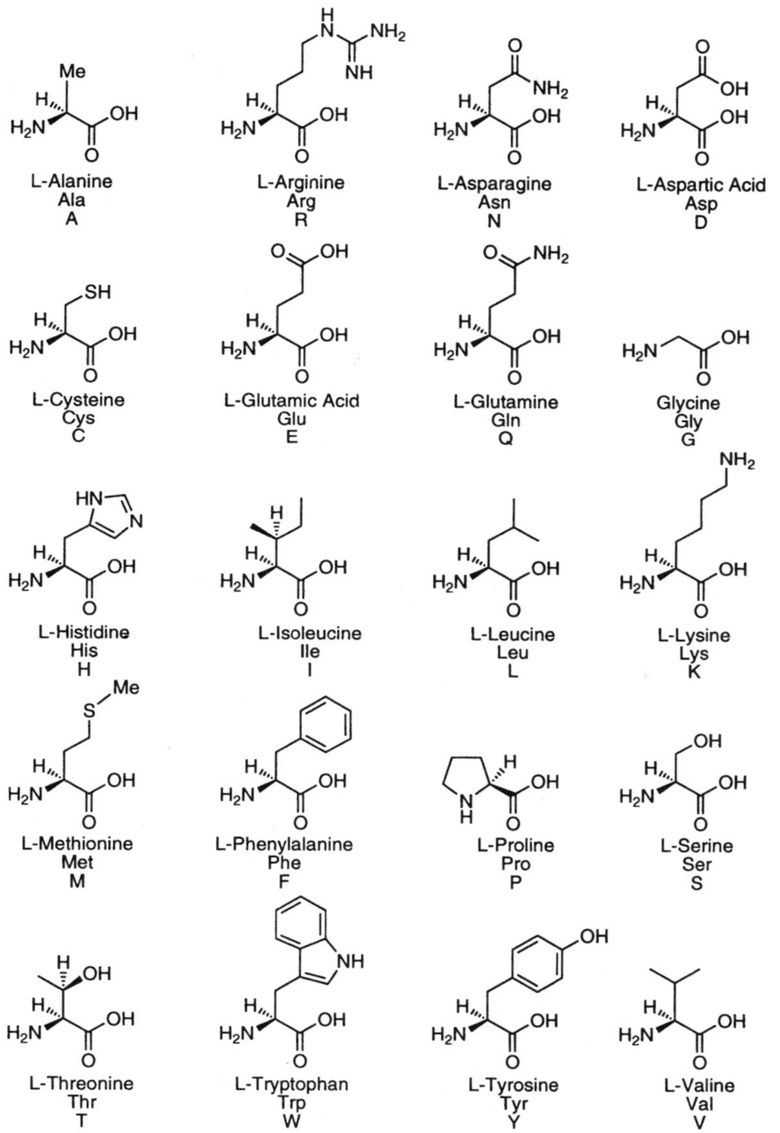

L-Alanine
Ala
A

L-Arginine
Arg
R

L-Asparagine
Asn
N

L-Aspartic Acid
Asp
D

L-Cysteine
Cys
C

L-Glutamic Acid
Glu
E

L-Glutamine
Gln
Q

Glycine
Gly
G

L-Histidine
His
H

L-Isoleucine
Ile
I

L-Leucine
Leu
L

L-Lysine
Lys
K

L-Methionine
Met
M

L-Phenylalanine
Phe
F

L-Proline
Pro
P

L-Serine
Ser
S

L-Threonine
Thr
T

L-Tryptophan
Trp
W

L-Tyrosine
Tyr
Y

L-Valine
Val
V

Note: The three-letter and one-letter codes for each amino acid are given below the chemical name.

to *activate* the carboxyl group of one of the amino acids so that nucleophilic attack by the amino group of the second can take place, forming the desired amide under mild conditions. In peptide chemistry, this process of amide bond formation between two amino acids is called *coupling*. Activation of the carboxyl group of an amino

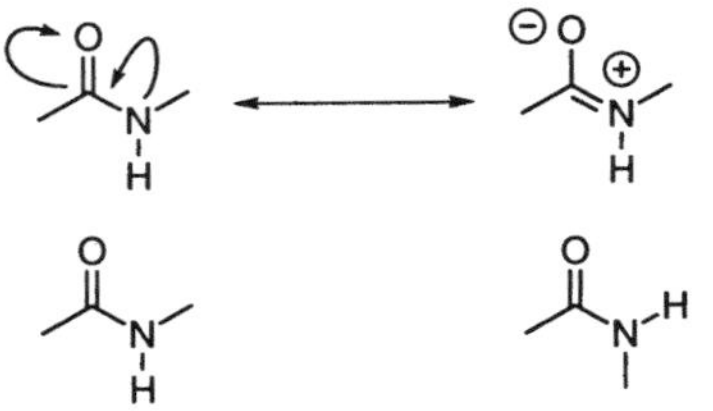

trans-amide   cis-amide

FIGURE 1.1   The amide or peptide bond.

acid may be brought about by chemical attachment of an electron-withdrawing leaving group (X in Scheme 1.1) to the acyl carbon of the carboxyl component. The activated carboxyl group can now undergo nucleophilic substitution by the amino group of the second amino acid, forming the amide bond as shown in Scheme 1.1.

There is a large choice of electron-withdrawing groups X, which can, in principle, be used for this purpose, but activation of the carboxyl groups of amino acids for peptide synthesis may be grouped into a few basic types. A brief outline is given here; coupling methods are treated in more detail in Sections 2.4.1 and 3.3.1.

1.2.1   Acid Azides, Acid Halides, and Anhydrides

The first derivatives used as reactive intermediates for the coupling of amino acids were the acid azides and acid chlorides employed by Curtius[2] and Fischer,[17] respectively, nearly a century ago. In spite of its antiquity, the acid azide method has been one of the most successful coupling procedures used in peptide synthesis and, although it has been superseded by newer methods, is still used occasionally today. The use of acid chlorides, or of other acid halides, on the other hand, did not find general acceptance. The reagents that were traditionally used to form these derivatives often had a deleterious effect on sensitive protected amino acid or peptide molecules. However, more recently acid chlorides[18] and acid fluorides[19] have been reevaluated and have proved their usefulness in cases where strong activation of an amino acid is required.

Anhydrides constitute a major class of activated carboxyl components and have been used extensively in peptide synthesis. There are two basic types, symmetrical anhydrides, formed from the relevant N^α-protected amino acids themselves, and mixed anhydrides, usually formed from the amino acid derivative and a carbonic or carboxylic acid.[20] When anhydrides are used as the activated carboxyl component, the leaving group is a carboxylate anion. Symmetrical anhydrides are relevant in

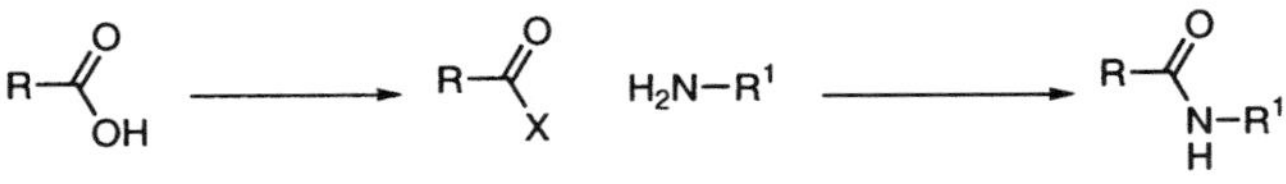

SCHEME 1.1   Generalized chemical formation of an amide bond.

modern peptide synthesis since they, along with other reactive intermediates, are formed *in situ* from the protected amino acid and a *coupling reagent* (see Section 1.2.3).

1.2.2 Active Esters

Simple alkyl esters undergo nucleophilic attack too slowly to be useful in peptide synthesis, but phenyl esters are more reactive and form the basis of the so-called *active ester* class of derivatives.[21] Among the most common are pentafluorophenyl, *p*-nitrophenyl, and 2,4,5-trichlorophenyl esters **7**, **8**, and **9**, in which the leaving group is a phenoxide ion.

They have the advantage of being stable, isolable species but are comparatively "underactivated" when compared with other derivatives such as anhydrides, so that coupling reactions tend to require more time to go to completion.

1.2.3 Coupling Reagents

The most common general coupling method in peptide synthesis today is to use a coupling reagent. This reacts with the free carboxyl group of an amino acid, generating a reactive species, which is not isolated and which is sufficiently reactive to allow amide bond formation to occur at room temperature or below. The most commonly used coupling reagents are the carbodiimides.[22] These have the general structure **10** and form the *O*-acylisoureas **11** as reactive intermediates, which can then undergo aminolysis by the amino component.

The most widely employed reagent of this type, without doubt, has been dicyclohexylcarbodiimide[23] **12**, known as DCC. Other reagents that have been used as coupling reagents in peptide chemistry include ethoxyacetylene[24] **13** and 1-ethoxycarbonyl-2-ethoxy-1,2-dihydroquinoline (EEDQ)[25] **14** and related compounds, although these have never enjoyed the popularity of DCC.

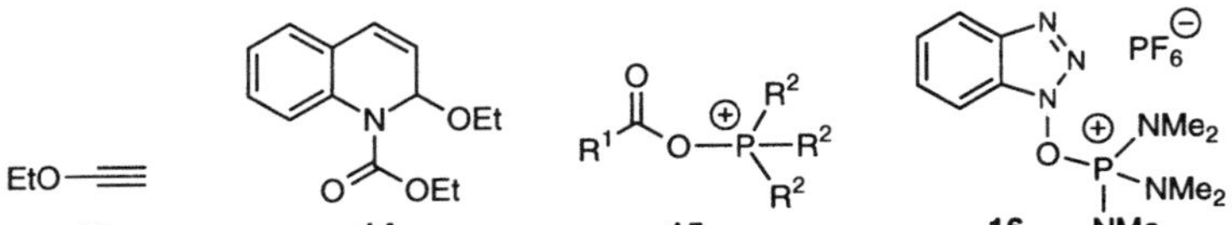

13   **14**   **15**   **16**

More recently, there has been a strong growth in the application of phosphonium salt-based coupling reagents. These are thought to form the reactive acyloxyphosphonium species **15** on reaction with carboxylate anions; **15** then reacts readily with nucleophiles. One of the most successful has been the benzotriazolyloxy-tris-(dimethylamino) phosphonium hexafluorophosphate (BOP) reagent **16**, developed by Castro.[26] Since then, many other, similar compounds have been described, all of which produce acyloxyphosphonium salts as the reactive intermediates. The newer generation of such reagents, based on uronium salts, is rapidly gaining acceptance in peptide synthesis (see Section 2.4.1.2).

1.3 PROTECTION SCHEMES FOR PEPTIDE SYNTHESIS

For the synthesis of even the smallest peptide in a controlled manner it becomes obvious that certain functional groups must be protected. For example, in the synthesis of a dipeptide, it is not sufficient simply to mix the two free amino acids in the presence of a coupling reagent. This would lead to the formation of a mixture of di-, tri-, and polypeptides, because there is no way to differentiate between the two amino groups or the two carboxyl groups of the respective amino acids. Coupling would take place in an uncontrolled fashion, as represented in Scheme 1.2.

In order to couple α-amino acids in such a way that is useful for the synthesis of complex peptides, the functional groups that are not directly involved in the amide bond-forming reaction must be protected or blocked. The N^α-amino group of one of the amino acids and the C-terminal carboxyl group of the other are both blocked with suitable protecting groups. Formation of the desired amide bond can now occur upon activation of the free carboxyl group. After coupling, peptide synthesis can continue by deprotection of the N^α-amino group of the dipeptide and coupling with

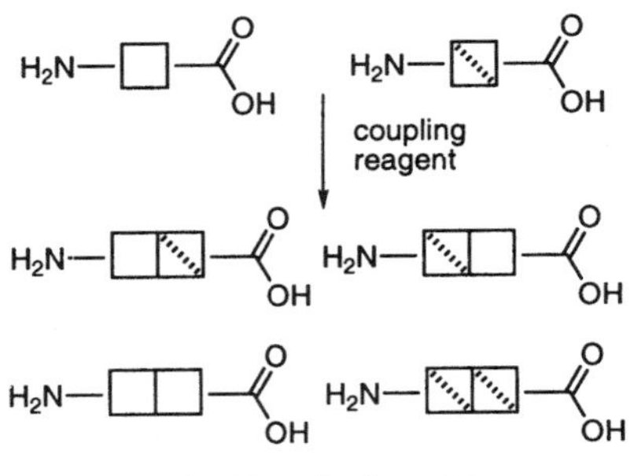

plus tri- and polypeptides

SCHEME 1.2   The uncontrolled coupling of two amino acids.

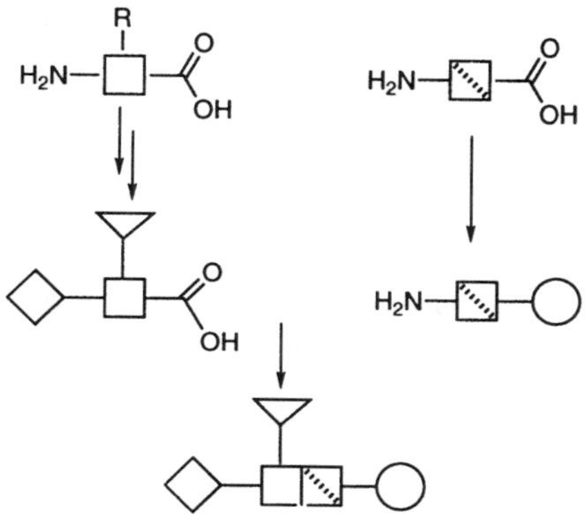

SCHEME 1.3   Controlled coupling of two amino acids. The *C*-terminus-protecting group is represented by a circle, the side chain-protecting group (if present) by a triangle, and the N^α-amino group by a diamond.

the free *C*-terminus of another protected amino acid or of a suitably protected peptide. At the end of the synthesis the protecting groups must be removed, giving the desired free peptide. Scheme 1.3 shows the formation of a protected dipeptide from two protected α-amino acids.

It is obvious, therefore, that the *protection scheme* adopted, that is to say, the combination of the various protecting groups required to block the different reactive functional groups of the amino acids or peptides under consideration, is of the utmost importance. It can, in fact, make the difference between success and failure in a synthesis. It must be possible to introduce, and more importantly to remove, all protecting groups under conditions that do not affect the integrity of the peptide (including its stereochemical purity) being synthesized. In addition, the yields for the introduction and removal of all protecting groups must be as high as possible. These are stringent requirements, but the field has been very extensively worked and nowadays a whole range of protecting groups useful in peptide synthesis is available.

In general, three types of protection are required (Figure 1.2). The functional groups of the amino acid side chains must be protected with groups that are stable to the repetitive treatments necessary both for removal of the N^α-amino-protecting group of the growing peptide chain and for repeated amino acid couplings. Such side chain-protecting groups are usually called "permanent" protecting groups. The N^α-amino function is normally protected with what is referred to as a "temporary" protecting group. The third type of protection is that of the *C*-terminus of the peptide. This must remain in place throughout the various synthetic processes required to elaborate the desired amino acid sequence. However, it may be necessary to remove it, in the presence of all other protecting groups, if the peptide is to be coupled at its *C*-terminus.

Merrifield's great contribution to peptide chemistry was to realize that the *C*-terminal-protecting group could in fact be a polymeric carrier and that, consequently, peptide synthesis could be carried out on an insoluble support. In SPPS the

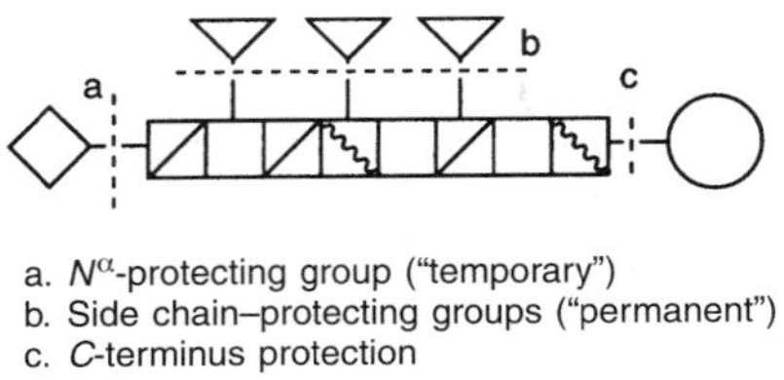

a. N^α-protecting group ("temporary")
b. Side chain–protecting groups ("permanent")
c. C-terminus protection

FIGURE 1.2 Generalized protection scheme for peptide synthesis.

C-terminal-protecting group may be considered to be the peptide–resin anchorage that attaches the growing peptide chain to the solid support. In solution synthesis, on the other hand, the C-terminus is protected with any one of a number of more conventional blocking groups.

1.3.1 N^α Protection

For the synthesis of peptides the N^α-protecting group is almost always a urethane (or alkoxycarbonyl) derivative, and there are several reasons for this. Urethane groups are easily introduced and, depending upon their structure, can be removed easily by alkyl–oxygen cleavage. This leads in the first instance to carbamic acids, which spontaneously decarboxylate generating the free amine of the N-terminal amino acid. In addition, and perhaps more important, the activation and coupling of amino acids with urethane N^α-protecting groups can be accomplished with minimal racemization of the α-stereogenic center. Epimerization and racemization are treated more fully in Section 3.3.2.

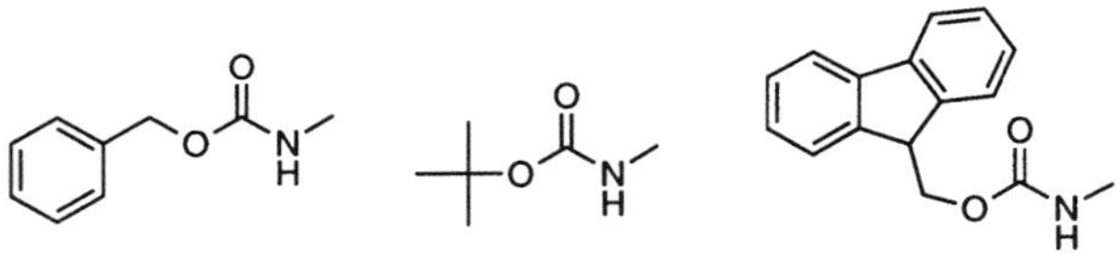

17   **18**   **19**

Structures are drawn to include the nitrogen atom of the amino acid.

Three urethane-based N^α-amino-protecting groups are, by far, the most commonly used in peptide synthesis. The benzyloxycarbonyl group[27] **17** is known as the Z group, in honor of its discoverer, Leonidas Zervas, and is removed by hydrogenolysis or by acidolysis with strong acid. The *tert*-butoxycarbonyl (Boc) group **18** is removed[28-30] by treatment with moderately strong acid, and the base-labile 9-fluorenylmethoxycarbonyl (Fmoc) group **19**, [31] by treatment with a solution of a secondary amine.

Many other N^α-protecting groups have been proposed, of which most, but not all, have been urethane or urethane-like structures. Some of these have found a more limited use in peptide synthesis.

1.3.2 Side Chain Protection

Since the different side chains of the DNA-encoded amino acids encompass the majority of the common functional groups in organic chemistry, several different types of side chain-protecting groups are required for peptide synthesis. However, these side chain-protecting groups are usually based on the benzyl (Bzl) or the *tert*-butyl (*t*Bu) group. For amino acids having alcohols (Ser and Thr) or carboxylic acids (Asp and Glu) in the side chain, these can be protected either as benzyl ethers or as benzyl or cyclohexyl esters. As an alternative they may be protected as *tert*-butyl ethers or esters. For other types of functional groups, such as the amino group of Lys, the thiol group of Cys, the imidazole of His, or the guanidino group of Arg, more specialized protection is required. Many different protecting groups have been described, but relatively few have found significant use in contemporary peptide synthesis. For certain amino acids, notably, His, Met, and Arg, completely satisfactory solutions to the protection of their side chains have not, as yet, been found.

Not all trifunctional amino acids require side chain protection on every occasion and, particularly in synthesis in solution, minimal protection strategies are sometimes used. This involves leaving the side chains of amino acids, such as Ser, Thr, Tyr, Asn, Gln, Trp, Met, and sometimes certain others, without protecting groups. The advantages of minimal protection are often to be found in the increased solubility of the various synthetic intermediates.

1.3.3 Protection of the *C*-Terminus

With regard to protection of the *C*-terminus, the question of whether or not the peptide is to be synthesized in solution or using solid-phase methods must be considered. For peptide synthesis in solution a range of *C*-terminal-protecting groups may be used. *tert*-Butyl esters **20**, benzyl esters **21**, and phenacyl esters **22**, in addition to simple methyl and ethyl esters, may all be used advantageously, depending upon the overall synthetic plan.

For SPPS the *C*-terminal-protecting group is a polymer, normally either a polystyrene- or a polyamide-based carrier. Merrifield's original resin was a chloromethylated polystyrene **23**, but since then many different types of resin have been introduced. The most important nonpolystyrene resins are the polyamide-based resins developed by Sheppard.[32,33] Throughout this book, polymeric solid supports will be represented by the ⬤ symbol.

1.3.4 Compatibility of Protecting Groups — Orthogonality

It is important that there be a high degree of compatibility between the different types of protecting groups such that one type may be removed in the presence of the others. An orthogonal[34,35] protecting scheme has been defined as one based on completely different classes of protecting groups such that each class of groups can be removed in any order and in the presence of all other classes of protecting group. Protecting schemes in peptide synthesis are usually not strictly orthogonal, but rather are based upon the difference in rate when the same chemical reaction (usually acidolysis) is used to remove different classes of protecting groups.

If, for example, the Boc group is to be used as N^α protection in combination with benzyl-based side chain protection groups, the C-terminus cannot be protected with any group that is appreciably labile to acid. If *tert*-butyl-based side chain protection is to be used, then this is obviously incompatible with the Boc group as N^α protection but fully compatible with the Fmoc group. If the Z group is to be used as N^α protection, then, obviously, neither the side chain-protecting groups nor the C-terminus can be protected with benzyl-based groups or groups that can be removed hydrogenolytically. Considerations such as these are the very essence of peptide chemistry, and the decisions made with respect to the choice of one protecting scheme over another may have profound implications for the outcome of any synthetic endeavor.

1.4 SYNTHESIS OF PEPTIDES

1.4.1 Synthesis in Solution

The controlled formation of a dipeptide in solution requires the coupling of two amino acids, one protected at its N-terminus and the other at its C-terminus, as represented in Scheme 1.3. After amide bond formation, the protected dipeptide is isolated, purified, and characterized. The N^α-protecting group of the purified product is then removed and, if necessary, the dipeptide can again be isolated, purified, and characterized. This dipeptide is then coupled with another amino acid protected at its N^α-amino group. A protected tripeptide is formed that is again isolated, and submitted to purification and characterization procedures. If peptide synthesis is now continued by repetitive N^α-amino group deprotection and amino acid–coupling steps, the synthetic strategy is said to be *linear*. If, on the other hand, protected peptide *segments* are coupled, then the strategy is said to be *convergent*. A typical convergent synthetic strategy is represented in Scheme 1.4.

Since in synthesis in solution the intermediate compounds can be isolated and characterized at every step,[36-38] it is quite easy to detect side reactions, including incomplete deprotection and coupling reactions. The unwanted side products can then be removed before proceeding. This approach has the advantage of allowing

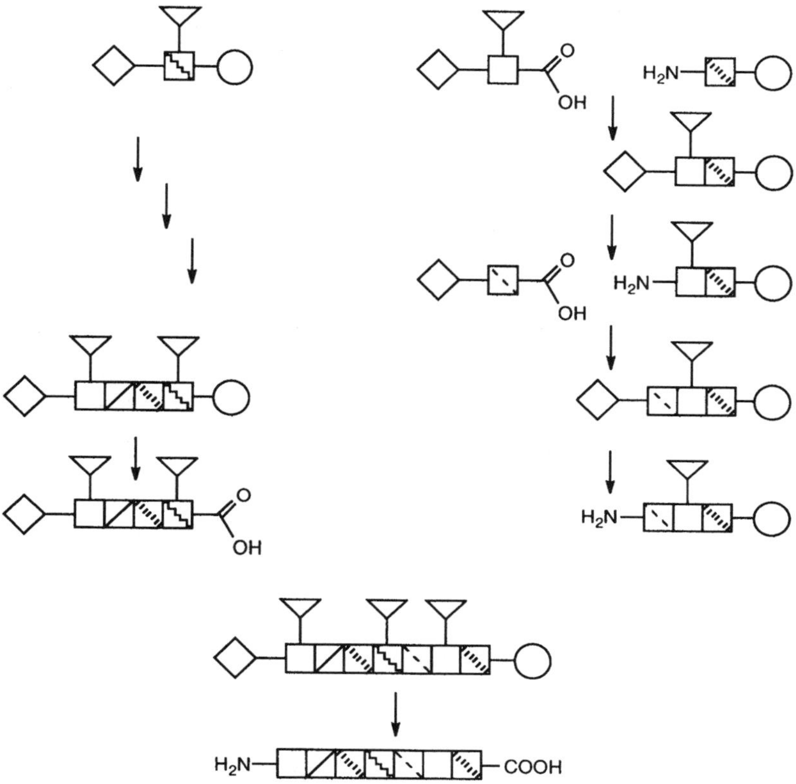

SCHEME 1.4   A convergent strategy for peptide synthesis.

chemists to know exactly which chemical species they are dealing with at any one point. However, the isolation and characterization of intermediates is time-consuming, and even relatively small peptides may require a considerable investment in time and energy in order to synthesize them. Apart from these considerations, peptide synthesis in solution suffers from the serious problem of the poor solubility of the larger protected intermediates. These can be so insoluble as to render their chemical reactions effectively impossible. Although many ingenious attempts were made to overcome the problems associated with classical peptide synthesis in solution, the most radical and successful was that of Merrifield with his invention of SPPS.

1.4.2   Solid-Phase Peptide Synthesis

In SPPS the peptide is not synthesized in solution but rather on an insoluble polymeric support.[39,40] These supports are resins that swell in the solvents used, forming open gel systems in which peptide synthesis takes place. This has two important consequences. First, the isolation of synthetic intermediates is now impossible, but is rendered superfluous if coupling yields can be pushed to completion. Second, the problem of poor solubility of the intermediates during the synthesis now disappears since the growing peptide chain remains attached to the insoluble solid support until the synthesis has been completed.

The first (C-terminus) amino acid of the peptide to be synthesized is protected at its N^α-terminus and its side chain if necessary. It is then *anchored* chemically, either by the formation of an ester or an amide bond, to a resin, typically a copolymer of styrene and 1% divinylbenzene. Peptide synthesis is then continued by repetitive cycles of deprotection of the N^α-protecting group of the peptide–resin and of coupling of the next protected amino acid. In this way, the desired peptide chain is built up on the support in a linear fashion, almost always from the C-terminus to the N-terminus (the $C{\rightarrow}N$ strategy). This is the reverse of the ribosomal synthesis of peptides in nature, where chain elongation takes place from the N-terminus. Since the peptide derivatives remain bound to the insoluble support during all synthetic operations, excess reagents and by-products can be removed simply by washing the peptide–resin with suitable solvents. Large excesses of the soluble N^α-urethane-protected amino acid derivatives can, therefore, be used in order to drive the coupling reactions to completion so that very high coupling yields are achieved with little, if any, racemization. This is the key to the success of the method.

When the desired peptide chain has been synthesized, the peptide must be separated from the solid support, purified, and characterized. *Cleavage* of the peptide from the resin is usually brought about by acidolysis with strong acid, often liquid hydrogen fluoride. This also removes all the amino acid-protecting groups at the same time, affording the crude free peptide that is subsequently purified and characterized. SPPS is represented in Scheme 1.5.

Merrifield's original concept survived early skepticism that was based in the main on two objections. First of all, since the growing peptide chain is attached to an insoluble polymer, the purification and characterization of the various synthetic intermediates is impossible. Characterization of the peptide can only take place at the end of the synthesis, when it has been detached from the polymer. Second, if N^α deprotection and amino acid-coupling yields are not quantitative, or very nearly so, peptides having amino acid compositions other than that of the desired target molecule will build up on the solid support. These may be either truncated sequences or peptides lacking one or more amino acids. The factors involved are discussed in more detail in Section 2.4.2.2. Deletion peptides are closely related structurally to the desired peptide and their presence in the final crude peptide mixture can make purification of the target molecule very difficult. The success of SPPS has demonstrated beyond any doubt that these problems can be overcome in practice and that the disadvantages inherent in the method are clearly outweighed by the advantages.

Solid supports are now being used more and more in other areas of synthetic organic chemistry, particularly for the preparation of mixtures, or libraries, of small molecules that might serve as lead compounds in drug development.[41] The controlled generation of chemical diversity is referred to as *combinatorial synthesis* and is currently being intensively investigated. The synthesis of peptide libraries is treated in more detail in Chapter 6.

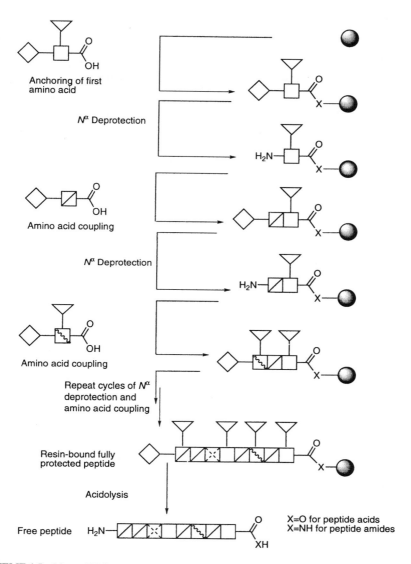

SCHEME 1.5   Linear SPPS. Amino acids are represented by squares, the protected α-amino group by a diamond, and the protected side chain functional groups by triangles. The *C*-terminus of the peptide is bound to an insoluble solid support.

1.4.3   SPPS vs. Synthesis in Solution

Several factors must be taken into account when deciding whether to synthesize a given peptide in solution or by using solid-phase methods. SPPS now dominates synthetic peptide chemistry, and today the majority of peptides are made by this method. This, of course, does not necessarily mean that it is always the best synthetic approach in all cases. For peptides composed of the standard DNA-encoded amino acids, SPPS is the method of choice unless there is some other factor, such as the scale at which the synthesis is to be carried out, that makes synthesis in solution more attractive. For milligram quantities of peptide, SPPS is unrivaled, but when larger quantities of peptide are required, although SPPS may still be used, synthesis in solution becomes more attractive. For the industrial synthesis of kilogram amounts of peptide, synthesis in solution is still the method of choice.

For peptides composed of uncommon or unusual amino acids, N-methylated amino acids, or hydroxy acids, however, synthesis in solution may well be the most promising option. The synthesis of peptides composed of these monomers has not been studied extensively using solid-phase methods, and the limited work that has been done highlights the difficulties of coupling them on solid supports.

REFERENCES

1. Wieland, T., in *Peptides. Synthesis, Structures and Applications*, Gutte, B., Ed., Academic Press, New York, 1995, 1.
2. Curtius, T., *Ber. Dtsch. Chem. Ges.*, 35, 3226, 1902.
3. Fischer, E., *Ber. Dtsch. Chem. Ges.,* 39, 530, 1906.
4. du Vigneaud, V., Ressler, C., Swain, J. M., Roberts, C. W., Katsoyannis, P. G., and Gordon, S., *J. Am. Chem. Soc.,* 75, 4879, 1953.
5. Meienhofer, J., Schnabel, E., Bremer, H., Brinkhoff, O., Zabel, R., Sroka, W., Klostermeyer, H., Brandenburg, D., Okuda, T., and Zahn, H. Z., *Naturforsch.,* 18b, 1120, 1963.
6. Katsoyannis, P. G., Fukuda, K., Tometsko, A., Suzuki, K., and Tilak, M., *J. Am. Chem. Soc.,* 86, 930, 1964.
7. Kung, Y.-T., Du, Y.-C., Huang, W.-T., Chen, C.-C., Ke, L.-T., Hu, H.-C., Jiang, R.-Q., Chu, S.-Q., Niu, C.-I., Hsu, J.-Z., Chang, W.-C., Cheng, L.-L., Li, H.-S., Wang, Y., Loh, T.-P., Chi, A.-H., Li, C.-H., Shi, P.-T., Yieh, Y.-H., Tang, K.-L., and Hsing, C.-Y., *Sci. Sin. Ser. B (Engl. Ed.),* 14, 1710, 1965.
8. Sieber, P., Kamber, B., Hartmann, A., Jöhl, A., Riniker, B., and Rittel, W., *Helv. Chim. Acta,* 57, 2617, 1974.
9. Yajima, H. and Fujii, N., *J. Chem. Soc. Chem. Commun.,* 115, 1980.
10. Yajima, H. and Fujii, N., *Biopolymers,* 20, 1859, 1981.
11. Merrifield, R. B., *J. Am. Chem. Soc.,* 85, 2149, 1963.
12. Hearn, M. T. W., *Methods Enzymol.,* 104, 190, 1984.
13. Cahn, R. S., Ingold, C., and Prelog, V., *Angew. Chem. Int. Ed. Engl.,* 5, 385, 1966.
14. Lipmann, F., *Science,* 173, 875, 1971.
15. Wenger, R. M., *Helv. Chim. Acta,* 66, 2308, 1983.
16. Rinehart, K. L., Kishore, V., Nagarajan, S., Lake, R. J., Gloer, J. B., Bozich, F. A., Li, K.-M., Maleczka, R. E., Todsen, W. L., Munro, M. H. G., Sullins, D. W., and Sakai, R., *J. Am. Chem. Soc.,* 109, 6846, 1987.
17. Fischer, E. and Fourneau, E., *Ber. Dtsch. Chem. Ges.,* 34, 2868, 1901.
18. Carpino, L. A., Chao, H. G., Beyermann, M., and Bienert, M., *J. Org. Chem.,* 56, 2635, 1991.
19. Carpino, L. A., Sadat-Aalee, D., Chao, H. G., and DeSelms, R., *J. Am. Chem. Soc.,* 112, 9651, 1990.
20. Meienhofer, J., in *The Peptides. Analysis, Synthesis, Biology,* Vol. 1, *Major Methods of Peptide Bond Formation,* Gross, E. and Meienhofer, J., Eds., Academic Press, New York, 1979, 263.
21. Bodanszky, M., in *The Peptides. Analysis, Synthesis, Biology,* Vol. 1, *Major Methods of Peptide Bond Formation,* Gross, E. and Meienhofer, J., Eds., Academic Press, New York, 1979, 105.
22. Rich, D. H. and Singh, J., in *The Peptides. Analysis, Synthesis, Biology,* Vol. 1, *Major Methods of Peptide Bond Formation,* Gross, E. and Meienhofer, J., Eds., Academic Press, New York, 1979, 241.

23. Sheehan, J. C. and Hess, G. P., *J. Am. Chem. Soc.,* 77, 1067, 1955.
24. Panneman, H. J., Marx, A. F., and Arens, J. F., *Recl. Trav. Chim. Pays-Bas,* 78, 487, 1959.
25. Belleau, B. and Malek, G., *J. Am. Chem. Soc.,* 90, 1651, 1968.
26. Castro, B., Dormoy, J. R., Evin, G., and Selve, C., *Tetrahedron Lett.,* 1219, 1975.
27. Bergmann, M. and Zervas, L., *Ber. Dtsch. Chem. Ges.,* 65, 1192, 1932.
28. Carpino, L. A., *J. Am. Chem. Soc.,* 79, 4427, 1957.
29. McKay, F. C. and Albertson, N. F., *J. Am. Chem. Soc.,* 79, 4686, 1957.
30. Anderson, G. W. and McGregor, A. C., *J. Am. Chem. Soc.,* 79, 6180, 1957.
31. Carpino, L. A. and Han, G. A., *J. Org. Chem.,* 37, 3404, 1972.
32. Atherton, E., Clive, D. L. J., and Sheppard, R. C., *J. Am. Chem. Soc.,* 97, 6584, 1975.
33. Arshady, R., Atherton, E., Clive, D. L. J., and Sheppard, R. C., *J. Chem. Soc. Perkin Trans. 1,* 529, 1981.
34. Barany, G. and Merrifield, R. B., *J. Am. Chem. Soc.,* 99, 7363, 1977.
35. Barany, G. and Albericio, F., *J. Am. Chem. Soc.,* 107, 4936, 1985.
36. Finn, F. M. and Hofmann, K., in *The Proteins,* Vol. 2, 3rd ed., Neurath, H. and Hill, R. L., Eds., Academic Press, New York, 1976, 105.
37. Sakakibara, S., *Biopolymers (Pept. Sci.),* 37, 17, 1995.
38. Kiso, H. and Yajima, H., in *Peptides. Synthesis, Structures and Applications,* Gutte, B., Ed., Academic Press, New York, 1995, 40.
39. Barany, G. and Merrifield, R. B., in *The Peptides. Analysis, Synthesis, Biology,* Vol. 2, *Special Methods in Peptide Synthesis,* Part A, Gross, E. and Meienhofer, J., Eds., Academic Press, New York, 1979, 1.
40. Merrifield, R. B., in *Peptides. Synthesis, Structures and Applications,* Gutte, B., Ed., Academic Press, New York, 1995, 93.
41. Ellman, J. A., *Acc. Chem. Res.,* 29, 132, 1996.

Chapter 2

Solid-Phase Peptide Synthesis

Merrifield's method for the synthesis of peptides on insoluble polymeric supports has been so successful that the great majority of peptides are now made using this technique.[1-4] The advantages of SPPS over classical synthesis in solution are those of simplicity and speed of execution. SPPS is amenable to mechanization and has led to the development of automated peptide synthesizers[5] that can be programmed to carry out the repetitive steps in the synthesis of a peptide. In favorable cases, quite complex peptides can be made in a matter of hours by machine-assisted synthesis. In less favorable ones, syntheses must be performed manually, but, even so, SPPS is still much faster than synthesis in solution. It is now the method of choice for the synthesis of peptides composed of the proteinogenic amino acids at the multimilligram scale.

The chemistry involved in SPPS has been refined to a degree that is unusual in synthetic organic chemistry. Most of the reactions take place repetitively and reproducibly in yields very close to 100% — indeed, for the method to be useful, near quantitative yields are a prerequisite. In order to attain such high performance all aspects of the chemistry, from the solid support itself to the protecting groups and coupling reagents used, must be rigorously optimized. Only those that present the highest levels of efficiency can be applied.

2.1 THE SOLID SUPPORT

For a solid support to be useful in SPPS, certain properties are essential. It must consist of particles of a convenient size and shape that are physically robust enough to permit easy manipulation and rapid filtration from liquids. At the same time, it must be chemically inert to all of the reagents and solvents used for peptide synthesis and must not interact physically, in any detrimental way, with any of these, nor with the peptide chain itself. Furthermore, it must be possible to modify the solid support chemically, in such a way that the first amino acid of the sequence to be synthesized can readily be attached by the formation of a covalent bond.

Few materials fulfill all of these requirements, but much investigation has been done and many materials have been evaluated with regard to their possible usefulness. The supports most commonly employed in SPPS all swell extensively in the solvents used, allowing reagents to penetrate throughout the particles, ensuring ready accessibility of the reactive sites to the reagents concerned. There are currently three basic types that have found lasting application in peptide synthesis.

2.1.1   Cross-Linked Polystyrene

The solid support chosen by Merrifield has stood the test of time and is currently the most popular type used in SPPS. It is composed of beads, of between 20 and 50 μm diameter, of a synthetic resin prepared by copolymerization of styrene with a small (typically 1%) amount of divinylbenzene. This resin has a mechanical stability sufficient to allow rapid filtration under vacuum but at the same time can swell to up to 8 ml/g in relatively apolar solvents, such as dichloromethane. This being so, compared with the solvent, the polymer matrix itself contributes only a small amount to the internal volume of the bead.[6,7] Friedel Crafts chloromethylation of the aromatic rings of the polystyrene matrix, using chloromethylmethyl ether in the presence of zinc chloride,[8] provides a way of functionalizing the polymer, and different resin types are usually produced by chemical modification of the chloromethyl groups. Aminomethyl groups may be introduced by Gabriel synthesis[9] and hydroxymethyl groups by displacement of chloride with acetate ion followed by hydrazinolysis or hydrolysis.[10,11] Polystyrene resins usually have a substitution level of 0.3 to 1.2 mmol of functional group per gram, and an ideal level for routine applications is of the order of 0.5 mmol/g.

2.1.2   Polyamide

In an effort to improve the efficiency of the fledgling solid-phase method, Sheppard[12-16] suggested that peptide synthesis would proceed more efficiently if the polymer support were chemically more similar to the peptide chain itself. Both would then be well solvated in dipolar aprotic solvents. This line of reasoning led to the development of polyamide resins as an alternative to polystyrene in SPPS. The most successful of this type of solid support consists of a cross-linked polydimethylacrylamide, incorporating sarcosine methyl ester side chains. These methyl esters can then be modified by treatment with ethylene diamine, providing a primary amino group as the point of attachment to the resin.[17] Polyamide resins swell up to ten times their dry volume in dimethylformamide and even more in water. On the other hand, they swell much less in dichloromethane, and, in this sense, their properties are the reverse of those of polystyrene resins. The few comparative studies that have been carried out between polystyrene and polyamide resins indicate that both give very similar results in routine SPPS.

The development of continuous-flow peptide synthesizers, in which the various solvents and reagents are pumped through a column of the functionalized resin, has given an impetus to the design of supports that can withstand the back pressures arising with this technique. Standard resins are not useful because they are compressed under the operating conditions, which in turn leads to higher back pressures and reduced flow rates. A solution to the problem is to embed the resin upon which peptide synthesis takes place, in a rigid but highly permeable matrix. Two such solid supports have been very successful and deserve special mention. The first consists of polyamide gel resin embedded within kieselguhr[18] and is known commercially as Pepsyn K™. The second consists of polyamide embedded within a highly cross-linked polystyrene,[19] and is sold as Polyhipe™. The former has functionalization levels of the order of 0.1 to 0.2 mmol/g, while Polyhipe has a higher substitution level, in the range of 0.3 to 1.8 mmol/g.

Other polyacrylamide resins developed for use in both batchwise and continuous flow synthesis include the polyacrylic support[20] Expansin™ and poly N-[2-(4-hydroxyphenyl)ethyl]-acrylamide.[21,22] This latter has a high capacity of the order of 5 mmol/g.

2.1.3   Polyethylene Glycol-Grafted Polystyrene

Polyethylene glycol-grafted polystyrene consists of an insoluble polystyrene matrix to which chains of polyethylene glycol have been attached. It was developed, as were the polyamide resins discussed above, in an attempt to improve the solvation of peptides bound to the solid support.[23,24] Polyethylene glycols of various chain lengths may be used, but the most successful in general application appear to be those with molecular weights between 2000 and 3000 Da. Such resins swell in a range of solvents and additionally have the mechanical stability necessary for synthesis under continuous-flow conditions. Polyethylene glycol-grafted polystyrene supports can be prepared either by the attachment of modified polyethylene glycol chains to standard amino-functionalized polystyrene resins[25,26] or by anionic copolymerization of ethylene glycol in the presence of polystyrene beads.[27-31] This latter procedure has been used to provide so-called *tentacle polymers*.

Polyethylene glycol-grafted polystyrene resins have been reported to give superior performance to standard polystyrene resins, both in batchwise and continuous-flow peptide synthesis. This is thought to be because they provide a better environment for peptide synthesis, since the polyethylene glycol chains act as a spacer between the polystyrene matrix and the first amino acid of the peptide. This separation is thought to make it less likely that the peptide adopt secondary structures in which the availability of the N^α-terminal amino group is decreased, thought to be one of the main factors involved in the phenomenon known as *difficult sequences*. (see Section 2.4.2.2.2). Typical functionalization levels for these resins are in the range of 0.15 to 0.3 mmol/g.

In addition to these three general types of solid supports, many others have been proposed and evaluated for use in peptide synthesis, without having established themselves, at least as yet. These include membranes,[32] cotton,[33] and other carbohydrates,[34,35] polyethylene-grafted polyamide,[36] controlled-pore silica glass,[37] linear polystyrene chain grafted onto Kel-F,[38-40] and polyethylene sheets.[41]

2.2   PROTECTION SCHEMES

There are very few protecting groups that have not been evaluated for use in SPPS. Indeed, peptide synthesis has been traditionally, and continues to be, a proving ground for the majority of protecting groups used in organic chemistry. Although many groups have been tried, only a small number are actually used in contemporary practice because of the strict limits of performance that the technique demands. The research and development effort in SPPS has led to the two major protection schemes that are used today. These are the so-called Boc/Bzl and Fmoc/tBu approaches to peptide synthesis. In the former, the Boc group is used for N^α-protection and the side chains of the various amino acids are protected with benzyl- or cyclohexyl-based protecting groups. In the latter the Fmoc group is used as N^α protection and the side chains are protected with tertiary carbon-based protecting groups such as the *tert*-butyl and the triphenylmethyl (trityl or Trt) groups.

2.2.1 N^α Protection

In SPPS, N^α protection is almost always either the Boc[42-44] group **1** or the Fmoc[45] group **2**. Both of these N^α-protecting groups are easily incorporated into amino acids. In the former case, di-*tert*-butyl dicarbonate,[46] (Boc)$_2$O, or less commonly 2-*tert*-butoxycarbonyloximino-2-phenylacetonitrile,[47] BocON, can be used. In the latter, treatment of the amino acid with 9-fluorenylmethyl succinimidyl carbonate[48,49] (FmocOSu), 9-fluorenylmethyl azidoformate (FmocN$_3$), or, less commonly, with 9-fluorenylmethyl chloroformate (FmocCl), usually gives good results. Care must be taken with the latter reagent, however, to avoid the formation of dipeptides.[50,51]

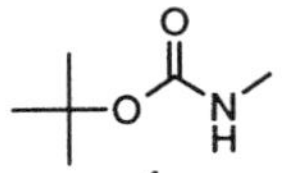

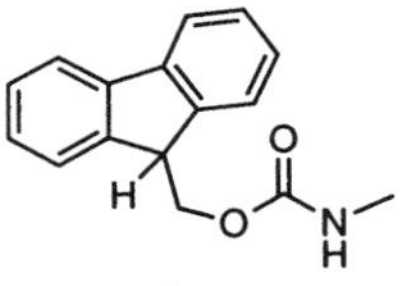

Structures are drawn to include the N^α-nitrogen atom of the amino acid.

In Boc/Bzl synthesis, all protecting groups are removed by acidolysis. The side chain-protecting groups are, however, stable to the repeated treatments with moderately strong acid solutions, such as 33% trifluoroacetic acid in dichloromethane, required to remove the Boc group. These deprotection conditions give rise to the trifluoroacetate salt of the N^α-amino group, as shown in Scheme 2.1.

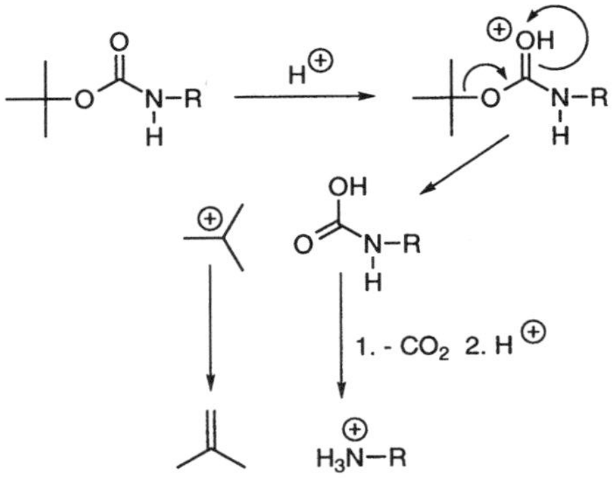

SCHEME 2.1 Acidolytic removal of the Boc group.

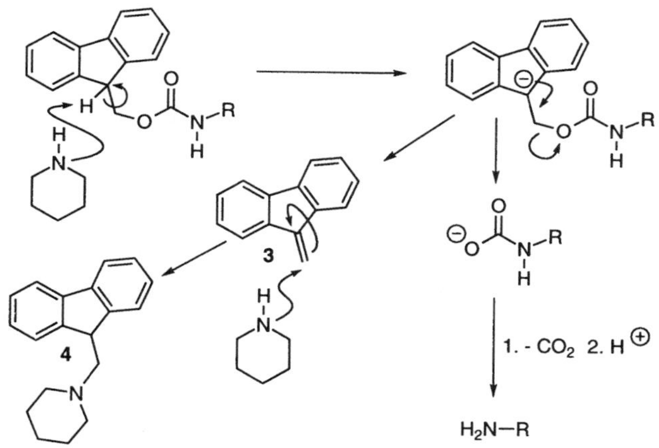

SCHEME 2.2 Removal of the Fmoc group with piperidine.

Once the peptide has been synthesized, the side chain-protecting groups are removed by treatment with strong acid, often liquid hydrogen fluoride or, less commonly, trifluoromethanesulfonic acid. This detaches the completed peptide chain from the solid support at the same time.[52] The final deprotection and cleavage step in SPPS is discussed more fully in Section 2.5.

In Fmoc/tBu synthesis, the Fmoc group is labile to a solution of a secondary amine, normally 20% piperidine in dimethylformamide. The mechanism of cleavage is known[53,54] to be E1cB, proceeding by initial proton abstraction to give the stabilized dibenzocyclopentadienide anion, which on elimination gives rise to dibenzofulvene, **3**. This latter is then trapped by reaction with excess piperidine giving adduct **4**. These processes are depicted in Scheme 2.2. Other reagents, such as 1,8-diazabicyclo[5.4.0]undec-7-ene[55,56] (DBU) or fluoride ion,[57] are also effective for Fmoc group removal.

The side chain-protecting groups and the peptide–resin anchorage are also labile to acid in the Fmoc/tBu approach. However, the use of strong acids such as liquid hydrogen fluoride is not necessary. In the majority of cases treatment with trifluoroacetic acid is sufficient to cleave the peptide from the solid support and to remove all protecting groups (see Section 2.5.1.2).

In common with other urethane N^α-protected amino acids, Boc and Fmoc amino acid derivatives are resistant to racemization under normal peptide synthesis conditions. (For a more detailed discussion of racemization and epimerization in peptide synthesis, see Section 3.3.2.1.)

2.2.2 Side Chain Protection of Individual Amino Acids

Global protection strategies are normally used in SPPS. That is to say, those amino acids with side chain functional groups have them protected, although there are certain exceptions. Met, Asn, and Gln, for example, are not always protected at their

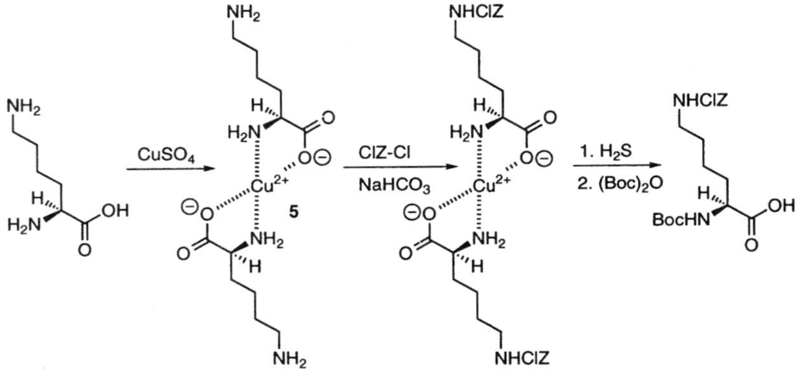

SCHEME 2.3 Protection of the Lys side chain by chelate formation.

side chains in standard SPPS. Under certain circumstances, other amino acids such as Thr, Ser, and Tyr may also be incorporated without side chain protection.

2.2.2.1 Lysine and Ornithine

Protection of the N^ε-amino group of Lys is mandatory in peptide synthesis; otherwise its acylation will lead to branching of the peptide chain. Selective protection of the N^ε-amino group requires differentiation of the two amino groups of Lys, and classical methods are available for achieving this. A useful procedure takes advantage of the tendency of the N^α-amino and carboxyl groups of Lys to form chelates, such as **5**, with copper ions,[58,59] leaving only the N^ε-amino group free to react. After its protection, destruction of the chelate then frees the N^α-amino group for protection with a complementary group, as shown in Scheme 2.3.

Alternatively, the greater basicity and nucleophilicity of the N^ε-amino group compared with the N^α-amino group allows it to be protected selectively. Reaction of Lys with benzaldehyde can be made to proceed so that only the N^ε-amino group forms the Schiff base,[60] as in **6**. With the N^ε-amino group blocked, the N^α-amino function can then be protected and the N^ε-group unmasked and reprotected in a complementary manner.[61] These processes are shown in Scheme 2.4.

In Boc/Bzl synthesis, the N^ε-amino group is normally protected with the 2-chlorobenzyloxycarbonyl (ClZ) group, **7**. The benzyloxycarbonyl (Z) group **8** is less useful here, since the acidolytic conditions required for deprotection of the N^α-Boc group causes some N^ε-Z group removal and, consequently, branching of the peptide.[10] This can become severe in longer peptides, where many Boc-removal cycles must be performed. The increased acid stability (up to 60 times more stable) of the ClZ group is helpful in reducing this unwanted deprotection. In the Fmoc/tBu approach, the use of the Boc group as N^ε protection provides an optimum combination with the Fmoc N^α protection.

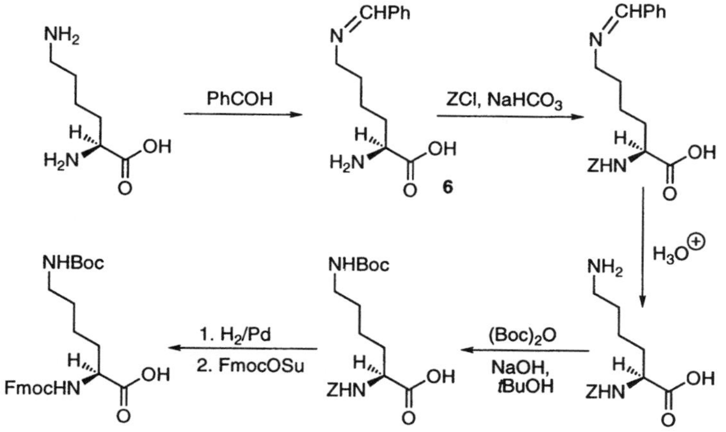

SCHEME 2.4 Protection of Lys by preferential Schiff base formation at the N^ε-amino group.

In some situations, orthogonal N^ε-amino protection is necessary. The solid-phase synthesis of branched peptides, for example, requires the independent removal of the Lys N^ε-amino protecting group, without affecting any other protecting groups present in the peptide. Alternatively, it may be necessary to maintain the N^ε-amino groups protected after deprotection and cleavage of the peptide from the solid support. Several possibilities exist for orthogonal protection of the Lys side chain, but among the most useful are the trifluoroacetyl **9** (Tfa), the 3-nitro-2-pyridine sulfenyl **10** (Npys), the 1-(4,4-dimethyl-2,6-dioxocyclohexylidene)ethyl **11** (Dde), and the allyloxycarbonyl (Alloc) **12** groups. N^ε-Tfa protection is stable to acidolysis and is therefore compatible with Boc N^α protection. A separate treatment with piperidine solutions is required for its removal,[62-64] which, of course, means that it is removed under conditions similar to those used for the Fmoc group. N^ε-Npys protection[65] is compatible with Boc/Bzl synthesis and is removed on treatment with

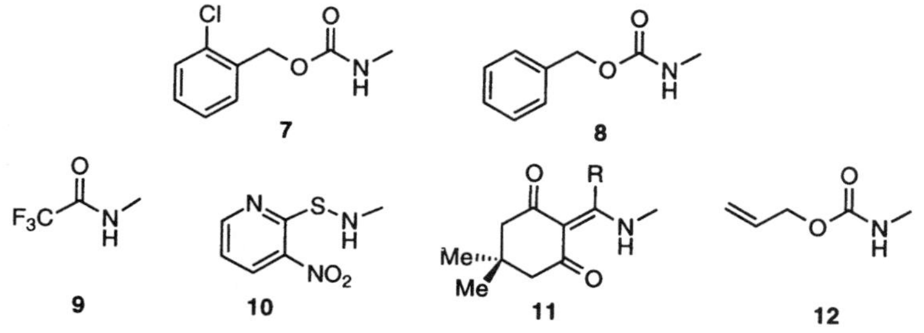

Structures are drawn to include the nitrogen atom of the Lys side chain.

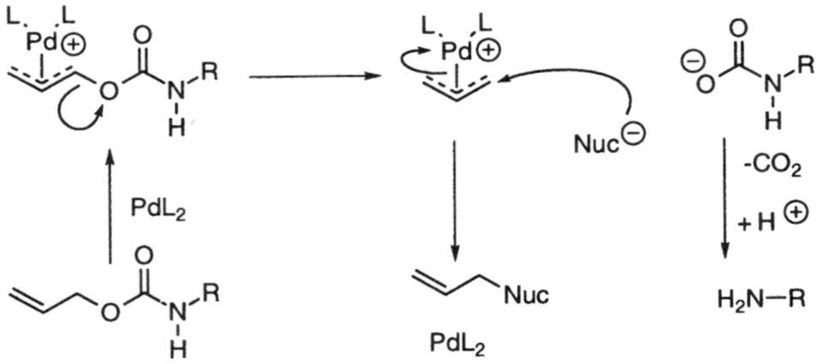

SCHEME 2.5 Palladium-assisted removal of the Alloc group.

triphenylphosphine or 2-pyridinethiol 1-oxide.[66] The Dde group, compatible with the Fmoc/tBu approach, can be removed by treatment with dilute hydrazine solutions.[67]

Unlike groups **9** to **11**, which are only compatible with one or the other of the Boc/Bzl or Fmoc/tBu approaches, N^ε-Alloc protection **12** is stable to the conditions used in both. It can be smoothly removed, however, by treatment with a suitable nucleophile in the presence of a palladium catalyst.[68-72] The mechanism for this process is thought to be similar to that shown in Scheme 2.5.

Ornithine has one fewer carbon atom in its side chain than Lys, and similar considerations apply to its protection.

2.2.2.2 Arginine

The δ-guanidino group of Arg is strongly basic (pK_a = 12.5) and protonation provides a simple method of protection that was often used in peptide synthesis in solution.[73] However, the poor solubility associated with the intermediates and the problems arising as a result of the protonated peptides behaving like ion-exchange resins, causing them to change anions during the course of a synthesis, prompted the search for a more satisfactory method for protecting the side chain of this amino acid.

A side reaction occasionally observed in peptides containing Arg is the formation of the corresponding Orn peptide during synthesis. Such transformations have been observed on ammonolysis,[74] or after strong acidolysis with HBr or HF.[75] The main side reaction associated with this amino acid, however, is its cyclization, upon activation of the *C*-terminus, to form δ-lactams **13**, derivatives of 2-piperidone, as shown in Scheme 2.6. This can be severe when the side chain is left unprotected under basic conditions and is not always fully suppressed by the incorporation of a protecting group at the ω-position of the guanidino function. Protection of the δ-amino group in δ, ω-diprotected derivatives,[76-79] on the other hand, does completely eliminate such cyclization.

The nitro group[80,81] found favor for a while as side chain protection for Arg. However, although generally speaking it performs reasonably well, its removal is problematic. Forcing acidolytic conditions that can be detrimental to other sensitive functionalities in a peptide are often required. Its removal by catalytic hydrogenolysis

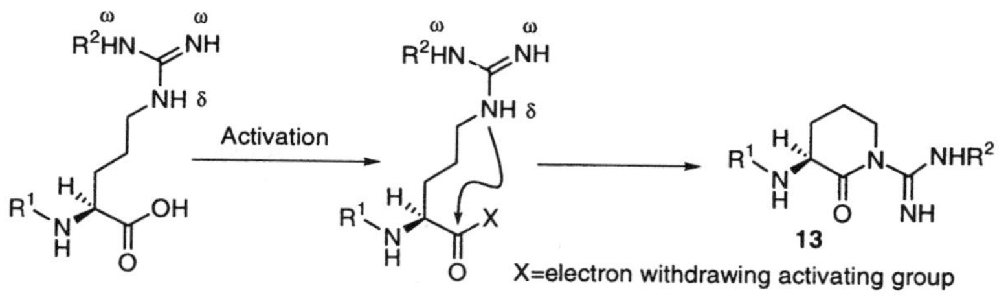

SCHEME 2.6 δ-Lactam formation in Arg derivatives on coupling.

is not always straightforward — long reaction times may be necessary, and stable, partially reduced products are often formed.[82]

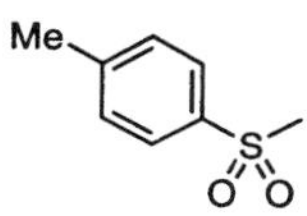

Arylsulfonyl-based protecting groups are currently the most commonly applied. Although such protection does not completely suppress δ-lactam formation on activation of the C-terminus for coupling, the use of additives such as 1-hydroxybenzotriazole is normally sufficient to prevent it occurring to an unacceptable extent. For Boc/Bzl SPPS, the tosyl group[83] **14** is adequate. It is stable to the repetitive Boc-deprotection cycles but can be removed at the end of the synthesis by treatment with liquid hydrogen fluoride. For Fmoc/tBu synthesis the 4-methoxy-2,3,6-trimethylbenzenesulfonyl (Mtr) group[84,85] **15** has been superseded by both the 2,2,5,7,8-pentamethylchroman-6-sulfonyl (Pmc) group[86] **16** and the closely related 2,2,4,6,7-pentamethyldihydrobenzofuran-5-sulfonyl (Pbf) group[87] **17** that are more easily removed in practice. Arylsulfonyl-based groups **15**, **16**, and **17** are sufficiently labile to trifluoroacetic acid to permit Arg side chain deprotection during the cleavage reaction at the end of the synthesis. Irrespective of whether the Boc/Bzl or the Fmoc/tBu approach to peptide synthesis is used, incomplete deprotection of the Arg residues may be observed in peptides that are rich in this residue.[88]

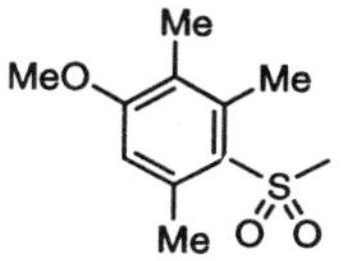

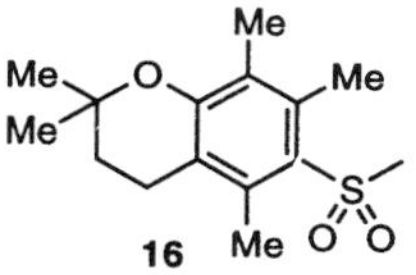

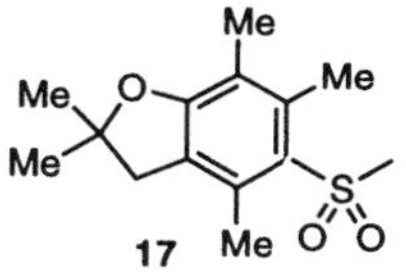

Incorporation of arylsulfonyl groups at the ω-position of the amino acid can be carried out as shown in Scheme 2.7.

A quite different solution to the problem of Arg protection is to carry out the synthesis with suitably protected Orn derivatives and then to convert these into Arg at the end.[89-92] This approach has been used successfully on more than one occasion in synthesis in solution,[93] and similar protocols have been used[92,94,95] in SPPS.

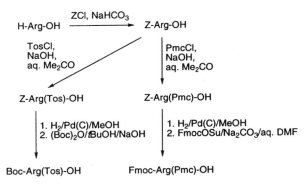

SCHEME 2.7 Synthesis of Boc-Arg(Tos)-OH and Fmoc-Arg(Pmc)-OH.

2.2.2.3   Histidine

The protection of the side chain of His was problematic for many years, and, even today, it has still not been resolved in a completely satisfactory manner. Difficulties arise because the imidazole ring has two nonequivalent, but similarly reactive nitrogen atoms, designated π and τ, as in **18**. In the earliest attempts at His protection, which of the nitrogens was blocked either was not considered or was often incorrectly assumed.

The location of the protecting group is important, however, since it is intimately related to the issue of His racemization in peptide synthesis. This amino acid is unique in that it can racemize easily on activation and coupling, even if its N^α-amino group is urethane protected. This takes place via mechanisms involving the His side chain, and the two main pathways for the process are thought to be those shown in Scheme 2.8. In the first of these, the π-nitrogen acts as nucleophile, attacking the activated carbonyl group intramolecularly, forming cation **19**. Enolization then leads to loss of optical activity. In the second, the π-nitrogen acts as base leading directly to the zwitterionic enolate **20**. Since, in both, it is the π-nitrogen that is involved, the suppression of His racemization on activation hinges upon the effective blocking or deactivation of this position. (His can, of course, also racemize by the mechanisms that are common to all the other chiral DNA-encoded amino acids. These are discussed in more detail in Section 3.3.2.1.)

Regiospecific protection of the π-nitrogen will prevent racemization but is not easy to achieve because of the comparable reactivity of the two nitrogens of the heterocycle. Unhindered alkylating agents tend to give mixtures of mono- and

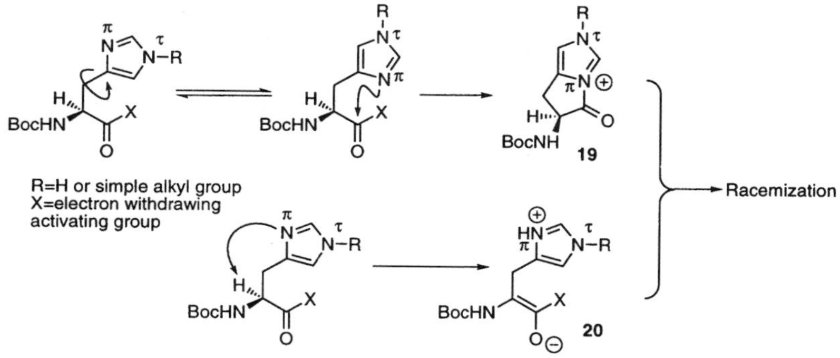

SCHEME 2.8 Mechanisms for the racemization of His on activation.

disubstituted products, because monoalkylation blocks one nitrogen without deactivating the other. Reagents with a larger steric requirement give more of the τ-protected derivatives, which are, generally speaking, very prone to racemization. In any case, simple π-alkyl substitution will not diminish the basicity and nucleophilicity of the τ-nitrogen atom so that other side reactions may occur during peptide synthesis. Alkylation of the imidazole ring with the benzyl group, one of the first methods attempted,[96] proceeds to give predominantly the N^τ-protected product, which is completely ineffective at preventing racemization on coupling.

An alternative to regiospecific blocking of the π-position is to attenuate its reactivity by reduction of the electron density. This can be done by introducing an electron-withdrawing group into one of the nitrogens of the imidazole. Even if it is at the τ-nitrogen, the inductive effect transmitted across the ring also deactivates the π-nitrogen. Electron-withdrawing groups therefore, provide more secure protection than simple alkyl substitution. Furthermore, problems with the formation of disubstituted products are avoided because once one of the nitrogens is protected the nucleophilicity of the second is suppressed.

For Boc/Bzl synthesis, the question of how to protect the His side chain still awaits a definitive answer so that a number of groups are currently used. The 2,4-dinitrophenyl-protected derivative[97,98] **21** is perhaps the most useful and may be removed by thiolysis.[99,100] In multiple His-containing peptides, however, incomplete deprotection may sometimes be observed. Furthermore, it presents the disadvantage that it is labile to other nucleophiles including possibly N^α-amino groups,[101] such that it may be lost progressively during peptide synthesis. Furthermore, at least one report[102] places its effectiveness at preventing racemization in doubt. The tosyl derivative[103-105] **22** is also quite widely used. The tosyl group may be removed at the end of the synthesis by treatment with liquid hydrogen fluoride. Unfortunately, it is not stable in the presence of N^α-amino groups[106] or 1-hydroxybenzotriazole,[107] a widely used additive in peptide synthesis, and this reduces its usefulness. The protecting groups in both the 2,4-dinitrophenyl and tosyl derivatives **21** and **22** are normally located at the N^τ-nitrogen, as a consequence of the steric requirements of the reagents used for their preparation.

Another group used for His protection in the Boc/Bzl approach is the π-benzyl-oxymethyl (Bom) group[108,109] **23**, which prevents racemization and can be removed by hydrogen fluoride.

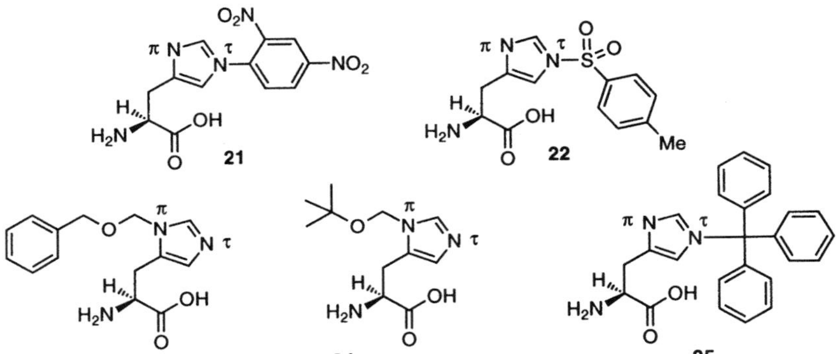

The problem associated with this derivative is that the Bom group generates formaldehyde on strong-acid-mediated cleavage from the solid support. This can lead to the irreversible blocking or "capping" of amino groups or to the modification of Cys residues, unless precautions are taken[110-113] (see Section 2.5). Regiospecific incorporation of the Bom group at the π-position may be carried out as shown in Scheme 2.9

For Fmoc/tBu synthesis the situation is clearer, although still not wholly convincing. The N^{π}-tert-butoxymethyl (Bum) group[114] **24** can be used; it is as effective as the Bom group at preventing racemization, although its synthesis is relatively involved[115] and, again, the formation of formaldehyde on removal may be problematic.[113] However, for this approach to peptide synthesis, protection as the N^{τ}-triphenylmethyl derivative[116] **25** is currently the method of choice. It is completely stable to the basic conditions required to remove the Fmoc group and is smoothly removed by mild acidolysis. Alkylation of suitably protected His derivatives with trityl chloride blocks the τ-position regiospecifically, which would suggest that the trityl group might not be effective in preventing racemization. However, racemization of N^{τ}-trityl-protected His is not normally a problem under standard Fmoc/tBu SPPS conditions. A comparison of the amounts of racemization observed when trityl- and benzyl-N^{τ}-protected His derivatives were coupled with H-Pro-OtBu in solution, however, showed similar amounts in each case.[117] The possibility of appreciable

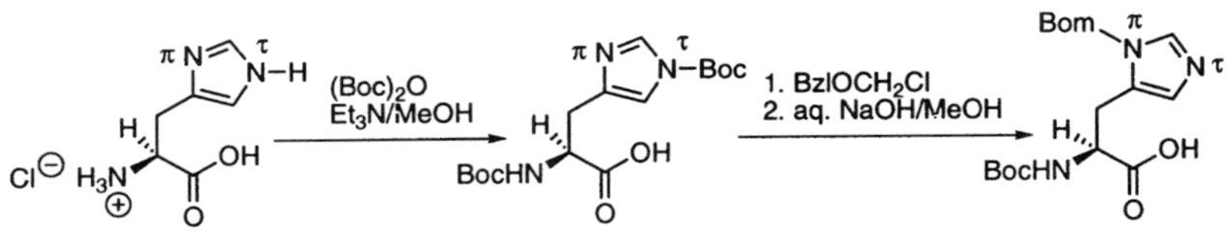

SCHEME 2.9 Synthesis of Boc-His(Bom)-OH.

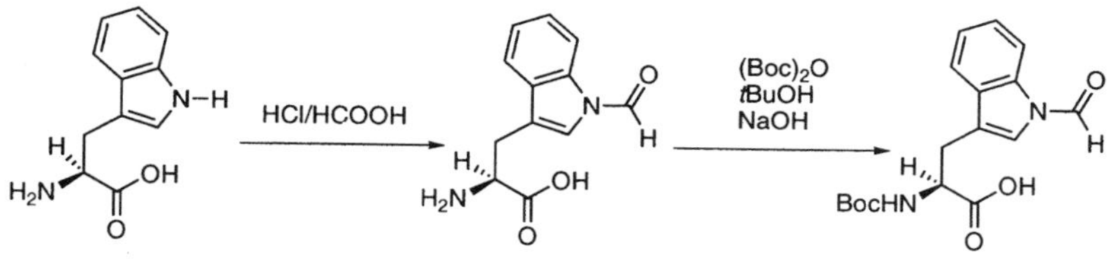

SCHEME 2.10   Synthesis of Boc-Trp(For)-OH.

racemization of N^τ-trityl-protected His in solid-phase couplings that are sluggish for whatever reason cannot, therefore, be discounted.

2.2.2.4   Tryptophan

The indole ring of tryptophan undergoes facile electrophilic aromatic substitution at more than one position.[118-124] In addition, this amino acid is also susceptible to oxidative degradation by atmospheric oxygen in acidic media. This means that the integrity of Trp is at risk either on removal of Boc N^α protection or, especially so, during the final acidolytic cleavage of the peptide from the solid support. Peptides containing Trp can be very troublesome to handle during this process and the use of suitable scavengers is essential (see Section 2.5). Nevertheless, the side chain of Trp is not always protected in peptide synthesis[125] although the more common approach to the management of this residue is to incorporate an electron-withdrawing group at the indole nitrogen atom. This reduces the electron density of the indole nucleus and with it the amount of unwanted alkylation.

The most popular choice for the protection of the Trp side chain in Boc/Bzl synthesis is the N-formyl (For) group.[126] Its introduction is straightforward, as shown in Scheme 2.10. Its removal can be accomplished by treatment with a solution of piperidine in water or with hydrazine or hydroxylamine.[127] A drawback, however, is that under basic conditions the N^{im}-formyl group can migrate to N^α-amino groups, in this way capping them[128] (see Section 2.4.2.2.1). The problem can normally be controlled by careful choice of reaction conditions.

For Fmoc/tBu peptide synthesis the Boc group may be used for protection of the indole nitrogen atom. The synthesis of Fmoc-Trp(Boc)-OH is straightforward,[129,130] as shown in Scheme 2.11. N^{im}-Boc protection has been reported to reduce alkylation of Trp residues on cleavage of the peptide from the solid support.[131]

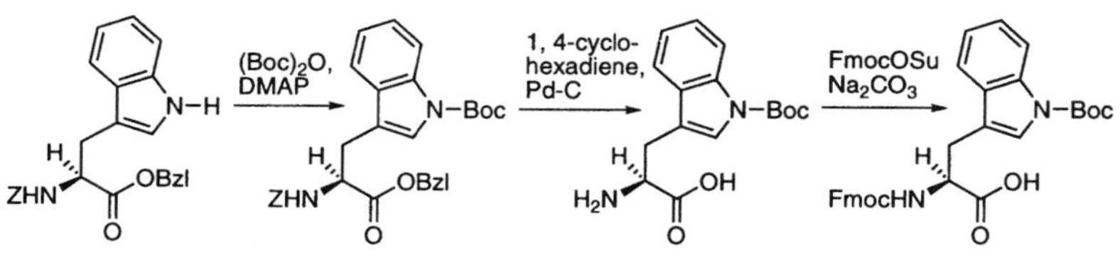

SCHEME 2.11   Synthesis of Fmoc-Trp(Boc)-OH.

2.2.2.5 Asparagine and Glutamine

The amide side chain functional groups of Asn and Gln are often left unprotected in SPPS. The main problem that this can cause, unless precautions are taken against it, is amide dehydration to the corresponding nitrile **26**, on coupling.[132,133] The mechanism by which it takes place is shown in Scheme 2.12.

A consequence of this mechanism is that dehydration should only occur on activation of the C-terminus of Asn or Gln for coupling, and not once the amino acid has been incorporated into the growing peptide chain. This accords with observed experimental data, and once Asn and Gln have been coupled their side chains are safe. In carbodiimide-mediated couplings, dehydration of Asn and Gln with unprotected side chains can be avoided by adding 1-hydroxybenzotriazole[134,135] (see Section 2.4.1.1). However, the use of this additive does not suppress nitrile formation when BOP reagent is used.[135] Dehydration-free incorporation of Asn and Gln without side chain protection can also be achieved if the corresponding amino acid active esters are used for coupling (see Section 2.4.1.3).

Activated Asn and Gln derivatives having unprotected side chains may also undergo intramolecular cyclization reactions, as in Scheme 2.13, forming succinimides **27** and glutarimides **28**, respectively.[136] Furthermore, Gln can also cyclize to give pyroglutamic acid derivatives **29**, a reaction that is catalyzed by weak acids.[137] This can be troublesome in SPPS when the Boc group is removed from N-terminal Gln-containing peptides,[138] and this is discussed in more detail in Section 2.4.2.1.3. Although none of these side reactions is especially severe, or at least can usually be avoided by judicious choice of the reaction conditions, it is clear that there are advantages in side chain amide protection. This not only effectively eliminates side reactions, but also improves the solubility of the amino acid derivatives themselves. Improved solubility on side chain protection is especially noticeable in Fmoc N^α-protected Asn and Gln, which are otherwise only poorly soluble in dichloromethane and dimethylformamide. An additional advantage of side chain protection of these residues, especially in Fmoc/tBu SPPS, is that it may also be helpful in reducing the hydrogen bonding that stabilizes secondary structure formation and that leads to reduced coupling rates and even difficult sequences (see Section 2.4.2.2.2). Factors such as these explain why the protection of the side chains of Asn and Gln during peptide synthesis has been gaining in popularity recently.

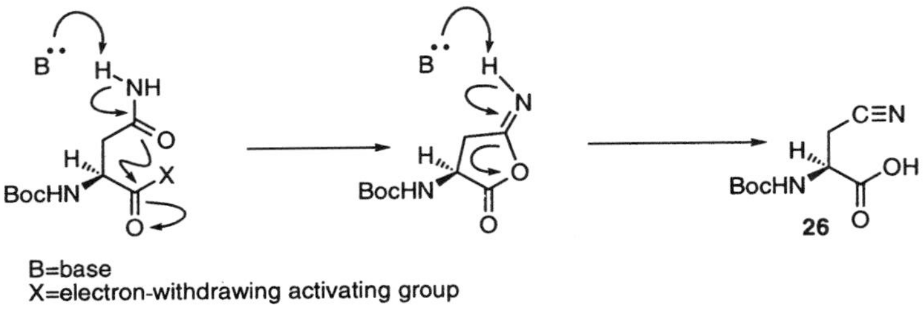

B=base
X=electron-withdrawing activating group

SCHEME 2.12 Mechanism for the dehydration of Asn on activation. A similar mechanism applies in the case of Gln.

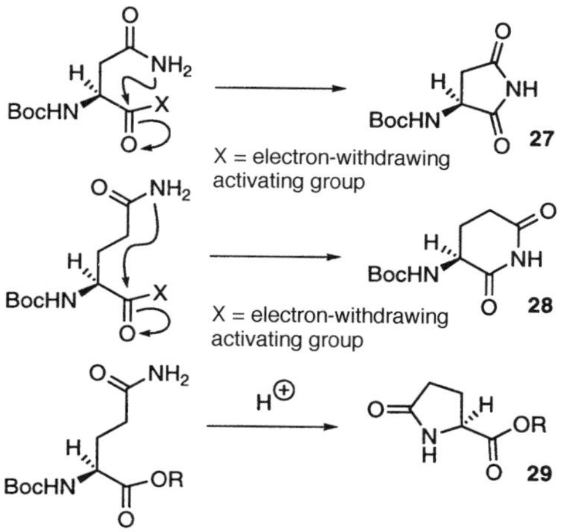

SCHEME 2.13 Side reactions involving Asn and Gln.

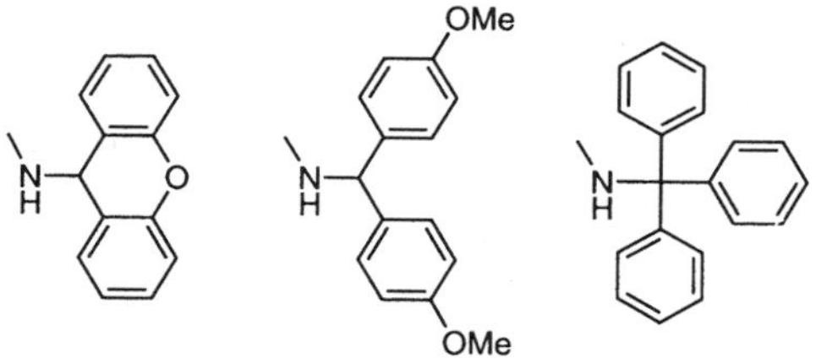

30	31	32

Structures are drawn to include the nitrogen atom of the Asn or Gln side chain

For Boc/Bzl synthesis, the xanthenyl[139](Xan) **30** or the 4,4'-dimethoxybenz-hydryl[140,141] (Mbh) **31** groups may be used to protect the side chain during incorporation of Asn or Gln. Neither is stable to the acidolytic cleavage conditions necessary for removal of the Boc group so that both Xan and Mbh side chain protection is progressively removed once the residue has been incorporated into the peptide chain. Both groups do, however, provide protection against nitrile formation when Asn or Gln are coupled and allow phosphonium or uronium reagents to be used for this. For Fmoc/tBu synthesis, the trityl group[142] **32** provides permanent side chain protection for these amino acids.

The preparation of Boc-Asn(Xan)-OH and Fmoc-Gln(Trt)-OH is shown in Scheme 2.14. Other protected derivatives may be prepared using similar methods.

2.2.2.6 Aspartic and Glutamic Acids

The carboxylic acid side chains of Asp and Glu must be protected during peptide synthesis; otherwise their activation will lead to amide bond formation and branching of the peptide chain. The most commonly used protecting groups are cyclohexyl

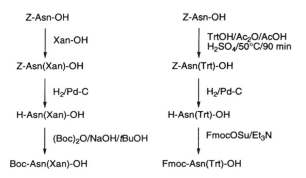

SCHEME 2.14 Synthesis of Boc-Asn(Xan)-OH and Fmoc-Asn(Trt)-OH.

esters in combination with Boc N^α protection or *tert*-butyl esters for the Fmoc/tBu approach. Cyclohexyl esters[143,144] are now favored over the more classical benzyl esters because they are more effective at preventing aspartimide formation[145] during synthesis. Aspartimides are cyclic, Asp-derived compounds that can be produced at various stages in peptide synthesis. They are discussed in more detail in Section 2.4.2.1.2.

Selective protection of the side chains of Asp and Glu requires differentiation of the two carboxylic acid groups, in each case, and may be achieved in several ways. The simplest is direct acid-catalyzed esterification of the free amino acid. Protonation of the amine under these conditions reduces the reactivity of the α-carboxylic acid, allowing the side chain carboxyl group to react selectively. This is shown in Scheme 2.15 for Glu; Asp reacts similarly.

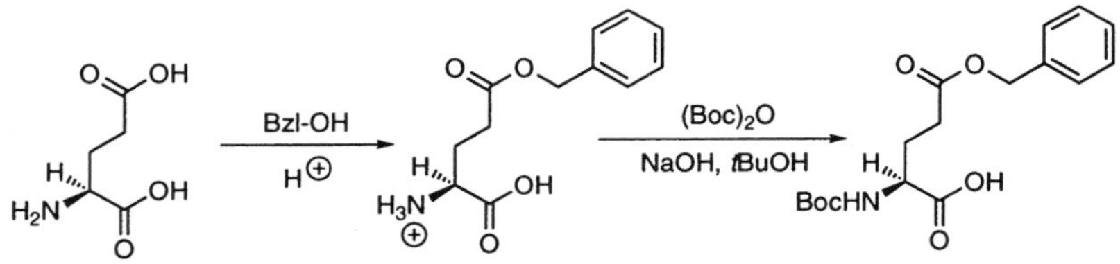

SCHEME 2.15 Acid-catalyzed selective esterification of the side chain of Glu.

Alternatively, the carboxyl groups of both Asp and Glu can be distinguished chemically by formation of the intramolecular anhydride (**33** in the case of Glu). The α-carboxyl group in each case is then rendered more electrophilic by the electron-withdrawing nitrogen derivative, allowing its selective esterification. The side chain carboxyl group can then be protected in a complementary manner, as shown in Scheme 2.16 for Glu. A similar reaction sequence can be carried out with Asp.

Chelate formation of the type already discussed for the protection of Lys (see Scheme 2.3) can also be used to protect the side chains of Asp and Glu. Here again, the copper chelate of the amino acid **34** has only the side chain carboxyl group free for reaction with the appropriate alkyl halide.[146,147] Destruction of the chelate then

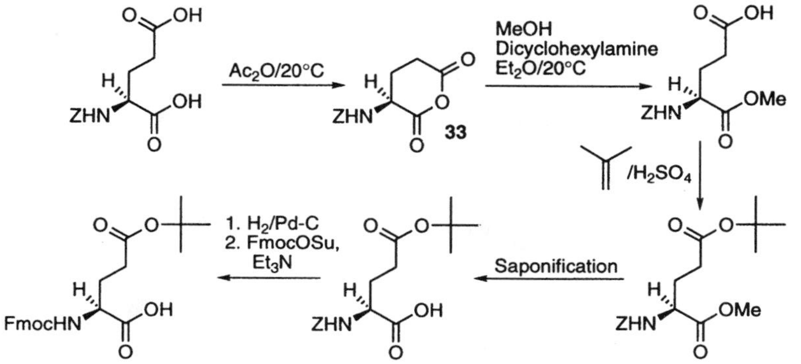

SCHEME 2.16 Synthesis of Fmoc-Glu(OtBu)-OH.

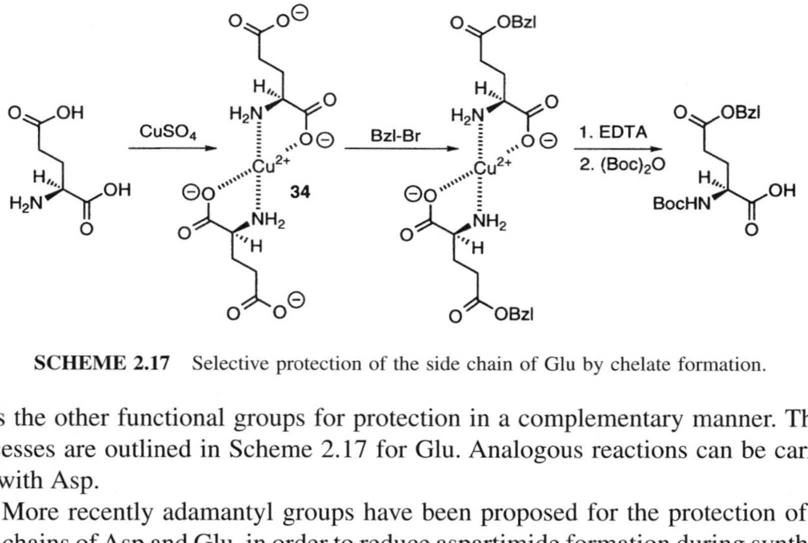

SCHEME 2.17 Selective protection of the side chain of Glu by chelate formation.

frees the other functional groups for protection in a complementary manner. These processes are outlined in Scheme 2.17 for Glu. Analogous reactions can be carried out with Asp.

More recently adamantyl groups have been proposed for the protection of the side chains of Asp and Glu, in order to reduce aspartimide formation during synthesis (see Section 2.4.2.1.2). For Boc/Bzl synthesis the 2-adamantyl group has been proposed, while for the Fmoc/tBu approach the 1-adamantyl group is recommended.[148,149] If orthogonal protection of the side chains of Asp and Glu is required, then this can be realized in Boc/Bzl synthesis using the base labile 9-fluorenylmethyl (Fm) group.[150] Allyl esters[68,69,151] removed by palladium-catalyzed transfer to a suitable nucleophile (see Alloc N^α protection and Scheme 2.5) provide orthogonal protection of Asp and Glu side chains in both Boc/Bzl and Fmoc/tBu synthesis.

2.2.2.7 Serine and Threonine

The hydroxyl groups of these amino acids can react with acylating reagents, more readily in the case of Ser than in Thr, owing to steric effects. In SPPS these amino acids are usually protected as benzyl ethers in Boc/Bzl synthesis or as *tert*-butyl

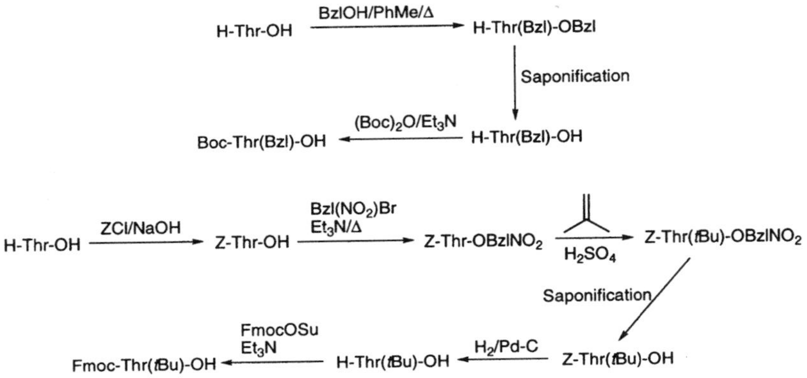

SCHEME 2.18   Preparation of Boc-Thr(Bzl)-OH and Fmoc-Thr(*t*Bu)-OH.

ethers for the Fmoc/*t*Bu approach. Benzyl-based groups are removed on treatment with liquid hydrogen fluoride, whereas trifluoroacetic acid treatment is sufficient for liberating hydroxyl groups protected as *tert*-butyl ethers. The protection of Ser and Thr is relatively straightforward, although often several steps are required. The preparations of Boc-Thr(Bzl)-OH and Fmoc-Thr(*t*Bu)-OH using typical methods[152-157] are shown in Scheme 2.18. Protection of the hydroxyl group of Ser is usually easier, on account of it being less hindered; similar routes to those outlined in Scheme 2.18 serve to provide the necessary derivatives.

2.2.2.8   Tyrosine

In common with the hydroxyl groups of Ser and Thr, the phenolic group of Tyr can also react with acylating agents. In SPPS it is usually incorporated in a protected form in order to avoid this. *O*-Acylation can, in fact, be more of a problem with Tyr, especially under basic conditions where the formation of the phenolate anion generates a potent nucleophile. Tyr is also susceptible to electrophilic aromatic substitution of the phenol ring, particularly at the positions *ortho* to the hydroxyl group. In peptide synthesis such alkylation usually occurs on acidolytic removal of protecting groups. Originally, the hydroxyl group of Tyr was protected as a benzyl ether.[158] Unfortunately, significant amounts of 3-benzyltyrosine were observed on acidolysis,[159,160] and this stimulated the search for protecting groups that give rise to less of this type of side product.

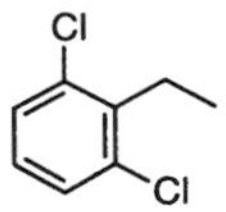

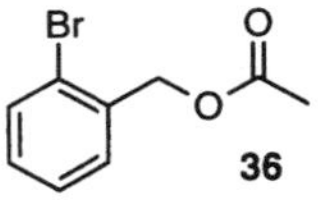

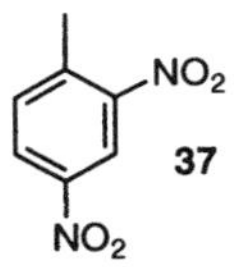

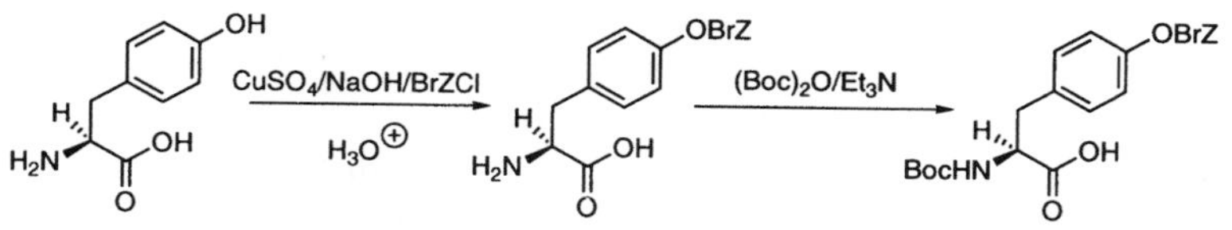

SCHEME 2.19   Preparation of Boc-Tyr(BrZ)-OH.

In contemporary Boc/Bzl SPPS, Tyr is normally protected using either the 2,6-dichlorobenzyl (Dcb) **35** or the 2-bromobenzyloxycarbonyl (BrZ) **36** groups, neither of which give rise to significant amounts of 3-alkylation.[160,161] Their incorporation may be accomplished by forming the copper chelate of the amino acid, (see Schemes 2.3 and 2.17), leaving the phenolic hydroxyl group free to react as shown in Scheme 2.19, for incorporation of the BrZ group.

In the Fmoc/tBu-approach, the phenol side chain is protected as the *tert*-butyl ether.[157] Acidolytic removal of the *tert*-butyl group gives rise to very little 3-alkylation product. The synthesis of the necessary derivative requires a longer, but still relatively straightforward route,[152] as shown in Scheme 2.20.

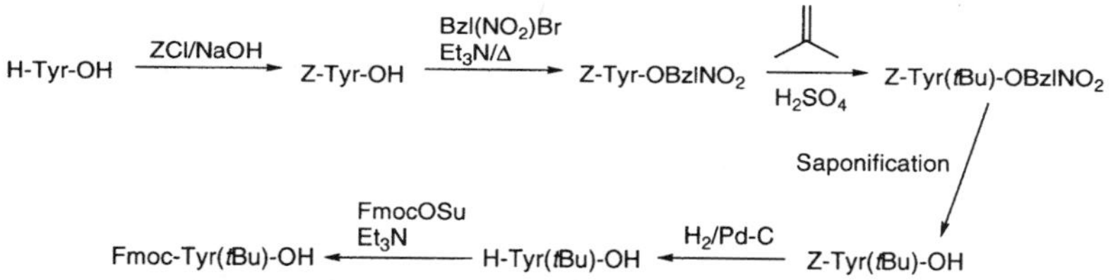

SCHEME 2.20   Preparation of Fmoc-Tyr(tBu)-OH.

The 2, 4-dinitrophenyl (Dnp) **37** group can be used to protect the phenol of Tyr, and it provides an orthogonal protection scheme for both Boc/Bzl and Fmoc/tBu synthesis. Dnp protection can be removed by thiolysis, and this type of Tyr derivative can be useful in the synthesis of peptide conjugates.[162]

2.2.2.9   Methionine

The major side reactions associated with unprotected Met are alkylation and oxidation of the thioether side chain. Oxidation to the sulfone takes place only under quite drastic conditions, but sulfoxides can be formed simply on prolonged exposure to air.[163,164] Although the reaction is reversible, partial oxidation of the thioether can complicate the handling of a peptide. The sulfoxide is chiral and, consequently, the mixture now contains three components — the unoxidized thioether and the two diastereomers corresponding to the sulfoxide. Alkylation of Met with carbocations proceeds rapidly under acidic conditions to give the relatively stable sulfonium salts.[165] This can take place either during repetitive acidolytic N^α-deprotection steps or in the final acidolytic cleavage (see Section 2.5.2.1). It is sometimes, but not always, suppressed by the addition of scavengers.

There are two general approaches to the treatment of Met in peptide synthesis, one option being to use derivatives in which the thioether is not protected, and to attempt to minimize oxidation and alkylation where possible. Alternatively, the thioether may be oxidized to the sulfoxide in a controlled manner and the synthesis carried out with derivatives of methionine-sulfoxide [Met(O)], such as **38**. This does not undergo oxidation or alkylation under normal peptide synthesis conditions.[166] There is no clear consensus on which of these general procedures gives the best results in practice, and examples of the use of both the protected and unprotected amino acid are to be found.

In susceptible substrates incorporating unprotected Met, complete suppression of oxidation is rarely possible, but alkylation can sometimes be avoided or at least diminished. The peptide S-*tert*-butylmethioninesulfonium salts **39** are substantial by-products whenever unprotected methionine is used in peptide synthesis along with *tert*-butyl-based protecting groups removable by acidolysis. However, since the *tert*-butylation of Met is reversible, this alkylation is normally not a severe problem.[167] The benzylation of Met residues can occur when the Z group is removed acidolytically on treatment with hydrogen bromide or when other benzyl-based protecting groups are removed in the cleavage of the peptide from the solid support. Such alkylation may be reduced by addition of diethylphosphite or ethyl methyl sulfide[168,169] in the case of Z group removal or by appropriate scavengers in the final cleavage reaction (see Section 2.5.2.1).

Protection of the thioether by controlled oxidation to the sulfoxide is a more secure way of avoiding the problems caused by undesired oxidation and alkylation. Reduction then permits regeneration of the normal Met residue at a later stage in the synthesis; several methods for accomplishing this have been reported.[170-174] Met protection as the sulfoxide may be advisable, not only for the avoidance of unwanted side reactions, but also for the beneficial effect that the extra polarity can have in the purification of Met-containing protected peptide segments. See Section 4.1.2.1.1. A drawback is that the oxidation of methionine does not give rise to an optically pure sulfoxide, but rather to a mixture of two diastereomers. Although these can be separated by fractional crystallization,[175] this is rarely done and consequently diastereomeric peptides are formed. These are often, but not necessarily always, resolved under standard high-performance liquid chromatographic conditions.

2.2.2.10 Cysteine

Protection of the side chain of Cys is obligatory in peptide synthesis since sulfhydryl groups are good nucleophiles and can compete effectively with amines in acylation reactions. Additionally, they can be oxidized, even by air, to give disulfides. The protection of cysteine has been extensively studied because of the importance that disulfide bridge formation has in peptide chemistry, something that is dealt with in more detail in Chapter 5 of this book.

Originally, the S-benzyl[176] thioether **40** was used as protection for the side chain of Cys. However, the unsubstituted benzyl group can only be removed with sodium in liquid ammonia or by liquid hydrogen fluoride at 25°C. Such harsh conditions are incompatible with sensitive functionalities so that nowadays protection with the unsubstituted benzyl group is seldom applied. In order to increase acid lability so that removal might be effected under less drastic conditions, substituted benzyl thioethers, such as the S-p-methylbenzyl (Meb) **41**, S-p-methoxybenzyl (Mob) **42**, S-trimethoxybenzyl (Tmob) **43**, or S-triphenylmethyl (Trt) **44**, were investigated and evaluated for complex peptide synthesis.[160,177-182]

Structures are drawn to include the sulfur atom of Cys.

The cleavage of Meb and Mob thioethers still requires treatment with liquid hydrogen fluoride, albeit at a lower temperature than that required by the unsubstituted parent compound. The Tmob and Trt groups, on the other hand, are labile to trifluoroacetic acid. This being so, the former two are compatible with Boc/Bzl SPPS while the latter two can be used in the Fmoc/tBu approach. Unusually, S-tert-butyl protection **45**, is very stable to acid and even survives treatment with liquid hydrogen fluoride at 0°C, although it is removed at 20°C. Other significant cysteine protection in peptide synthesis includes the acetamidomethyl (Acm) **46** and the Fm **47** groups.[183-185] The former is compatible with both Boc/Bzl and Fmoc/tBu SPPS and can be removed by oxidative treatment with iodine or with thallium trifluoroacetate. The latter can be used in the Boc/Bzl approach and is removed by treatment with secondary amines.

A different approach to cysteine protection is based on mixed disulfides such as the *S-tert*-butylmercapto (S*t*Bu) **48** or *S*-Npys **49** groups.[186-188] These are useful in Boc/Bzl peptide synthesis and may be removed by a variety of methods including thiolysis. A particularly useful aspect of the Npys group is that it can be displaced by the free sulfhydryl group of another Cys residue, allowing the directed formation of disulfide bridges. (See Chapter 5 for a more detailed discussion.)

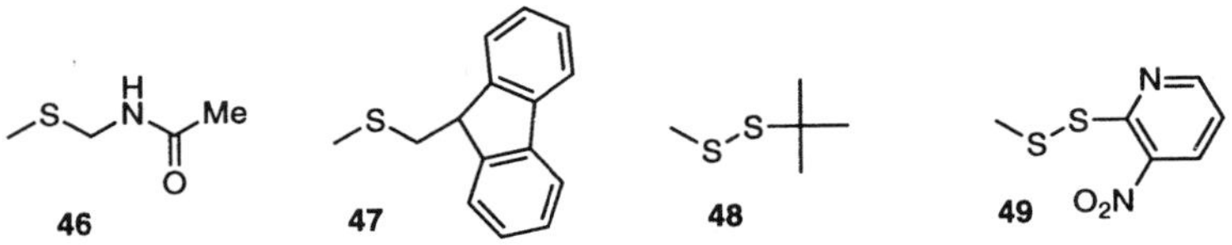

46   **47**   **48**   **49**

Structures are drawn to include the sulfur atom of Cys.

Incorporation of protecting groups into Cys is straightforward because of the pronounced soft nucleophilicity of the sulfhydryl function.[189] Treatment of free Cys with either the relevant alkyl halides under basic conditions or the requisite alcohols, with acid catalysis, normally leads to the formation of the desired *S*-protected derivatives in high yield. Mixed disulfides may be formed by treatment of cystine with the relevant thiol under basic conditions. Typical reactions are shown in Scheme 2.21.

Protected Cys is prone to a number of side reactions. Oxidation and alkylation of the thioether can occur, as with Met, but these appear to be less severe in the case of Cys.[2,190,191] Under basic conditions, *S*-protected Cys can also undergo β-elimination leading to the formation of dehydroalanine,[192] although the extent of elimination depends upon the *S*-protecting group used. The problem is not severe for the Cys derivatives normally used in the Fmoc/*t*Bu approach.[16] Elimination to give dehydroalanine derivatives has also been observed under acidic conditions.[193] The formation of piperidinyl alanine has been observed[194] in Fmoc/*t*Bu synthesis.

Certain protected Cys derivatives are particularly prone to racemize on coupling[195,196] and on attachment to the solid support. The amount observed appears to depend both upon the methods used for anchoring and upon the *S*-protecting group.[183,197,198] Judicious choice of these usually allows the problem to be minimized (see also Section 2.3).

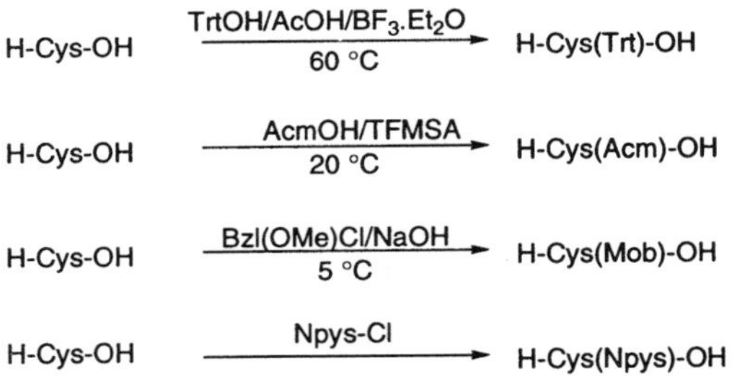

SCHEME 2.21   Methods for the protection of the side chain of Cys.

2.3 ATTACHMENT OF THE FIRST AMINO ACID TO THE SOLID SUPPORT

The union between the peptide and the resin (often called the peptide–resin anchorage) must be stable to the repetitive treatments necessary for removal of the N^α-amino protecting group and for amino acid coupling. At the same time, however, it must be possible to cleave the peptide–resin bond at the end of a synthesis so that the peptide is detached without damage to any of the sensitive structural features it might possess. The peptide–resin anchorage is created when the first amino acid is attached (or "loaded" or "anchored") to the solid support, making this one of the key steps in SPPS. Loading is always accomplished by the formation of either an ester or an amide bond between the solid support and the first amino acid. In the former case, peptide C-terminal acids are ultimately produced and, in the latter, peptide C-terminal amides.

The simplest approach, represented as Method A in Scheme 2.22, is to attach the first amino acid directly to a suitably functionalized polymer. Merrifield's original procedure was to reflux a suspension of the N^α-protected amino acid and a chloromethylpolystyrene resin in ethanol in the presence of a tertiary amine. This has now been superseded by another method[199] that involves treatment of a suspension of the chloromethyl resin in dimethylformamide with the cesium carboxylate of the amino acid. In both cases, an ester bond is formed between the first amino acid and the solid support. For the synthesis of peptide C-terminal amides, the N^α-protected first amino acid is anchored by amide bond formation to an amino-functionalized polystyrene resin.

A more sophisticated alternative is to incorporate a handle (sometimes called a linker or, less commonly, a linkage agent or spacer) between the resin and the first amino acid of the peptide, as in Methods B, C, and D in Scheme 2.22. A handle is a bifunctional compound that allows the first amino acid to be attached to the resin in two discrete steps.[2] This provides more flexibility, since the handle itself may be designed in such a way as to modify the properties of the peptide–resin anchorage, which can be made more (or less) labile to certain reagents, depending upon the type of synthesis that is to be carried out on the polymer. Handles are not usually amino acids, although the use of dehydroalanine residues has been investigated.[200,201] Generally speaking, there are two ways in which a handle can be used in SPPS, depending upon whether the bond between it and the first amino acid is formed on the resin or in solution. In the first case, represented as Method B in Scheme 2.22, the handle is joined chemically (usually by amide bond formation) to the polymeric resin and the first amino acid of the peptide linked to it either by esterification or amidation. Both of the bonds involving the handle are formed on the solid support in this variation. An alternative procedure, however, is to attach the first amino acid to the handle in solution, again either by ester or amide formation. This amino acid handle unit is now called a preformed handle[202] and is incorporated into the resin, again usually by amide bond formation. This is represented as Method C in Scheme 2.22. Here only one bond to the handle is formed under solid-phase conditions. Synthesis with preformed handles is preferred since they can be purified and

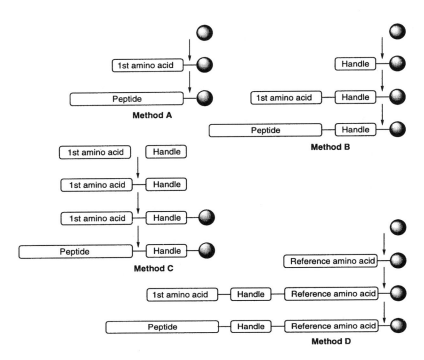

SCHEME 2.22   Different methods for anchoring the first amino acid to a solid support.

characterized prior to attachment to the polymer, allowing the formation of a chemically well-defined peptide–resin anchor.

A further advantage of handles is that they allow an internal reference amino acid to be used.[13,203] This is an amino acid that is directly attached to the solid support and to which the handle is then joined. The first amino acid is now attached to the handle, either on the solid phase or by linking a preformed handle to the internal reference amino acid; peptide synthesis then takes place under normal conditions (Method D in Scheme 2.22). The internal reference amino acid should, ideally, not be present in the peptide to be synthesized, although this is not always possible in longer peptides, unless some non-DNA-encoded residue such as norleucine is used. The amino acid-coupling yields and final cleavage steps in the synthesis can now be monitored more efficiently by amino acid analysis[204] (see Section 2.4.3.3), since the amount of each amino acid incorporated into the peptide can be compared with the amount of reference amino acid, which remains constant throughout. After the cleavage step, the amount of peptide (if any) remaining on the resin can be compared with the amount of internal reference amino acid, allowing yields to be calculated.

2.3.1   Peptide Acids

By far the most common peptide–resin anchorage used for the synthesis of peptide *C*-terminal acids is the substituted benzyl ester. Originally, the peptide was attached to the resin by the benzyl ester formed from the functionalized polystyrene matrix

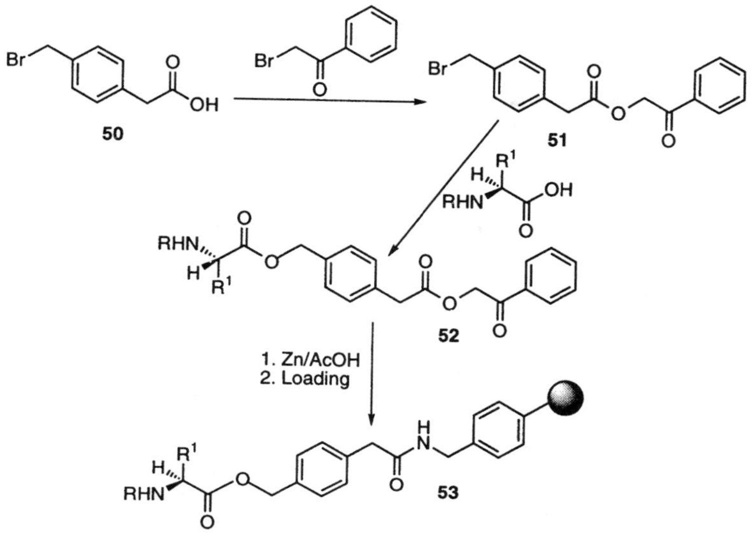

SCHEME 2.23 Preparation of an amino acid Pam resin.

itself. However, these are not sufficiently stable to the repetitive acidolytic treatment necessary to remove the Boc group. Each acidolysis leads to a 1.1% loss of peptide chain from the support.[205] This loss, although unimportant in the synthesis of shorter peptides, becomes a serious problem in the synthesis of larger ones. It can be avoided by increasing the stability of the peptide–resin anchor to the acidic conditions used in synthesis. The most successful approach to accomplishing this has been to use the 4-(hydroxymethyl)phenylacetic acid handle **50** which, when incorporated into an aminomethyl polystyrene solid support, gives rise to the phenylacetamidomethyl resin (commonly known as Pam resin).[205-207] The presence of the electron-withdrawing carboxamidomethyl group in the *para* position of the peptide benzyl ester increases the acid stability of Pam resin approximately 100 times compared with that of standard polystyrene. This, in turn, reduces the loss of peptide per deprotection step to levels that permit the synthesis of large peptides; at the same time the peptide–resin bond of Pam resin may be cleaved by treatment with liquid hydrogen fluoride at 0°C. The Pam linker has become one of the most widely used handles in SPPS. Scheme 2.23 shows the preparation of an amino acid Pam resin.

4-(Bromomethyl)phenylacetic acid **50** is made to react with bromoacetophenone giving the phenacyl ester **51**. The first protected amino acid of the peptide is then esterified onto **51**, giving preformed handle **52**. After removal of its carboxyl protecting group, this is then loaded onto an amino-functionalized solid support, giving the amino acid Pam resin **53**. This procedure is that of Method C in Scheme 2.22. An alternative preparation of **53** can be carried out following Method B in the same scheme. In this case, reaction of a carboxyl-activated derivative of **50** with an amino-functionalized resin attaches the handle. Reaction of this handle–resin with the cesium carboxylate of a protected amino acid then furnishes **53**. Of the two procedures, the former is preferred since it forms an unambiguous product and does not give rise to unreacted hydroxyl groups on the resin, which can cause problems throughout the synthesis.

The Pam resin was designed for the synthesis of free peptide C-terminal acids since the strong acidolytic conditions necessary for cleavage of the peptide also bring about the removal of most, if not all, of the side chain-protecting groups. If the objective is the synthesis of protected peptide C-terminal acids, other handles are available that allow cleavage of the peptide from the resin under milder conditions that allow side chain protection to be maintained. The solid-phase synthesis of protected peptide segments is discussed in more detail in Section 4.1.1.

For the Fmoc/tBu approach, several handles and resins are available for the synthesis of free peptide acids. Among the most commonly used are the 4-alkoxy-benzyl alcohol[208] resin **54** (known as the Wang resin) and the 4-hydroxymethylphe-noxyacetic acid[209] (HMPAA) handle, **55**. An alternative to both is the 3-(4-hydroxy-methylphenoxy)propionic acid[210] handle **56**. In all these cases, the peptide is anchored to the solid support by a benzyl ester that has an alkoxy electron-donating substituent in the *para* position. This has the effect of allowing peptide cleavage to occur under milder conditions. Since in the Fmoc/tBu approach the conditions for the construction of the peptide chain are basic, premature cleavage of the peptide from the resin by acidolysis does not occur. The peptide–resin anchorage can, therefore, be designed in such a way that cleavage can take place under much milder acidolytic conditions than in Boc/Bzl synthesis. Several so-called "super acid labile" handles are available for the synthesis of protected peptide segments using the Fmoc/tBu approach (see Section 4.1.1.1.1).

The anchoring of an amino acid by esterification is somewhat more difficult than by amide bond formation because of the less favorable reaction rates. It must therefore be carried out under conditions that maximize yields but at the same time do not give rise to unacceptable amounts of racemization.[211] The methods currently available offer reasonable compromises between these two factors. Generally speaking, for hydroxymethyl-functionalized handles and supports, the activation of the N^α-protected amino acid with carbodiimides in the presence of catalytic amounts of dimethylaminopyridine[212] or of hydroxybenzotriazole[213] gives good results. However, several other procedures have also been reported. These include the use of amino acid active esters,[214,215] of amino acid mixed anhydrides,[216] of amino acid chlorides[217] or fluorides,[218,219] and of urethane-protected amino acid N-carboxyan-hydrides.[220] For halomethyl-functionalized handles and supports, esterification with the cesium carboxylate of the N^α-protected amino acid concerned[199,221] usually proceeds in satisfactory yield with little, if any, racemization. Fmoc-His and Fmoc-Cys derivatives are, however, particularly difficult to anchor efficiently and can racemize to an unacceptable extent, although the problem is less serious for His derivatives protected at the π-nitrogen. These amino acids can be anchored as their cesium carboxylates,[222] but care must be exercised since cesium carbonate can promote the removal of the Fmoc group. Minimal racemization in the esterification of N^α-Fmoc-protected His and Cys derivatives has also been reported using dicyclohexylcarbo-diimide in the presence of 1-hydroxybenzotriazole[197] or 2,4,6-mesitylene-sulfonyl-3-nitro-1,2,4-triazolide in the presence of 1-methylimidazole.[223]

2.3.2 Peptide Amides

Many naturally occurring peptides have a C-terminal amide, which is often essential for biological activity. The biosynthesis of peptides with this structural characteristic requires the two-step posttranslational enzymatic modification of glycine C-terminal peptides.[224] There is much interest in methods for the chemical synthesis of peptide amides since they cannot be produced by the currently available biotechnological techniques.[225] Furthermore, the production of peptide C-terminal N-alkylamides is also useful for the investigation of structure–activity relationships and for confor-mational studies.

In SPPS, peptide amides have been obtained by ammonolysis of peptide–resins having ester anchorages to the solid support. This is unsatisfactory, however, because of the often low cleavage yields, especially for peptides having sterically hindered C-terminal residues such as Val or Ile. Prolonged exposure to ammonia can also lead to epimerization at one or more stereogenic centers. A much more useful approach is the acidolysis of peptide–resins having a C-terminal amide bond to the solid support. Carboxamides are appreciably more stable to anhydrous acids than esters, so that premature acidolytic loss of peptide chains from the resin in Boc/Bzl synthesis is not a problem. The most common solid supports for the synthesis of peptide amides are benzhydrylamide derivatives[226] such as the 4-methylbenzhydryl-amine resin **58**. This may be synthesized[227] from polystyrene **57** as shown in Scheme 2.24.

The attachment of the first amino acid to **58** is accomplished by amide formation; peptide synthesis is then continued using standard Boc/Bzl SPPS procedures. Cleav-age of the completed peptide amide from the resin can be brought about, normally in good yield, by treatment with hydrogen fluoride. The 4-methylbenzhydrylamine-resin **58** is the most commonly used solid support for the synthesis of peptide C-terminal amides using the Boc/Bzl approach.

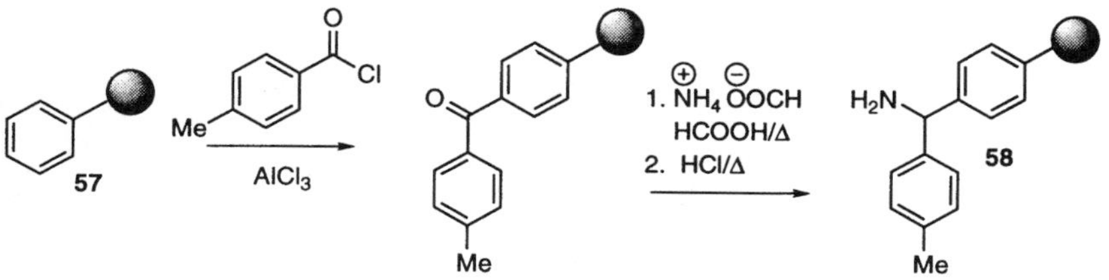

SCHEME 2.24 Preparation of p-methylbenzhydrylamine polystyrene.

For Fmoc/tBu synthesis a wider range of solid supports is available. Several, such as the 4-(2′, 4′-dimethoxyphenylaminomethyl)phenoxymethyl resin **59**, the 4-(4′-methoxybenzhydryl)phenoxyacetic acid **60** and 3-(amino-4-methoxybenzyl)-4, 6-dimethoxyphenylpropionic acid **61** handles, and the 4-succinylamino-2, 2′, 4′-trimethoxybenzhydrylamine (SAMBHA) resin **62**, are benzhydrylamine derivatives.[228-232] The most widely used of these is **59**, which is now commercially available.

Two other options also exist for the Fmoc/tBu synthesis of peptide amides. The 5-(4-aminomethyl-3,5-dimethoxyphenoxy)valeric acid (PAL) **63** and 5-(9-aminox-anthen-2-oxy)valeric acid (XAL) **64** handles[233-235] both give excellent results in practice. The former is commercially available and is perhaps the most useful of all handles for the solid-phase synthesis of peptide C-terminal amides. The latter is more acid labile and is a very promising alternative. For all the resins and handles **59** to **63**, the completed peptide amide can be cleaved on treatment of the peptide–resin with mixtures (typically 50 to 95%) of trifluoroacetic acid in dichloromethane. Peptide amides may be cleaved from resins incorporating the XAL handle **64** by treatments with a 5% solution of this acid in dichloromethane. Several other handles and supports for the Fmoc/tBu synthesis of peptide amides have also been described.[236-238]

2.3.3 Side Chain and Backbone Amide Anchoring

Although the first amino acid of a peptide is usually anchored to the resin through its α-carboxyl group, an alternative in certain cases is attachment at the side chain. For peptides having Asp or Glu at the C-terminus this can easily be done either by amide or ester bond formation to the side chain rather than to the α-carboxylic acid. This is illustrated in Scheme 2.25 where a suitably protected Glu derivative is anchored at its side chain to a resin **65** that may be amino- or hydroxyl-functionalized

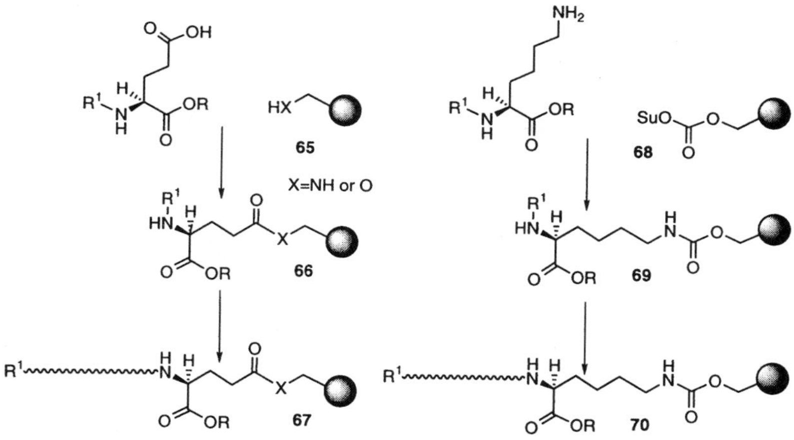

SCHEME 2.25 Side chain anchoring of Glu and Lys.

(X = NH or O, respectively), giving amino acid resin **66**. If the protection for the Glu α-carboxyl group is chosen so as to be stable to the synthesis conditions, then normal chain elongation can take place at the N^α-terminus, giving peptide–resins **67**. This provides a way of making head-to-tail cyclic peptides on the solid support,[239-244] since deprotection of the C-terminus at some suitable point allows on-resin cyclization by amide bond formation with the deprotected N-terminus. Once cyclization has taken place, the peptide is cleaved from the resin by acidolysis and purified. If the initial side chain anchorage of Glu was brought about by amide bond formation, then cleavage leads to a peptide incorporating a Gln residue. If, however, it was brought about by esterification, then a Glu-containing peptide is produced on cleavage. An analogous procedure can, of course, be carried out for side chain anchoring with Asp.

This strategy for the solid-phase synthesis of cyclic peptides is not limited to attachment through Asp or Glu side chains. Modification of the solid support, as, for example, in **68**, allows the side chain of Lys or Orn to serve as attachment points, giving the carbamate amino acid resin **69**. Here again, chain elongation can take place at the N^α-terminus giving peptide–resins **70**, from which head-to-tail cyclic peptide can again be elaborated on the solid support.[245]

A related idea is anchorage of the peptide to the resin through one of its amide bonds. This can be accomplished using a backbone amide linker[246] (BAL). Imide formation of an amino acid ester with aldehyde **70**, followed by reduction and attachment to the solid support, gives amino acid–resin **71**, which is shown in Scheme 2.26.

Chain elongation is then carried out at the N^α-alkylated terminus of **71**. Once the desired sequence has been elaborated, on-resin cyclization allows the production of head-to-tail cyclic peptides. An added feature of backbone amide anchoring is that it allows the facile preparation of a range of different C-terminal functionalities, such as peptide thioacids, thioesters, alcohols, substituted amides, and aldehydes.

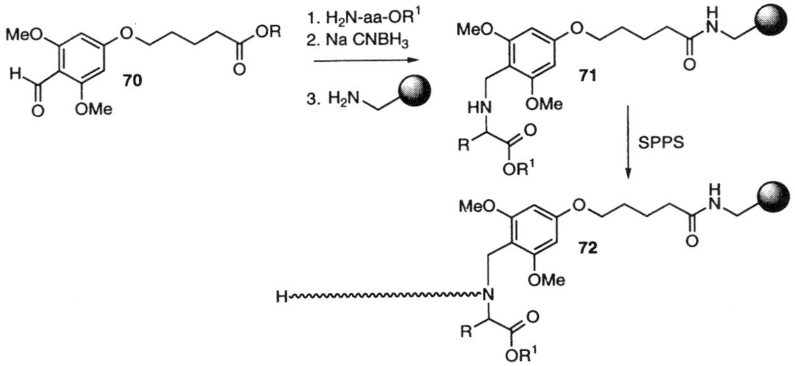

SCHEME 2.26 Backbone amide anchoring.

2.4 CHAIN ELONGATION

Once the first amino acid has been loaded onto the resin by esterification or amidation, the peptide chain is built up by repetitive N^α deprotection and amino acid-coupling reactions. One of the great advantages of the solid-phase method is that these steps, which represent a great part of the drudgery associated with peptide synthesis in solution, can be carried out efficiently by machines. That is not to say that amino acid couplings will always proceed without problems or that side reactions will not occur, but in many cases even quite complex peptides can be synthesized automatically. The coupling methods used in solid-phase chain elongation together with some of the most common side reactions that can intervene are now considered.

2.4.1 Amino Acid Coupling

Most of the chemical methods for coupling amino acids in solution have been evaluated in SPPS. The most popular method in contemporary practice, owing no doubt to its simplicity and rapidity, is to use coupling reagents. Amide bond formation is complete in less than an hour in the majority of cases and is often significantly faster. Other methods for coupling amino acids are notably less popular in SPPS so that preformed mixed anhydrides, symmetrical anhydrides, or active esters are not normally used, unless there is a particular synthetic reason for doing so (see Section 3.3). These species may, however, be formed as transient intermediates in the mixture of compounds that coupling reagents generate.

2.4.1.1 Carbodiimides

Dicyclohexylcarbodiimide (DCC) **73** was introduced[247] into peptide chemistry in 1955 and continues to be the most widely used carbodiimide. Diisopropylcarbodiimide (DIPCDI) **74**, however, is equally effective as a coupling reagent[248] and forms a more soluble urea by-product.

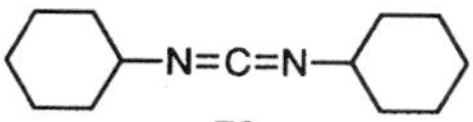

73 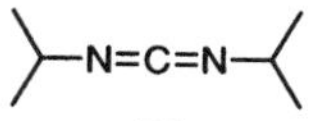   **74**

Addition of the carboxyl group of the N^α-protected amino acid to one of the N=C bonds of the carbodiimide gives the O-acylisourea intermediate **75**, the first active species formed in the coupling reaction.[249-251] Although carbodiimide reagents can form guanidine derivatives with amines, this reaction is usually too slow to compete effectively with the formation of O-acylisoureas.[252,253] The highly reactive **75** can then undergo aminolysis by the amino component, leading to the formation of an amide **76** and the dialkylurea by-product **78**. Alternatively, another molecule of carboxylic acid can react with the O-acylisourea, forming the amino acid symmetrical anhydride **77**. This is another potent acylating species that can also react with the amino component, again leading to amide bond formation.[252,254] These reactions are shown below in Scheme 2.27 for DCC; other carbodiimides function in an analogous fashion.

Although several mechanistic studies have been carried out, it is not clear how much each pathway contributes to any one coupling reaction, but the symmetrical anhydride may well be the main intermediate[255] in SPPS. At the practical level, couplings are usually carried out with ratios of N^α-protected amino acid to DCC of 2:1 or 1:1. If the former stoichiometry is used and the mixture allowed to equilibrate, a solution of the symmetrical anhydride is formed,[252,254] which can then be added to the amino component. If the amino acid-to-DCC ratio is 1:1 then, it has been argued, the symmetrical anhydride will still form because the carboxyl group of the N^α-protected amino acid will react more rapidly with the O-acylisourea than with DCC itself.[256,257] The latter ratio of reactants is, however, less desirable because DCC is basic and can catalyze the decomposition of symmetrical anhydrides. On the other

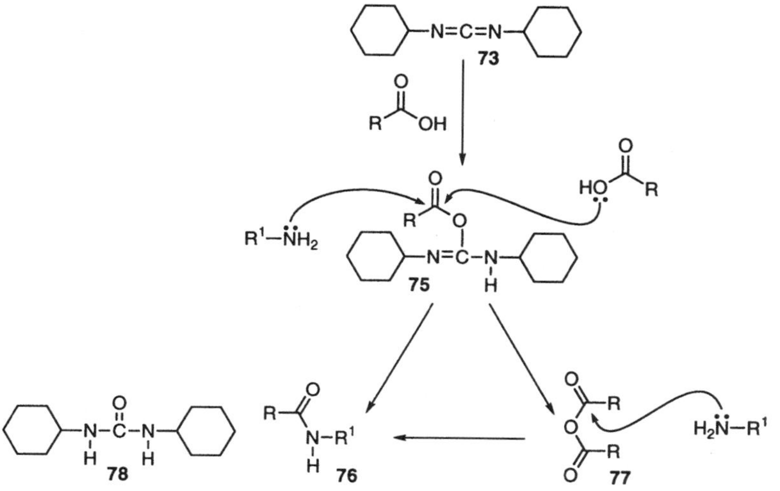

SCHEME 2.27   Mechanism of DCC-mediated coupling.

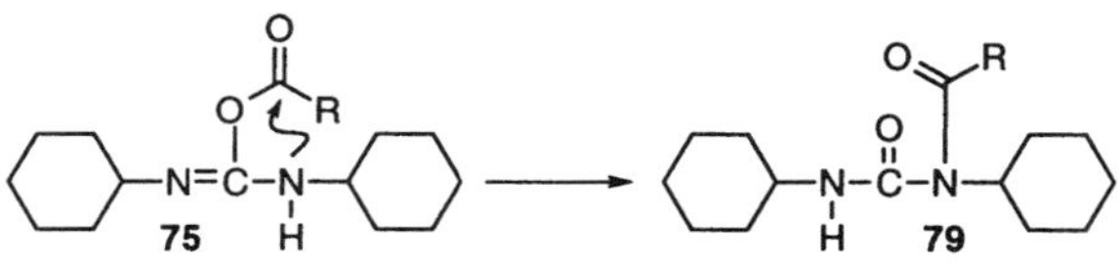

SCHEME 2.28 *N*-Acylurea formation.

hand, a different set of experiments indicates that, in chloroform at least, a 1:1 ratio of protected N^α-amino acid to carbodiimide gives rise to a solution in which only the *O*-acylisourea can be detected, even after prolonged periods of time.[258,259] It should be emphasized that the solvent may well play a key role in determining which mechanistic pathway is favored and the situation obtaining for those couplings carried out in dichloromethane may be quite different from that in dimethylformamide.

Several side reactions can occur when carbodiimide reagents are used, particularly if the coupling reaction is slow for whatever reason. The dehydration of the side chain amide groups of Asn and Gln to the corresponding nitriles has already been discussed (see Section 2.2.2.5). The much less reactive *N*-acylurea derivative **79** can also form,[260-263] either via a four-membered transition state[264] as shown in Scheme 2.28, in which the $O \rightarrow N$ acyl transfer has been reported to be a solvent-independent process,[265] or, alternatively, by reaction of dicyclohexylurea with the amino acid symmetrical anhydride. This latter reaction occurs only slowly in dichloromethane or acetonitrile but more rapidly in dimethylformamide[256,258,259] so that greater amounts of *N*-acylurea are formed when this solvent is used.

Another important side reaction that can occur when carbodiimides are used as coupling reagents is the formation of 5(4H)-oxazolones **80**. These are produced on intramolecular cyclization of the N^α-protected amino acid *O*-acylisourea **75** or anhydride **77**. Such oxazolones[266,267] are themselves acylating agents and can undergo aminolysis, providing an additional alternative pathway for amide bond formation,[268] as shown in Scheme 2.29.

However, the formation of such species in coupling reactions is undesirable because they can lose a proton forming a resonance stabilized anion **81**. This can re-protonate on either side of the plane of the ring leading to racemization of the amino acid, as shown in Scheme 2.30. In the solid-phase coupling of urethane N^α-protected amino acids under normal conditions, racemization by this mechanism

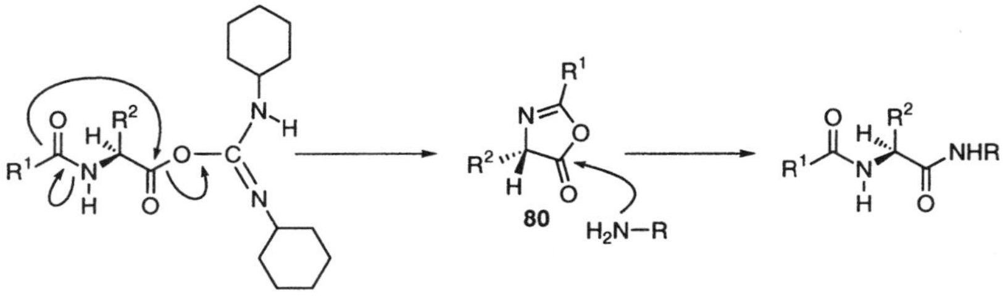

SCHEME 2.29 Coupling via oxazolone formation.

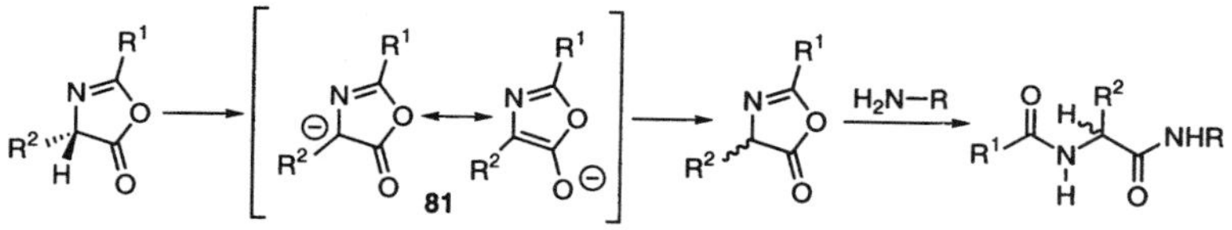

SCHEME 2.30 Racemization as a consequence of oxazolone formation.

is maintained at very low levels. The situation is, however, very different when protected peptide segments are coupled, since epimerization of the C-terminal amino acid can occur readily. Racemization and epimerization in peptide synthesis are treated more fully in Section 3.3.2.1.

Many of the side reactions that occur when activation is carried out with carbodiimides alone can be avoided when *additives* are introduced into the reaction mixture. These compounds intercept the highly reactive O-acylisourea and symmetrical anhydride intermediates, forming an acylating agent of lower reactivity (an active ester, see Section 2.4.1.3), which is still potent enough to allow rapid amide bond formation. The first additive proposed for peptide synthesis[269,270] was N-hydroxy-succinimide **82**. Although it is still used occasionally, it has been superseded by 1-hydroxybenzotriazole **83**, known as HOBt, probably the most widely used compound of this type.[271,272]

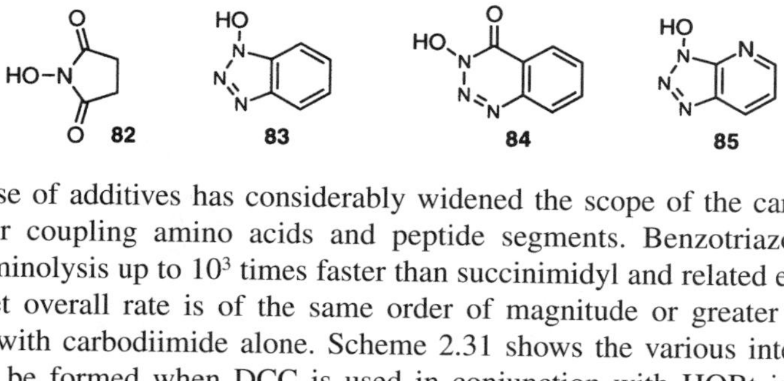

The use of additives has considerably widened the scope of the carbodiimide method for coupling amino acids and peptide segments. Benzotriazolyl esters undergo aminolysis up to 10^3 times faster than succinimidyl and related esters,[271,273] and the net overall rate is of the same order of magnitude or greater as that of couplings with carbodiimide alone. Scheme 2.31 shows the various intermediates thought to be formed when DCC is used in conjunction with HOBt in coupling reactions.

For N^α-protected amino acid benzotriazolyl esters in solution, it has been proposed that the normally represented structure **86** is in equilibrium[271] with the N-oxides **87** and **88**, as shown in Scheme 2.32. With regard to their structure in the solid state, amino acid benzotriazolyl esters have been reported[274-277] to crystallize either as **86** or **88**. The X-ray crystal structure of the benzotriazolyl ester of N-trityl methionine **89** has been determined.[278] The unusually high reactivity of benzotriazolyl esters, compared with esters of phenols having similar pK_a values, has been attributed to intramolecular base catalysis,[272] operating through a six-membered cyclic transition state, as in **90**. Firm evidence for this, however, is somewhat elusive.[274,279]

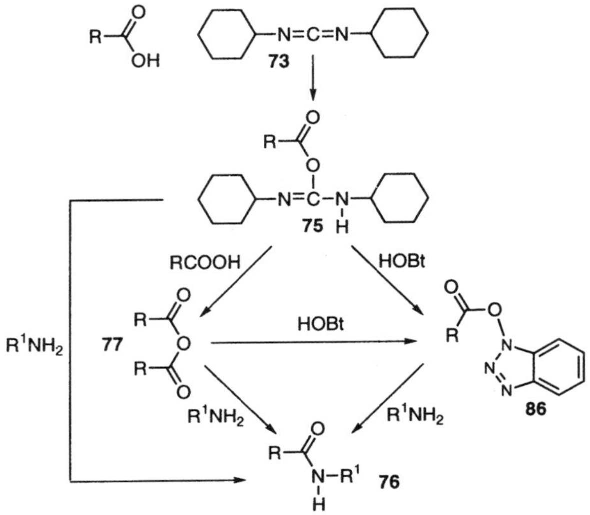

SCHEME 2.31   Mechanism of DCC-mediated coupling in the presence of HOBt.

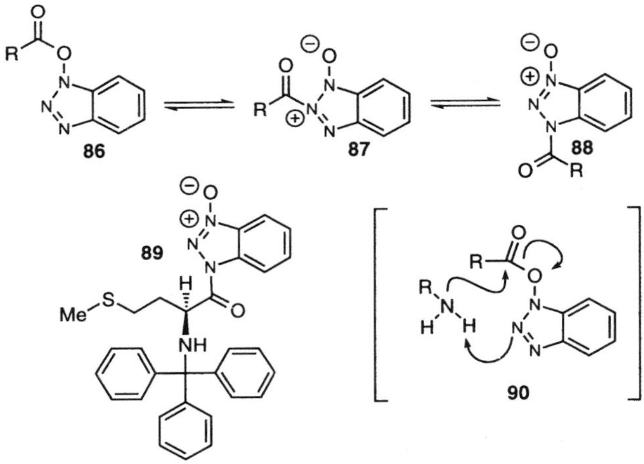

SCHEME 2.32   Structure and reactivity of amino acid benzotriazolyl esters.

Another additive, 3-hydroxy-4-oxo-3,4-dihydro-1,2,3-benzotriazine (HODhbt) **84** was originally reported to be more efficient than HOBt in reducing racemiza-tion.[273,280] Unfortunately, it can give rise to an azide by-product, which reacts readily and irreversibly with N^α-amino groups, in this way blocking or capping them. This reduces its usefulness and it has not achieved the popularity of HOBt, although its effectiveness in reducing racemization in susceptible substrates is beyond doubt,[281] and this makes it useful in certain cases. 4-Oxo-3,4-dihydro-1,2,3-benzotriazinyl esters of N^α-protected amino acid can also be used as preformed species. In this case the azide by-product is removed in the purification process and amine capping in the coupling reactions is avoided.[282] The use of these preformed esters in peptide synthesis is discussed below in Section 2.4.1.3.

A promising new addition to the list of additives is 1-hydroxy-7-azabenzotriazole **85**, known as HOAt. The structure **85** incorporates the components of HOBt and a tertiary base within the same molecule. This compound, used in conjunction with a range of coupling reagents, has been reported[283-285] to accelerate reaction rates significantly, to provide optimal coupling yields, and to reduce racemization levels in susceptible substrates.

Despite the growth in popularity of phosphonium and uronium reagents (see Section 2.4.1.2), carbodiimides continue to be among the most widely used coupling reagents in peptide chemistry. They are equally useful in SPPS and in synthesis in solution, whether used alone or with additives. Active research into the development of new carbodiimides for peptide coupling continues apace.[286,287]

2.4.1.2 Phosphonium and Uronium Reagents

The use of acylphosphonium salts as reactive intermediates for the formation of amide bonds was first postulated by Kenner[288] in 1969. Several procedures that generate these species were subsequently suggested as coupling methods for peptide synthesis,[289-291] including the oxidation–reduction condensation, which is discussed in more detail in Section 3.3.1.5. However, compounds that promoted peptide couplings by the formation of acylphosphonium salts did not become widely used until Castro[292-294] developed the 1*H*-benzotriazol-1-yloxy-tris(dimethylamino) phosphonium hexafluorophosphate (BOP) reagent **91**. Over the last decade the use of phosphonium salts in peptide chemistry has increased dramatically and several such reagents are now commercially available.

91   **92**

The BOP reagent has been used in many syntheses since its introduction; it is easy to use and promotes rapid coupling. A drawback, however, is that it produces the toxic hexamethylphosphorotriamide **94** as a side product, and this has stimulated the search for other compounds that are less hazardous.[295] The most promising of these is the 1*H*-benzotriazol-1-yloxytris(pyrrolidino)phosphonium hexafluorophosphate (PyBOP) reagent **92**, which presents all of the advantages of BOP but which forms a less noxious by-product.

The main mechanistic pathway in coupling reactions with BOP and related compounds is probably that shown in Scheme 2.33. Since the reagent does not react with N^α-amino groups it can be added directly to the amino and carboxyl components that are to be coupled. A tertiary amine is normally used as base to form the carboxylate

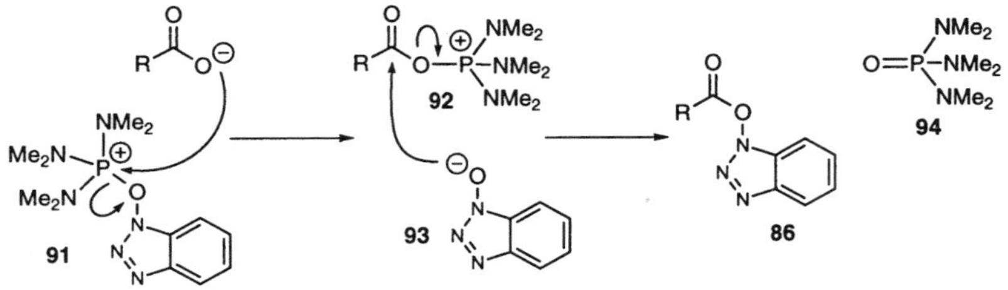

SCHEME 2.33 Mechanism of BOP-mediated coupling.

ion of the carboxyl component. Initial attack of this on the phosphonium salt **91** leads to the acyloxyphosphonium salt **92**. This is extremely reactive and is attacked by the oxyanion of 1-hydroxybenzotriazole **93**, forming the benzotriazolyl ester **86**, thought to be the predominant species suffering aminolysis.[268,294,296] Formation of the amino acid symmetrical anhydride by attack of a second carboxylate on **92** has also been postulated.[268] Whether or not acyloxyphosphonium salts **92** themselves intervene to any appreciable extent in reactions involving these reagents is in some doubt. One study[297] claims that at $-20°C$ **92** is the main reactive intermediate, whereas another[298] affirms that it is always predominantly **86**.

The related uronium salts have also gained acceptance as coupling reagents in SPPS since the first of these was introduced[299,300] in 1978. By analogy with phosphonium salts, such as the BOP reagent, this compound was initially called O-(benzotriazol-1-yl)-1,1,3,3-tetramethyluronium hexafluorophosphate (HBTU), and was assumed to have the structure **95**. Several other uronium reagents were introduced subsequently,[301] one of the most popular being **96**, named O-(benzotriazol-1-yl)-1,1,3,3-tetramethyluronium tetrafluoroborate, and known as TBTU. More recently, a new generation of phosphonium and uronium coupling reagents, based upon HOAt **85** have been developed,[285,302] such as the aza analogue of HBTU, known as HATU. Again, by analogy with the BOP reagent, its structure was initially formulated as **97**. The increased popularity of uronium salts in SPPS has led to a more thorough investigation of their properties, particularly with regard to their structures. X-ray crystallography[303] has shown that both HBTU and HATU crystallize as the guanidinium N-oxides **98** and **99**, respectively, and it is likely that other uronium salts follow suit. This being so, their names must be reformulated as N-[(1H-benzotriazol-1-yl) (dimethylamino)methylene]-N-methylmethanaminium hexafluorophosphate N-oxide in the case of **98**, and as N-[(dimethylamino)-1H-1,2,3-triazolo[4,5-b]pyridin-1-ylmethylene]-N-methylmethanaminium hexafluorophosphate N-oxide for **99**. The acronyms HBTU and HATU are, however, maintained. The structure of these salts in solution may well be subject to the same considerations as outlined above in Section 2.4.1.1 for amino acid benzotriazolyl esters.

Mechanistically, uronium salts are thought to function in a similar way to their phosphonium analogues, as outlined in Scheme 2.33. Unlike carbodiimides or phosphonium salts, however, they do react with amino groups under the conditions commonly employed in peptide synthesis, forming tetramethylguanidinium derivatives.[135,304] Irreversible blocking of the N^α-amino group can therefore take place if the reagent is added directly to the amino component. In order to avoid this, the amino acid to be coupled should be preactivated with the uronium reagent before addition to the component with the free N^α-amino group. The use of excess uronium salt in couplings should also be avoided.

Phosphonium salt- and uronium salt-based coupling reagents are a powerful addition to the peptide chemist's arsenal. They are especially useful for the coupling of sterically hindered amino acids[305-308] and in machine-assisted peptide synthesis where they allow coupling times to be reduced.

2.4.1.3 Preformed Active Esters

Simple alkyl esters undergo aminolysis at rates that are too slow to be useful in peptide synthesis. Esters incorporating good leaving groups as their alcohol component *(active esters)*, however, react much more rapidly and allow amide bonds to be formed between amino acids at room temperature or below. Although very many active esters have been tried in peptide synthesis,[309] only relatively few have been adopted into general practice. These fall into two main categories: substituted aryl esters and *O*-acyl hydroxylamine derivatives.

Substituted aryl esters of N^α-protected amino acids were the first general type of active ester to find widespread use in classical peptide synthesis in solution. The rate of aminolysis of phenyl esters in organic solvents is dependent upon the leaving ability of the ester group which, for such processes, is the rate-limiting step: esters having better leaving groups as their phenol component give faster coupling reactions.[310] The most frequently used are the *p*-nitrophenyl **100**, 2,4,5-trichlorophenyl **101**, and pentafluorophenyl **102** esters, which have proved their worth time and again in peptide synthesis in solution. Their preparation is straightforward[311-313] and is normally done by condensation of the N^α-protected amino acid with the appropriate phenol in the presence of DCC. The compounds are usually easily isolable, stable, crystalline solids. Because they are at a lower level of activation compared with most of the other methods for the formation of peptide bonds discussed so far, active esters generally give rise to fewer or less extensive side reactions during coupling. This lower level of activation, of course, means that slower rates are observed than when other methods are used for carbonyl group activation.

O-acyl hydroxylamine derivatives constitute the second major class of active esters and, generally speaking, are somewhat more reactive than substituted aryl esters. The most common are the succinimidyl **103**, the oxobenzotriazinyl **104**, and benzotriazolyl **86** esters. In addition to being used as preformed species, prepared in a similar way to their substituted aryl counterparts, these esters are also produced *in situ* when the parent alcohols, **82**, **84**, or **83**, respectively, are used as additives in carbodiimide-mediated coupling reactions (see Section 2.4.1.1). The benzotriazolyl active ester **86** is also formed *in situ* when phosphonium- or uronium-based reagents such as BOP, HBTU, or TBTU are used (see Section 2.4.1.2). It is, consequently, particularly important in modern peptide chemistry.

Preformed active esters are not widely used in Boc/Bzl SPPS because of the relatively long coupling times required, compared with those necessary when carbodiimides or phosphonium or uronium reagents are used. However, the preformed esters[314] **102** and **104** have been applied in automated continuous-flow Fmoc/*t*Bu synthesis. In this mode, active esters of the type **104** give rise to bright yellow-colored anions, providing a visual indication of the extent of coupling.[282,315] In the case of esters **102**, their reactivity can be increased[316,317] by the addition of HOBt to the reaction mixture.

Even when carbodiimide or phosphonium or uronium reagents are used for the synthesis of a given peptide, there may be certain types of coupling where active esters are to be recommended. The active esters of Asn and Gln provide an example, since they allow the dehydration-free incorporation of these residues, without the need for side chain protection.[134,135] Once incorporated, the synthesis can be continued with coupling reagents, if preferred. Active esters can also be used for the attachment of the first amino acid to the solid support where this involves an esterification reaction. Although reaction times are slower than with carbodiimides in the presence of dimethylaminopyridine catalysis, the risks of racemization are greatly reduced.[214,215] The milder reaction conditions associated with the use of active esters also allow certain amino acids such as Ser, Thr, and Tyr to be incorporated without side chain protection, if so desired.[318,319]

2.4.1.4   Preformed Anhydrides

Preformed symmetrical anhydrides (PSAs) are highly reactive and may be generated[127,320,321] from the corresponding N^α-protected amino acid and DCC in a

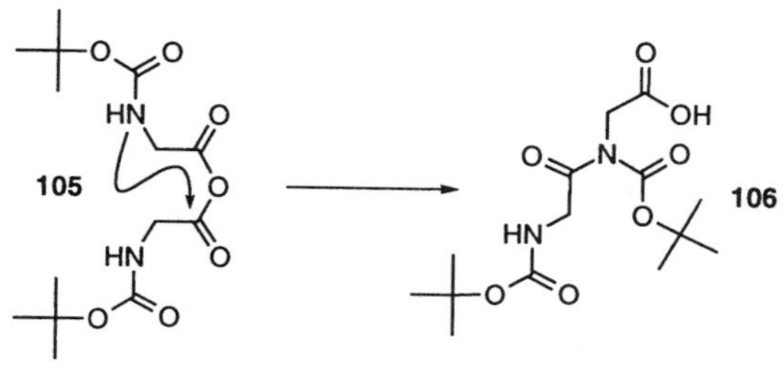

SCHEME 2.34 Formation of dipeptide derivatives from Boc-N^α-protected amino acid PSAs.

ratio of 2:1 (see Section 2.4.1.1). The mixture is allowed to stand, the urea removed by filtration, and the resulting solution of the anhydride can then be added directly to the peptide–resin. Alternatively, the solvent (usually dichloromethane) can be removed and the anhydride redissolved in dimethylformamide if preferred. PSAs promote rapid and clean coupling and are quite widely used. They are a good alternative to coupling reagents or active esters, but, since they are highly activated species, care must be exercised when using them systematically throughout a synthesis.

The method is inherently wasteful since one equivalent of N^α-protected amino acid derivative is not available for acylation. The ready availability of Boc- and Fmoc-protected amino acids today, however, means that this factor is not as important as it once was. More importantly, certain amino acids undergo side reactions, sometimes to an unacceptable extent when coupled as PSAs. In the case of Boc or Fmoc Gly or Ala, dipeptide derivatives can be incorporated into the growing peptide chain.[322-324] These are formed as shown in Scheme 2.34. The symmetrical anhydride of N^α-Boc-protected glycine **105** rearranges to dipeptide **106**, which then undergoes coupling with the amino component. Fortunately, this side reaction is only important for amino acids having the least-bulky side chains.

Other residue-specific side reactions that can occur when PSAs are used include δ-lactam formation with Arg, racemization of N^τ-protected His derivatives, and dehydration of the unprotected side chains of Asn and Gln. A further drawback in Fmoc/tBu synthesis is that the solubility of certain Fmoc-N^α-protected symmetrical anhydrides is poor, being even less soluble than the parent amino acid.[325] These problems are much less important when symmetrical anhydrides are generated *in situ* by coupling reagents, especially if additives are used. Along with 1-hydroxybenzo-triazole active esters, they are among the most important activated derivatives in modern peptide synthesis.

Other preformed anhydrides, such as mixed carboxylic or mixed carbonic anhydrides, are not commonly used in SPPS. They are more frequently applied in synthesis in solution and are discussed more fully in Section 3.3.1.2. However, N^α-urethane-protected N-carboxyanhydrides (see also Section 3.3.1.2), known as UNCAs, have been applied in solid-phase synthesis and are potentially very useful.[326] The advantage of these compounds is that they are highly active species that produce only carbon dioxide as a by-product. The preparation of the Boc **107** and Fmoc **108** derivatives is reasonably straightforward.[327-330] The compounds are, generally speaking, crystalline solids that are stable in the absence of water for extended periods

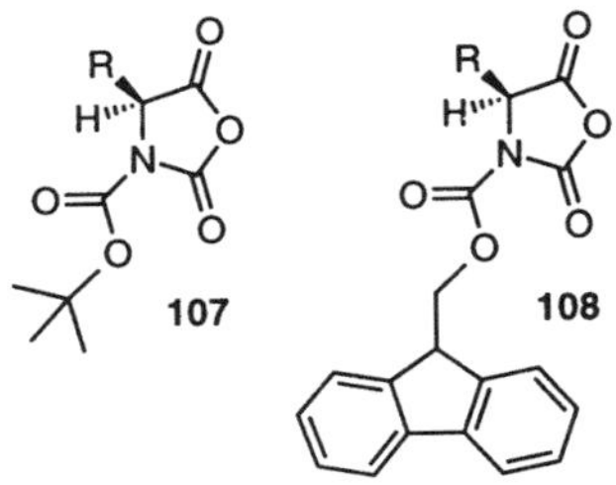

107   **108**

of time at low temperatures.[331] UNCAs have been used in several solid-phase syntheses, including acyl carrier protein (65-74) decapeptide,[332] and an LH-RH analogue.[333] Their use for the solid-phase synthesis of sterically hindered peptides has also been described.[334] Solid-phase coupling with UNCAs, with respect to reaction rates and to the amount of racemization observed, has been shown to be comparable with couplings carried out with the BOP or TBTU reagents.[335] UNCAs are very promising reagents that may well become more popular in the future.

2.4.1.5 Amino Acid Halides

The use of N^α-protected amino acid halides (see also Section 3.3.1.6) for coupling is attractive because the carboxyl group is strongly activated and, consequently, coupling reactions are rapid. However, amino acid chlorides cannot be applied in either Boc/Bzl or Fmoc/tBu SPPS because the *tert*-butyl- and trityl-based protecting groups are not stable to the conditions required for their preparation. However, since the Fmoc group withstands even thionyl chloride treatment, Fmoc-N^α-protected amino acid chlorides **109** can be used in peptide synthesis,[336,337] as long as the side chains are not protected with tertiary carbon-based protecting groups. This constraint does, of course, limit the usefulness of such derivatives in SPPS.

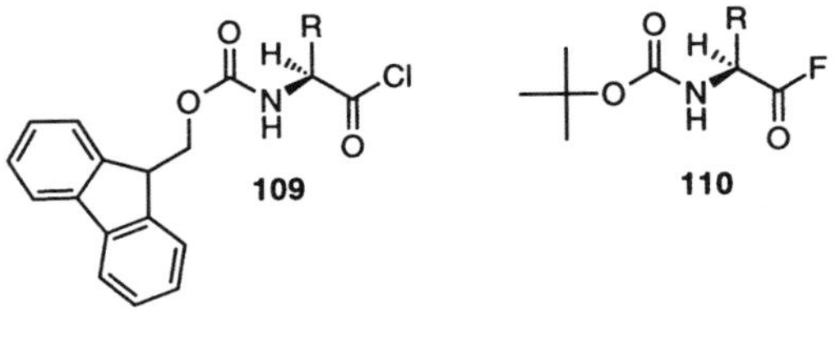

109   **110**

More recently, however, amino acid fluorides **110** have been developed as reactive intermediates. Unlike N^α-protected amino acid chlorides, their use is compatible with both the Boc/Bzl and Fmoc/tBu approaches.[338,339] Amino acid fluorides may be prepared by treatment of suitably protected derivatives with cyanuric fluoride,[340] or with (diethylamino)sulfur trifluoride.[341] Alternatively, the amino acid fluoride-forming coupling reagent tetramethylfluoroformamidinium hexafluorophosphate (TTFH) reagent may be used to generate them *in situ*.[342] Couplings proceed rapidly with very little racemization, even in the absence of tertiary amine base,[343] and with sterically hindered amino acids.[344,345] The high activation provided by these compounds, together with their excellent solubility and stability in dimethylformamide, also makes them useful in multiple peptide synthesis.[346,347]

2.4.2 Problems During Chain Elongation

Several residue- or sequence-specific side reactions can occur during SPPS even though global protection schemes are used. Different side reactions may occur to different extents, depending upon the peptide to be synthesized and upon whether Boc/Bzl or Fmoc/tBu synthesis is applied. The most important problems that can occur during chain assembly are considered below.

2.4.2.1 Undesired Cyclizations

2.4.2.1.1 Formation of Diketopiperazines

The solid-phase synthesis of peptide C-terminal acids sometimes requires special protocols for the coupling of the third amino. In the absence of these, the free amino group of the resin-bound dipeptide **111** can attack the peptide–resin anchorage intramolecularly,[348] giving rise to cyclic dipeptides, **112**, known as diketopiperazines.[349] Their formation is kinetically and thermodynamically favored and causes the loss of resin-bound dipeptide together with low incorporation of the third amino acid. Additionally, hydroxyl sites are generated on the solid support **113**, and these can subsequently cause other side reactions during synthesis (see Section 2.4.2.2.1). The formation of diketopiperazines is shown in Scheme 2.35.

Several factors govern the extent of the formation of diketopiperazines in SPPS. The most important are the type of peptide–resin anchorage present, the nature of amino acids of the dipeptide, and the coupling method used. Intramolecular cyclization will be favored by the presence of good leaving groups at the peptide–resin anchorage. When the peptide is bound to the resin as a benzyl ester, as in **111**, this side reaction can be quite severe in certain cases. The presence of electron-withdrawing groups on the aromatic ring, as in **114**, will tend to exacerbate the problem. On the other hand, diketopiperazine formation is suppressed when the C-terminal

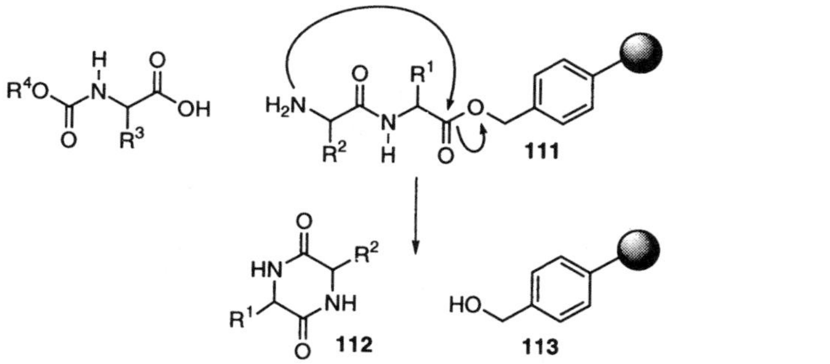

SCHEME 2.35 Formation of diketopiperazines on incorporation of the third amino acid.

peptide–resin anchorage has a large steric requirement, such as, for example, in the case of trityl-based **115** or *tert*-butyl-based **116** resins.[350-352] In the synthesis of peptide *C*-terminal amides, on the other hand, very little diketopiperazine formation occurs, because benzylamines are such poor leaving groups.

The nature of the first and second amino acids of the peptide also has an important influence on the amount of diketopiperazine formed. Amino acids that can easily adopt an amide bond having the *cis*-configuration, in either the first or second positions, will cause greater problems. The presence of Pro, of Gly, or of *N*-alkyl amino acids can, therefore, lead to substantial amounts of diketopiperazine formation. If such amino acids are at both the first and second positions, then cyclization can be severe.[353,354] Another unfavorable combination is to have one L- and one D-amino acid in the dipeptide, as in **117**. Stereochemical considerations now place the amino acid side chains on opposite sides of the plane of the transition state leading to the formation of the six-membered ring. There is, consequently, less steric hindrance to cyclization,[348,355-358] which gives the diketopiperazine **118**, as shown in Scheme 2.36. The nature of the incoming third amino acid is less important, but any effect, such as, for example, steric hindrance, which acts as a brake on the coupling reaction, will tend to favor intramolecular aminolysis over the desired intermolecular process.

The formation of diketopiperazines is catalyzed by weak acids[348,359] or by bases.[353,354,357,360,361] However, except in the most severe cases, the side reaction can be controlled by adopting special protocols either for the removal of the N^α-protecting group of the dipeptide–resin, or for coupling of the third amino acid, or both. In Boc/Bzl synthesis, acid-catalyzed diketopiperazine formation may be provoked on contact between the neutralized dipeptide–resin and the carboxyl group of the third N^α-protected amino acid. This can be minimized by addition of the carbodiimide coupling reagent to the resin, before addition of the third amino acid, a procedure known as *reverse addition*.[348] Alternatively, for more severe cases, diketopiperazine formation can be suppressed using coupling methods in which neutralization of the dipeptide–resin is done *in situ* during coupling of the third amino acid, rather than in a separate wash step with tertiary amine. Several such protocols have been reported, in which the N^α-protected amino acid, tertiary amine

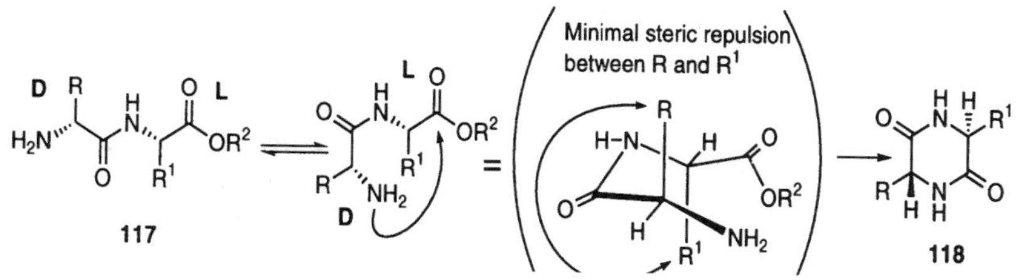

SCHEME 2.36 Mechanism of diketopiperazine formation.

base, and coupling reagent are added together to the unneutralized dipeptide–resin.[358,362,363]

In Fmoc/tBu synthesis the basic conditions used throughout chain assembly mean that coupling with *in situ* neutralization is impossible and appreciable loss of cyclic dipeptide from the resin on coupling of the third amino acid can sometimes occur. Solutions to the problem tend to concentrate on modification of the method for Fmoc group removal at the dipeptide stage. When piperidine in dimethylformamide is used, a shorter reaction time[357] can alleviate the problem. Fmoc group removal with tetrabutylammonium fluoride quenched with methanol has also been recommended.[57] Furthermore, in the Fmoc/tBu approach, special peptide–resin anchors, such as **115** and **116**, are available whose bulk impedes intramolecular cyclization of the resin-bound dipeptide.

In the most severe cases (the Pro-Pro sequence being a particularly troublesome example) intramolecular aminolysis cannot be avoided and the only recourse is to incorporate the second and third amino acids as a protected dipeptide.[312,348,353,354,364-366] The critical N^α deprotection, neutralization, and coupling steps are then avoided. This does, however, require that the requisite dipeptide be synthesized in solution and purified, making it more time-consuming. Additionally, the risks of epimerization at the *C*-terminal amino acid of the dipeptide on coupling are substantially greater, and care must be taken to reduce this hazard. Epimerization and racemization in peptide synthesis, together with preventative methods, are treated more fully in Section 3.3.2.

2.4.2.1.2 Formation of Aspartimides

The cyclization of aspartic acid residues to form aspartimides is a common side reaction in peptide synthesis. Fortunately, it is only troublesome when Asp-Gly, Asp-Ala, and Asp-Ser sequences are present.[144] With other amino acids the steric hindrance to cyclization usually impedes the reaction, although the Asp-Asn sequence is problematic[367,368] in Fmoc/tBu SPPS. An analogous cyclization can, but does not often, occur with Glu, leading to the formation of the corresponding glutarimides.[369,370] Both cases are undesirable because subsequent hydrolysis of the imide-containing peptide leads to a mixture of the desired peptide and a product (sometimes called a β-peptide) in which the side chain carboxyl group forms part of the peptide backbone. These processes are shown in Scheme 2.37, in which an Asp-Gly-containing peptide **117** undergoes aspartimide formation. On hydrolysis, aspartimide **118** gives a mixture of both the α- and β-peptides, **119** and **120**, respectively.

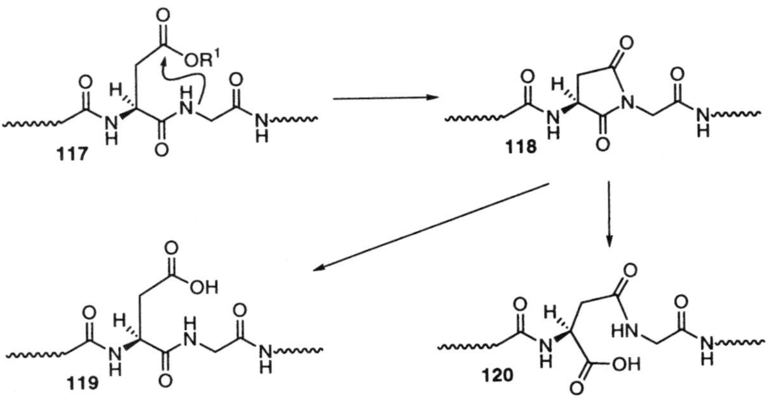

SCHEME 2.37   Aspartimide formation and its consequences in peptide synthesis.

The formation of aspartimides can occur under acidic or basic conditions. Acid-catalyzed aspartimide formation is one of the problems encountered when peptides are cleaved from the resin by strong acid treatment at the end of a synthesis (see Section 2.5). Aspartimide formation during chain elongation can be minimized in the Boc/Bzl approach by using cyclohexyl esters[144,371] as protecting groups for Asp. In Fmoc/tBu synthesis the Asp tert-butyl ester normally provides sufficient protection, although occasionally aspartimide formation can be considerable. Attack of piperidine, used for Fmoc group deprotection, on the aspartimide **118** can then lead to the formation of peptide piperidides, **121** and **122**.

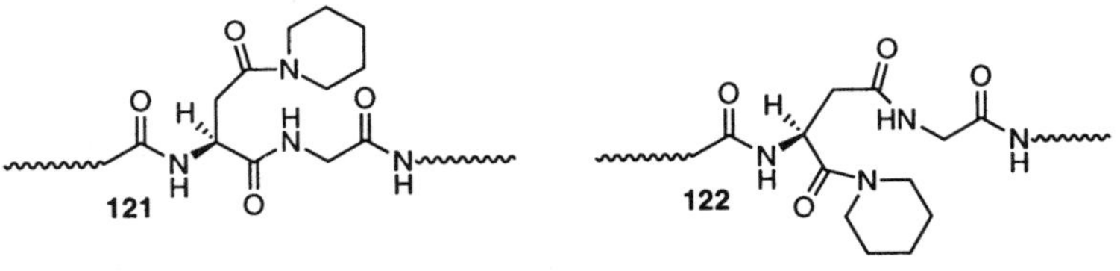

The Asp(tBu)-Asn(Trt) sequence is particularly prone to suffer aspartimide formation under standard Fmoc/tBu synthesis conditions, and significant amounts of piperidides can be produced during chain elongation unless precautions are taken.[367,368,372] The addition of HOBt or 2,4-dinitrophenol to the piperidine solution used for Fmoc group removal can alleviate the problem. A surer method for avoiding the side reaction altogether is to protect the aspartyl amide bond with the 2-hydroxy-4-methoxybenzyl (Hmb) group,[373] as in **123**. This amide bond protection is easily introduced into a peptide by incorporation of an amino acid derivative such as **124**.

Aspartimide formation is now suppressed by the steric hindrance to nucleophilic attack at the Asp side chain, caused by the presence of the bulky amide bond protection. Further uses of Hmb amide bond protection are discussed in Section 2.4.2.2.2 and in Section 2.6.4.

2.4.2.1.3 Formation of Pyroglutamic Acid

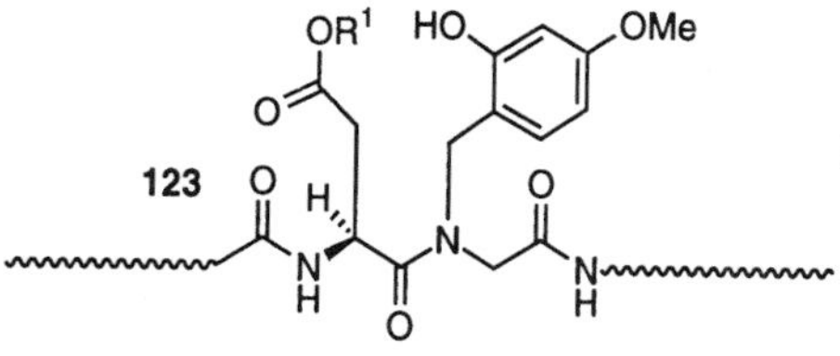

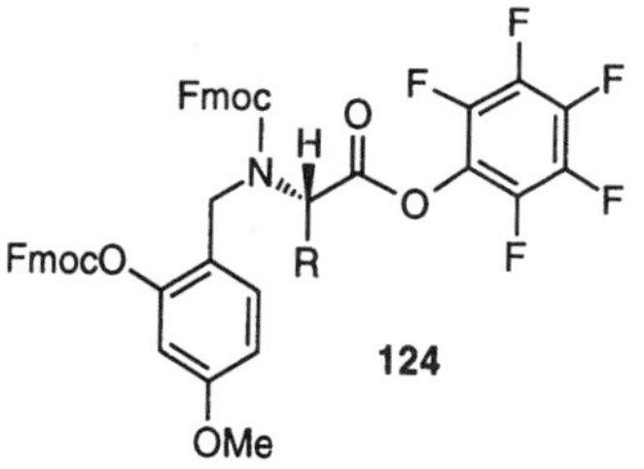

Intramolecular cyclization of *N*-terminal Gln can lead to the formation of pyro-glutamic acid residues. This cyclization is catalyzed by weak acids[374,375] and can occur on Boc group removal with trifluoroacetic acid. However, it is more commonly observed on coupling of the next N^α-protected amino acid to Gln. 1-Hydroxyben-zotriazole should not be used in such couplings since it, together with the N^α-protected amino acid itself, provides the weakly acidic species necessary to catalyze cyclization. Pyroglutamyl formation is shown in Scheme 2.38; the *N*-terminal Gln-containing peptide **125** gives rise to the pyroglutamyl peptide **126**. This peptide is now effectively capped and can take no further part in chain elongation.

Such cyclization can usually be avoided if hydrogen chloride acid in dioxane is used for the removal of the Boc group, followed by reverse addition[348] of the coupling reagent and the N^α-protected amino acid to the peptide–resin. Other pos-sibilities are to incorporate the next amino acid of the sequence by coupling it as the PSA or preformed active ester.[374] Alternatively, coupling protocols relying upon *in situ* neutralization of the peptide–resin, after removal of the Gln Boc group, can

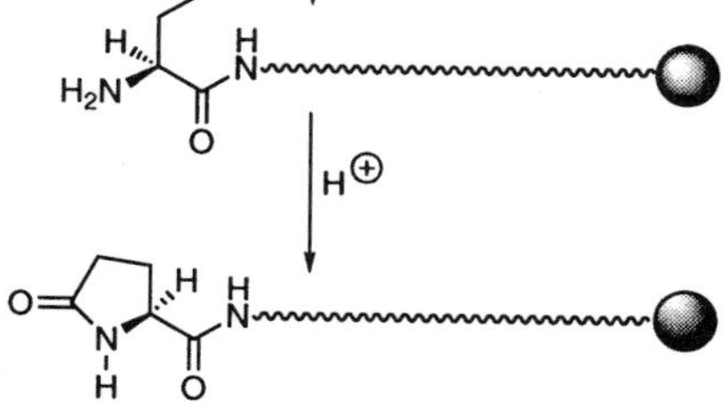

SCHEME 2.38 Capping by pyroglutamyl formation in SPPS.

also be effective. Pyroglutamyl formation is not normally an issue in Fmoc/*t*Bu synthesis, since basic conditions are maintained throughout chain elongation.

2.4.2.2 Incomplete N^α Deprotection and Amino Acid Coupling Reactions

2.4.2.2.1 *Deletion Peptides and Truncated Sequences*

The failure of N^α-deprotection reactions or of amino acid–coupling steps to proceed in optimal yield, either because the N^α-protecting group is not removed from all of the N^α-termini or because not all of the liberated N^α-amine couples with the next amino acid, causes the same effect: some of the resin-bound peptide chains do not incorporate a given amino acid. However, the *N*-termini of these chains may subsequently undergo both deprotection and coupling reactions normally, with the net result that peptides lacking in one, or more, amino acid residues are formed. Such peptides are called deletion peptides.

If the N^α-termini of some peptide chains become irreversibly blocked, further elongation is prevented, and a truncated sequence results. Apart from the deliberate capping of any unreacted amino groups that might remain after coupling reactions, irreversible blocking of N^α-amino groups can occur in other ways, one of the best known being by trifluoroacetylation. This can occur if free hydroxyl sites (generated if esterification of the first amino acid to hydroxymethyl resins does not go to completion, or if the acidolysis of some ester peptide–resin anchorages occurs on Boc group removal) are present on the solid support (see also Section 2.3). These hydroxyl groups are easily trifluoroacetylated and transfer of the trifluoroacetyl groups to the *N*-terminus of nearby peptide chains, as in **127**, can then occur, when the peptide–resin is neutralized after Boc-group removal.[376] These trifluoroacetylated peptide chains are effectively capped and the hydroxymethyl group is regenerated, as in **128** in Scheme 2.39.

Irrespective of their modes of formation, the truncated sequences and deletion peptides formed during SPPS remain bound to the resin throughout the synthesis and are cleaved, along with the target peptide, in the final acidolysis. In the early days of SPPS these problems were held to be insurmountable. It was argued that peptide synthesis on a solid support would always produce mixtures of peptides since it would be impossible to completely avoid the premature termination of some of the chains or to achieve quantitative yields in all N^α-deprotection reactions and amino acid–coupling steps. The deletion peptides, so the reasoning went, would be impossible to separate from the target peptide because their chromatographic properties would be nearly identical to those of the desired peptide. The success of the solid-phase method has clearly demonstrated that this view of events was unduly pessimistic, and there are two main reasons why this is so. First, it has been possible to refine the chemistry involved in SPPS to a superlative degree so that truncated sequences can often be avoided or at least reduced to a minimum. Furthermore, both the N^α-deprotection steps and the amino acid–coupling reactions can usually be made to proceed in yields which, to all intents and purposes, are quantitative. Second,

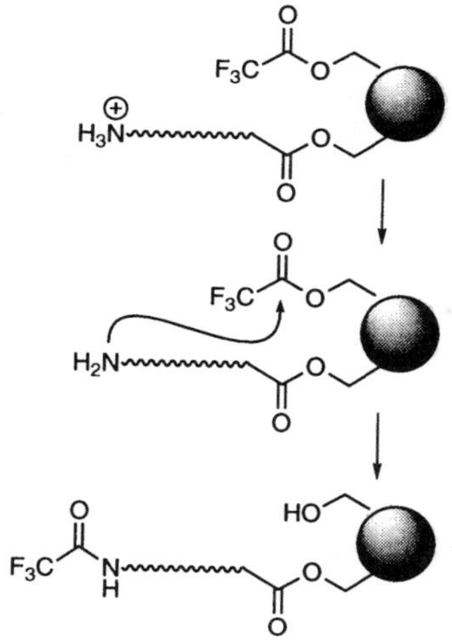

SCHEME 2.39 Capping by trifluoroacetylation in SPPS.

the enormous advances in liquid chromatographic techniques, particularly in high-performance liquid chromatography, which have taken place in parallel with the development of SPPS, mean that peptide chemists can be reasonably confident of isolating the target molecule as a single, homogeneous species.

2.4.2.2.2 Difficult Sequences

N^α-Deprotection steps and amino acid–coupling reactions that do not go to completion can, however, cause more intractable problems. The remedy for sluggish N^α-deprotection reactions can often be the use of longer reaction times and/or increased reagent strength. Nevertheless, in some stubborn cases, these measures are not sufficient and the N^α-protecting group cannot be removed or is removed in low yield. Similarly, even though deprotection may be achieved, coupling of the next amino acid of the sequence cannot be brought about satisfactorily, even by longer reaction times or by increased excess of acylation agent. The failure of these reactions is often sequence related and has been described[377] as the most serious potential problem facing contemporary SPPS. When N^α-deprotection reactions and/or amino acid–coupling steps do not go to completion, or proceed in low yields, for a series of amino acids in a peptide, then these amino acids are said to constitute a difficult sequence.[378-381]

The cause is thought to be the on-resin intermolecular association of the protected peptide intermediates.[382] The available evidence, mainly from gel-phase nuclear magnetic resonance spectroscopy[383,384] of peptide–resin suspensions,[7,385,386] suggests that, for some sequences, the peptide chains are prone to undergo self-association by hydrogen bonding. This leads to aggregation and to the formation of β-sheet-like secondary structures that can render the N-terminal amino acid inaccessible to the reagents necessary to continue peptide synthesis. Peptides having such sequences, consequently, often cannot be satisfactorily synthesized by conventional linear SPPS, and severe cases may require a different type of synthetic strategy. This might involve attempting the synthesis by a convergent solid-phase approach (see Chapter 4) or even by segment coupling in solution (see Chapter 3). The occurrence of difficult sequences has, however, stimulated much research into ways in which the hydrogen bonding that stabilizes the secondary structures involved can be disrupted, improving the solvation of the polymer-bound protected peptide, so that linear SPPS can be applied.

Two general methods have been receiving attention. The first of these is to bring about a change in the environment in which peptide synthesis is carried out. This can be done by using solvents such as dimethylformamide, trifluoroethanol, hexafluoroisopropanol, or, alternatively, by adding chaotropic salts.[387-392] Improvements both in yield and in quality of the final product have been noted in such cases. The optimized synthesis protocols developed by Kent[307] for Boc/Bzl SPPS are also thought to provide a method for circumventing the difficult sequences problem. In conventional Boc/Bzl synthesis aggregation is thought to occur predominantly when the peptide–resin is neutralized in the separate base treatment prior to the coupling reaction. In Kent's method, neutralization is carried out *in situ,* once the amino acid derivative and the coupling reagent have been added to the peptide–resin after Boc group removal, and it has been speculated that the presence of charged ammonium salts on the resin disfavors aggregation.

A different approach is to bring about a structural change in the peptide itself by the protection of some of its amide bonds, giving rise to peptides containing tertiary amides at periodic intervals. This again has the effect of disrupting hydrogen bonding, leading to better solvation of the peptide chain and to more-efficient deprotection and coupling reactions. Several methods have been proposed[393-398] for such amide bond protection. The most promising, however, is the use of the Hmb group[399-401] that can be introduced into peptides synthesized using the Fmoc/*t*Bu approach, by incorporating the amino acid derivative **124** at strategic points in the peptide chain.

This is normally done with Gly (**124**, R=H), but other amino acids can also be incorporated as their *N*-(*O*-FmocHmb) derivatives, if necessary. Although the alkylated N^{α}-amino group of Gly or of other amino acids is somewhat more hindered than usual, coupling of the next Fmoc-N^{α}-protected amino acid can normally be made to proceed in good yield by using longer reaction times. Experimental observations demonstrate that Hmb protection is only necessary at approximately every seventh residue and that aggregation does not occur after about 21 residues[378] so that no further amide bond protection is necessary beyond this point. The Hmb group is removed, along with side chain-protecting groups, on acidolytic cleavage of the peptide from the resin using trifluoroacetic acid.

Finally, to place the difficult sequences problem in perspective, the improvements that have taken place in SPPS since its invention up to the present are nicely illustrated by the case of the acyl carrier protein (64-75) decapeptide. The synthesis of this peptide failed using early Boc/Bzl SPPS on polystyrene supports,[402] but it was later synthesized successfully on a polyamide resin using the Fmoc/tBu approach.[13] Later improvements in Boc/Bzl chemistry allowed this peptide to be synthesized on polystyrene,[403] and at present the synthesis of this sequence has been refined to such an extent that it has become a test peptide for SPPS methodological improvements,[401] demonstrating that what was once a difficult sequence is now a standard synthesis for automatic peptide synthesizers. Further improvements in synthetic methodology, as outlined above, can be expected to allow the routine synthesis of other difficult sequences.

2.4.2.3 Other Problems in Chain Assembly

Several other problems can occur during chain elongation, but they can usually be overcome by modification of the reaction conditions. Among those that might arise are the loss of peptide from the resin during synthesis. This can occur in Boc/Bzl synthesis if standard Merrifield polystyrene solid supports are used for long peptides. The repetitive acidolyses bring about a progressive detachment of the peptide from the resin, calculated to be of the order of 1% every cycle.[205] If the Pam resin (see Section 2.3) is used, this loss can be avoided. Resins that are not stable to nucleophiles (see Section 4.1.1.2) may also undergo similar loss of peptide chains on resin neutralization after Boc group removal, since the free N^α-amino group is a sufficiently potent nucleophile to provoke this. Such cases can normally be solved by appropriate choice of synthesis strategy and of solid support.

tert-Butylation of the aromatic side chain of Trp and of the sulfur atom in Met and Cys can occur during acidolytic Boc group removal, if side chain protection is lacking. In the case of unprotected Trp, alkylation can be minimized by the judicious addition of scavengers to the solution used to effect Boc group removal.[125,404] The *tert*-butylation of Met is reversible,[167] so that under normal circumstances it is not especially problematic, although it may become so in peptides containing several Met residues. The alkylation of Met and Trp does not occur during Fmoc/tBu synthesis, so that these amino acids can be incorporated without side chain protection. This has led some workers to favor this synthetic approach for Met- and Trp-containing peptides.

Although N^α-protected derivatives of Val and Ile do not undergo side reactions in themselves, steric hindrance, caused by the branching at the β-carbon atom, can retard couplings and consequently favor side reactions in other parts of the peptide under construction; similar effects are observed for Thr. In critical cases, larger excesses of coupling reagent and of these amino acids can be used in order to speed up the coupling reaction at the expense of the side reaction concerned.

2.4.3 On-Resin Monitoring of SPPS

In SPPS the isolation, purification, and characterization of intermediates is impossible. Furthermore, the monitoring of the progress of the synthesis by techniques such as thin-layer chromatography or high-performance liquid chromatography is also impossible, without first cleaving the peptide from an aliquot of the resin.

Methods for the real-time monitoring of SPPS are highly desirable, but at present it is only feasible in the continuous-flow Fmoc/tBu approach.[405,406] Here, both Fmoc group removal and amino acid–coupling reactions can be checked, as long as the latter are carried out using preformed ODhbt esters. However, although it gives valuable information on the progress of a synthesis, it is not accurate enough for detecting the very low levels of incomplete N^α-deprotection or incomplete amino acid coupling that can still produce serious separation problems when the peptide is cleaved from the resin. Furthermore, it gives no information on the majority of the side reactions that might occur. In Boc/Bzl synthesis, there is no effective way of monitoring the removal of the Boc group so that real-time synthesis control is even less developed.

It is, therefore, important to obtain as much information as possible by analysis of the resin after coupling steps. There is a series of measures that can, and indeed should, be adopted in order to evaluate the quality of a synthesis, before the peptide is cleaved from the solid support.

2.4.3.1   Peptide–Resin Mass

As amino acids are incorporated, the mass of the peptide–resin increases. If the initial degree of functionalization of the resin is known, then the theoretical weight of the peptide–resin after any given coupling step can be calculated and compared with its observed weight after drying. Although, in principle, this procedure could be carried out after each and every coupling reaction, this is rarely done. The dried peptide–resin is usually only weighed at the end of the synthesis, before the peptide is detached from the solid support. If it weighs appreciably less than the theoretical weight, it is indicative of the failure of one or more of the amino acid incorporation cycles to go to completion.

While the method gives a useful rough idea of the quality of a synthesis, an exact interpretation of the data is not necessarily straightforward. If the peptide–resin mass is much less than that expected, then this clearly indicates a problem, although it gives no information as to its nature. If the peptide–resin mass is approximately the expected value (75% of theoretical, or above, is considered "acceptable"), then this is an indication that the synthesis has proceeded according to plan. More difficult to interpret (and more common in practice) are values that fall somewhere between these two extremes. A value of 50% or less is symptomatic of problems in the synthesis, but does not mean that the desired product cannot be isolated. This will depend upon the type of problem and the efficiency of the purification procedures available. In any case, in order to judge more accurately the quality of the synthesis, additional information must be obtained.

2.4.3.2   The Ninhydrin Test

The ninhydrin test[407] can be carried out on small samples of the resin removed from the synthesis vessel after the incorporation of an N^α-protected amino acid. Treatment of the sample with the requisite reagents then provides what is normally a highly sensitive colorimetric test for the presence of unreacted amino groups on the solid support. Samples containing less then 0.5% of unreacted amine groups can usually be detected and a modified version of the test allows quantitative analysis.[408] However, in certain cases, especially for couplings onto Pro, and less so for couplings onto Asp, Ser, Cys, and Asn, the test does not achieve this level of sensitivity and care must be exercised in interpreting the results. Another drawback of the ninhydrin test is that it irreversibly consumes what can become significant amounts of peptide–resin if carried out after each coupling step in a long synthesis. Notwithstanding these inconveniences, the test is simple and quick and can provide very useful information during SPPS. In the event of a test being positive, the coupling reaction can be repeated or, alternatively, the resin can be capped by treatment with acetic anhydride.

2.4.3.3 Amino Acid Analysis

Amino acid analysis[204] can also be carried out on small resin samples taken from the reaction vessel after amino acid–coupling steps. The peptide resin must be hydrolyzed to provide the free amino acids, and the resulting hydrolysate is then treated in the requisite manner and analyzed. Modern amino acid analyzers are based upon ion-exchange chromatography and the *in situ* derivatization of the individual amino acids for detection. Analysis nowadays can routinely be carried out on 50 pmol quantities of amino acids, and the information given is the relative molar concentrations of each of the amino acids in the peptide chain up to the step at which the peptide–resin sample was removed.

Hydrolysis of peptide–resins is usually carried out using a 1:1 mixture of concentrated hydrochloric acid–propionic acid, in a sealed tube at 110°C for 24 h. Some amino acids, however, do not survive these conditions. Ser, Thr, Tyr, and Met are partially destroyed and usually give values less than theoretical. Asn, Gln, Cys, and Trp are usually completely destroyed. Tyr destruction can normally be reduced by adding 0.1% phenol to the hydrolysis mixture. Cys, Cis, and Met are usually determined after oxidation, as cysteic acid and methionine sulfone, respectively. Asn and Gln are converted into Asp and Glu, so that the values obtained for Asp and Glu in an analysis represent the Asp plus Asn and Glu plus Gln content respectively. The rate of hydrolysis may also depend to some extent on sequence, above all in peptides containing the β-branched amino acids Val and Ile adjacent to one another. Considerably longer reaction times may then be required for complete hydrolysis of the amide bond between them.

Amino acid analysis is best suited to shorter peptides rather than to large peptides or proteins because the inherent error in the method is of the order of 10% for analysis at the nanomole level and can reach 20% at the picomole level. Consequently, analyses of smaller peptides lead to less ambiguity and any significant deviation from the expected values in these cases may be assumed to arise from heterogeneity of the sample.

2.4.3.4 Solid-Phase Edman Degradation

The analysis of peptide–resins may also be carried out using a variation of the Edman degradation,[409-411] which is often referred to as *preview analysis*.[412] For such analysis, which must be carried out using an automated sequencer, a resin sample is required after deprotection of the N^{α}-protecting group. Very little sample is required because modern instruments can routinely analyze 10 to 100 pmol quantities of amino acids. The chemistry involved is outlined in Scheme 2.40.

The peptide–resin **129** is subjected to treatment with phenylisothiocyanate **130** at basic pH forming the phenylthiocarbamyl peptide–resin **131**. On acid treatment, this loses the *N*-terminal amino acid as its anilinothiazolinone derivative **132**, at the same time liberating the *N*-amino terminus of the next amino acid in the peptide **133**. Amino acid derivative **132** is then transformed into the phenylthiohydantoin **134**, which is the species detected and quantified by the sequencer. The procedure is then repeated until all the amino acids of the resin-bound peptide have been analyzed.

Certain amino acids such as Ser, Thr, Cys, and Met are difficult to analyze because they are destroyed to varying extents under the analysis conditions. Furthermore, the technique cannot be used to detect peptides that are chemically modified at their *N*-terminal amino acid, so that if capping has been carried out in a synthesis, the capped peptides will be invisible to sequence analysis. Solid-phase Edman degradation is best carried out on peptide–resins synthesized using the Boc/Bzl approach since, in Fmoc/*t*Bu peptide–resins, neither the peptide–resin anchorage nor the *tert*-butyl-based side chain-protecting groups are stable to the degradation conditions and the technique must be modified accordingly. However, in spite of this, Edman degradation is a very powerful technique for the analysis of either resin-bound or free peptides. It gives the actual sequence of the amino acids in a peptide, information that is impossible to obtain for resin-bound peptides by

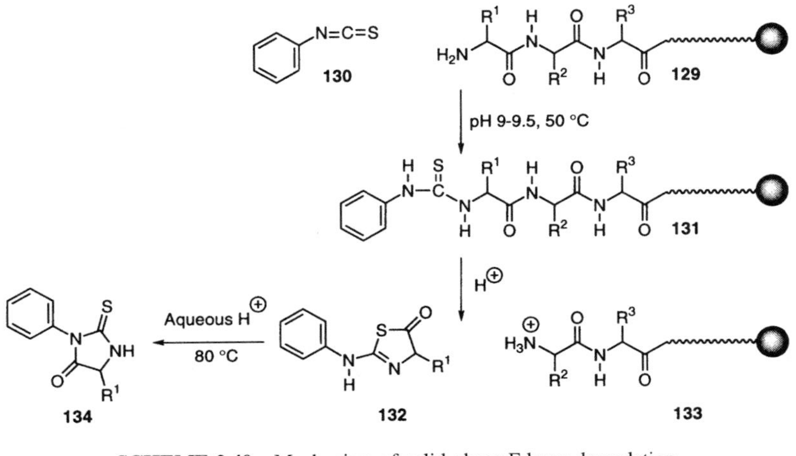

SCHEME 2.40 Mechanism of solid-phase Edman degradation.

any other means. Solid-phase Edman degradation is also a very efficient method for detecting the presence of deletion or addition sequences in a peptide.

2.5 CLEAVAGE OF THE PEPTIDE FROM THE RESIN

In many ways the final cleavage of the completed peptide from the solid support is the key step in SPPS. A very fine balance is necessary between conditions that are sufficiently vigorous to remove the peptide from the resin and, at the same time, are sufficiently mild to allow sensitive structural features to survive. These are stringent requirements but the success of SPPS is testament to the efficiency with which the cleavage can be carried out. The chemical methods used for detaching the peptide from the solid support depend upon the type of resin used, but, for the synthesis of free peptides, it is normally accomplished by acidolysis. Other methods of cleavage, such as photolysis, nucleophilic cleavage, and hydrogenolysis, are relatively unimportant in the synthesis of free peptides but are used to a greater or lesser extent in the synthesis of protected peptide segments (see Section 4.1.1).

If the target peptide contains disulfide bridges, then the formation of these will constrain, to some extent, the manner in which the final cleavage and deprotection reactions are carried out. Disulfide bridges can either be formed on the resin before the cleavage reaction, or in solution afterwards (see Chapter 5).

2.5.1 Acidolysis

In the majority of cases, the cleavage of free peptides from solid supports is achieved by acidolysis.[52] For peptides synthesized using the Boc/Bzl approach, strong acids, such as liquid hydrogen fluoride,[413,414] hydrogen bromide,[415] or trifluoromethane-sulfonic acid,[416] are required. Peptides synthesized using the Fmoc/tBu approach, on the other hand, can be cleaved using weaker acids, such as trifluoroacetic acid.[16] Whichever of the two approaches is used, the final cleavage reaction usually removes most, if not all, of the side chain-protecting groups of the peptide. As a consequence, the reaction medium is rich in potent electrophilic alkylating species. These can lead to the alkylation of susceptible residues, and, in order to reduce this as much as possible, cleavages are almost always carried out in the presence of scavengers, the most common being anisole or thiol derivatives.

2.5.1.1 Liquid Hydrogen Fluoride

Since its introduction into solid-phase synthesis,[417] liquid hydrogen fluoride[418] has become the most commonly used cleavage reagent for peptides synthesized using the Boc/Bzl approach. Originally, a 9:1 mixture of liquid hydrogen fluoride and anisole was used,[413,414,419,420] although other scavengers are often used at present. Generally speaking, such mixtures are applicable to a wide range of peptides and give high cleavage yields, even with relatively acid-stable anchoring groups such as

Pam resins (see Section 2.3.1) and difficult to cleave *C*-terminal amino acids such as Phe.

Reactions are conducted in a special all-fluorocarbon apparatus, since hydrogen fluoride attacks glass; the reagent is also highly toxic and extremely corrosive. Refinement of the original liquid hydrogen fluoride–anisole (9:1) cleavage mixture led to the development of the so-called "low-high" cleavage process.[421] This is a two-stage operation in which the peptide–resin is first treated with hydrogen fluoride diluted with large amounts (typically 70 to 80%) of dimethylsulfide. Under these conditions the more acid-labile benzyl–oxygen bonds are cleaved, under what are presumed to be S_N2-promoting conditions. In the second ("high") stage the peptide is treated either with a 9:1 mixture of liquid hydrogen fluoride and scavenger (usually anisole or *p*-cresol) or with neat hydrogen fluoride. This completes S_N1 cleavage from the resin and deprotects residues that survive the first stage, such as Arg(Tos) or Asp(OcHex).

2.5.1.2   Trifluoroacetic Acid

One of the advantages of Fmoc/*t*Bu synthesis over its Boc/Bzl counterpart is that the hazardous liquid hydrogen fluoride is not necessary for cleaving the peptide from the resin at the end of a synthesis. Milder conditions can be used[234,422] that avoid the need for special apparatus or working protocols. However, as in hydrogen fluoride-mediated cleavages, it is essential to add scavengers to the reaction mixture in order to minimize the unwanted alkylation of sensitive residues. Again, the most common are anisole or *p*-cresol and thiols, such as ethanedithiol or dimethyl sulfide, but silane derivatives may also be used.[423]

2.5.2   Side Reactions during Acidolytic Cleavage of Peptides from the Solid Support

Some of the side reactions that can occur during cleavage have already been discussed in Section 2.2.2. Amino acids that can undergo acid-catalyzed side reactions may suffer these to a greater or lesser extent in the cleavage reaction. However, there are factors that are specific to the cleavage reaction itself, and the most important points are considered below.

2.5.2.1   Alkylation

The free carbocations or other electrophilic alkylating species produced in the acidolytic cleavage reaction can attack sensitive residues such as Trp, Met, Cys, and Tyr, whether this is carried out with hydrogen fluoride or with trifluoroacetic acid. Trp is particularly troublesome in this respect and serious alkylation of this residue has been observed on hydrogen fluoride-mediated cleavage of peptides having it adjacent to Cys or Met. A possible solution is to add an excess of Trp to the cleavage mixture or during lyophilization.[424] Cleavage and deprotection of peptide–resins synthesized using the Fmoc/tBu approach results in the formation of *tert*-butyl cations and *tert*-butyl trifluoroacetate in the reaction medium. These can also cause significant *tert*-butylation of the indole ring.[119-123] Acidolysis of peptide–resins containing Arg residues protected with the Mtr or Pmc groups,[124,325,422,425] or of those containing Asn, Gln, or Cys residues protected with the Tmob group,[135,142] can also cause irreversible alkylation of Trp. A related problem is alkylation by resin-bound carbocations, or other potent alkylating species, produced in the cleavage reaction of some trifluoroacetic acid-labile ester and amide linkers.[234,316,426-428] This can lead to reattachment of the initially cleaved peptide to the resin at Trp residues, and the phenomenon can be mistaken for lack of initial peptide–resin cleavage; similar reattachment can occur at Tyr. Appropriate choice of reaction conditions and scavengers must be found for each individual case.

Alkylation of other residues, particularly of the thioether of Met and, to a much lesser extent, of the side chains of Tyr or of Cys, may also occur on cleavage, although the *tert*-butylation of Met is reversible.[167] Trifluoroacetic acid deprotection of Cys(Trt) is reversible in the absence of scavengers and the addition of silanes has been recommended in order to suppress realkylation.[429] Two cocktails have been developed[234,425] for the trifluoroacetic acid–mediated cleavage of peptides containing sensitive residues such as Trp, Met, Tyr, Thr, and Ser, namely, trifluoroacetic acid–phenol–thioanisole–ethanedithiol–water (82.5:5:5:2.5:5), known as *reagent K*, and trifluoroacetic acid–thioanisole–ethanedithiol–anisole (90:5:3:2), *reagent R*. Their use can lead to significant reductions in unwanted alkylations.

2.5.2.2 Undesired Cyclization Reactions

The γ-carboxyl group of Glu can become protonated in strongly acidic media and lose water, forming an acylium ion **135**. This can then be trapped either intramolecularly, forming a pyrrolidone **136**, or intermolecularly by a scavenger such as anisole,[420] giving the ketone **137**, as shown in Scheme 2.41. This potentially serious side reaction can be controlled by a reduction of the acid strength, using low hydrogen fluoride conditions and by maintaining the temperature of the medium at 0°C or below during acidolysis.[52,430]

Acidolytic cleavage from the solid support can provoke the cyclization of unprotected *C*-terminal Met. Alkylation of the sulfur atom by *tert*-butyl cations leads to the formation of the *C*-terminal *S-tert*-butylmethioninesulfonium salt peptide **138**. This can then cyclize,[431] forming the corresponding homoserine lactone peptide **139**, as shown in Scheme 2.42. This cyclization may be avoided by removal of all *tert*-butyl-based protecting groups before cleavage of the peptide from the resin[167] or by working with Met protected as the sulfoxide.[166]

Aspartimide formation (see Section 2.4.2.1.2) can also be troublesome on acidolysis of peptide–resins, particularly if the peptide contains Asp-Gly, Asp-Ala, or Asp-Ser sequences. Care must taken with regard to the length of acidolysis and the temperature of the medium[52,432] in order to avoid excessive amounts of by-product.

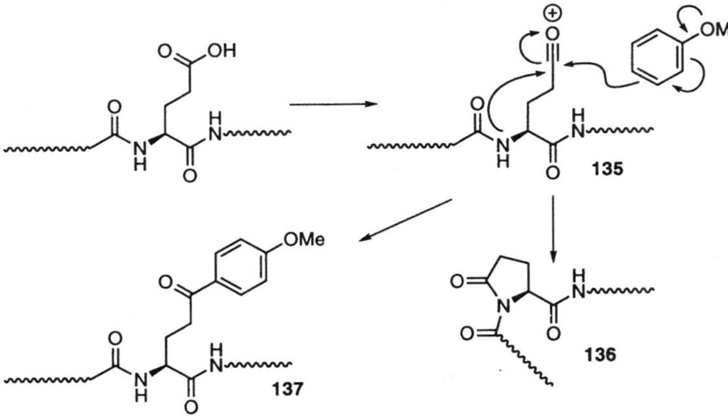

SCHEME 2.41 Side products produced as a consequence of acylium ion formation in acidolytic cleavage from the solid support.

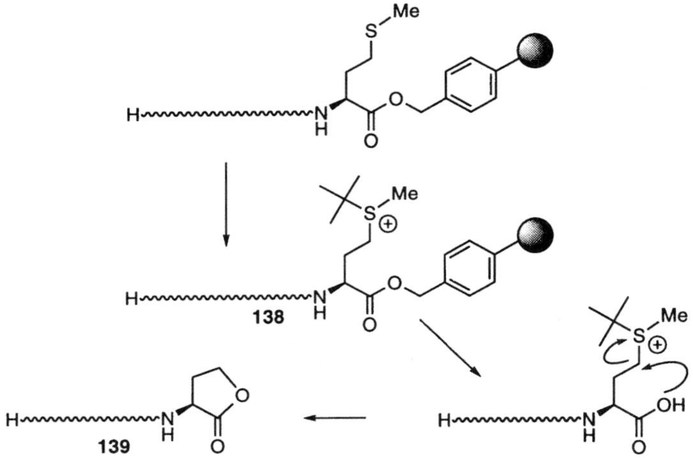

SCHEME 2.42 The production of homoserine lactone-terminal peptides as a consequence of Met alkylation during acidolytic cleavage from the solid support.

2.5.2.3 Other Side Reactions during Acidolytic Cleavage

The majority of peptide bonds are stable to hydrogen fluoride treatment, but there are some cases where its use can cause fragmentation of the peptide chain on cleavage. Peptides containing Met, but not Met(O), have been reported[433] to undergo such rupture at the C-terminus amide of this amino acid, on treatment with liquid hydrogen fluoride at 30°C for 48 h. These conditions are, however, rather drastic in comparison with modern cleavage protocols. Somewhat more troublesome are peptides containing the Asp-Pro sequence, which is appreciably more acid labile than other amides.[434] Fragmentation of the peptide at this point can occur[435] under the

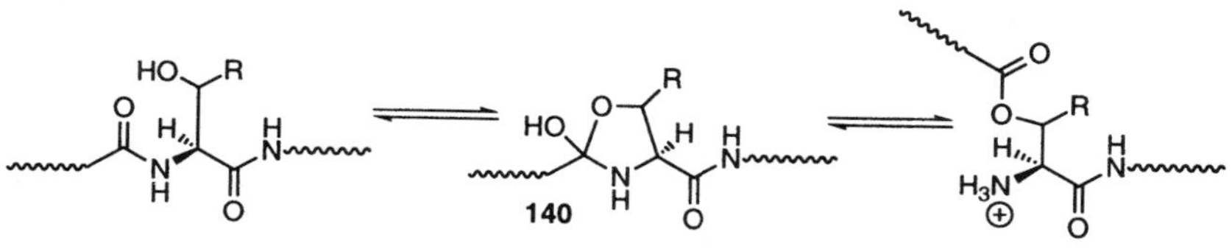

SCHEME 2.43 *N*- to *O*-acyl shifts during acidolytic cleavage from the solid support.

normal conditions used for peptide–resin cleavage with liquid hydrogen fluoride; in severe cases yields of the desired peptide can be low. A possible solution is to synthesize the peptide using the Fmoc/*t*Bu approach, in which the final acidolysis takes place under milder conditions.

Fragmentation at peptide bonds can also occur occasionally on strong acid treatment of peptides having Ser and Thr residues. In this case it takes place by *N*→*O* shifts,[436,437] as shown in Scheme 2.43. Such shifts proceed via intermediates of the type **140**, but are not normally serious since they can be reversed, in most cases, by treatment of the cleaved, deprotected peptide with aqueous base.[2] The problem may be more severe for peptides rich in Ser or Thr residues, however. Facile acidolytic amide bond rupture has also been reported for peptides containing *N*-methylated amino acids.[438]

Hydrogen fluoride cleavage of His(Bom)-containing peptides leads to the formation of formaldehyde that can lead to the methylation of susceptible side chains and to the cyclization of *N*-terminal Cys to a thiazolidine. Such side reactions may be suppressed by adding a formaldehyde scavenger such as resorcinol to the cleavage mixture.[110-112] Similar reactions can occur when the Bum group is used for Cys protection in Fmoc/*t*Bu synthesis[113] (see Section 2.2.2.3).

2.6 EXAMPLES OF PROTEIN SYNTHESIS BY LINEAR SPPS

2.6.1 Bovine Pancreatic Ribonuclease A

In 1969 a short paper titled "The Total Synthesis of an Enzyme with Ribonuclease A Activity," by B. Gutte and R. B. Merrifield appeared in the *Journal of the American Chemical Society*.[439] This communication announced the solid-phase synthesis of a biologically active protein molecule and, by any criterion, may be classed as epoch making. Although Gutte and Merrifield's synthesis did not provide the pure enzyme, the 0.4 mg of material that they were able to isolate exhibited a specific activity of 78% and a peptide map very similar to that of natural ribonuclease A. More importantly, it also provided conclusive proof that the native three-dimensional structure of a protein is determined by the amino acid sequence alone, something that had long been held to be true but that had, hitherto, not been possible to demonstrate.

The synthesis was quite controversial in its day, mainly because it was felt in some quarters that the classical chemical criteria for purity had not been respected. A mixture of compounds had almost certainly been produced, as a consequence of the *N*$^\alpha$-deprotection and amino acid–coupling reactions not going to completion in every case (See Section 2.4.2.2). This being so, it could not be known with certainty which of the peptides present in the mixture was in fact responsible for the observed biological activity; it could not, therefore, be claimed as a synthesis of an enzyme or protein. These criticisms must be taken seriously and touch upon a key issue in SPPS, namely, the problem of the production of closely related peptidic impurities, especially in the synthesis of longer peptides. The formation of these undesired peptides can only be avoided if yields in all of the chemical reactions involved are quantitative, something that, strictly speaking, is not possible to attain. At a more practical level, however, yields can often be pushed very near to quantitative by a rigorous optimization of the synthesis conditions so that, in favorable cases, even quite large peptides can be synthesized in a high state of purity by linear SPPS. Although, at the time, the solid-phase technique was still in its infancy, the preparation of ribonuclease A convincingly demonstrated that it was a viable (and indeed superior) alternative to classical synthesis in solution for the provision of proteins, paving the way for its current dominance in modern peptide chemistry. The synthesis is outlined in Scheme 2.44.

The solid support used was chloromethylated polystyrene cross-linked with 1% divinylbenzene. The Boc group was used for temporary protection of the amino acid N^α functions and benzyl groups were used for the protection of the side chains of

Ribonuclease A

H-Lys-Glu-Thr-Ala-Ala-Ala-Lys-Phe-Glu-Arg-Gln-His-Met-Asp-Ser-Ser-
Thr-Ser-Ala-Ala-Ser-Ser-Ser-Asn-Tyr-Cys-Asn-Gln-Met-Met-Leys-Ser-
Arg-Asn-Leu-Thr-Lys-Asp-Arg-Cys-Lys-Pro-Val-Asn-Thr-Phe-Val-His-
Glu-Ser-Leu-Ala-Asp-Val-Gln-Ala-Val-Cys-Ser-Gln-Lys-Asn-Val-Ala-Cys-
Lys-Asn-Gly-Gln-Thr-Asn-Cys-Tyr-Gln-Ser-Tyr-Ser-Thr-Met-Ser-Ile-Thr-Asp-
Cys-Arg-Glu-Thr-Gly-Ser-Ser-Lys-Tyr-Pro-Asn-Cys-Ala-Tyr-Lys-Thr-Thr-
Gln-Ala-Asn-Lys-His-Ile-Ile-Val-Ala-Cys-Glu-Gly-Asn-Pro-Tyr-Val-Pro-Val-
His-Phe-Asp-Ala-Ser-Val-OH

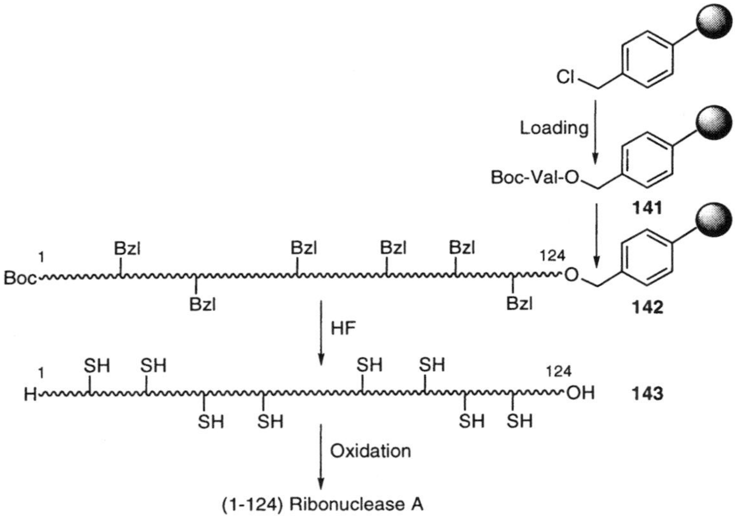

SCHEME 2.44 Synthesis of ribonuclease A.

Asp, Glu, Ser, Thr, Tyr, and Cys. The side chain of Lys was protected with the Z group and that of Arg with the nitro group; Met was introduced as the sulfoxide and His without side chain protection. The first amino acid was anchored by refluxing a mixture of the resin, Boc-Val-OH, and triethylamine in ethyl acetate for 50 h, giving an initial substitution level of 0.21 mmol g^{-1}, as in **141**.

Chain elongation was carried out using an early version of an automatic peptide synthesizer. Coupling reactions for all amino acids except Asn and Gln were performed using DCC in dichloromethane, except where solubility problems required the addition of dimethylformamide; Asn and Gln were coupled as their preformed *p*-nitrophenyl esters, in dimethylformamide. Once chain assembly had been completed, cleavage of peptide–resin **142**, with concomitant removal of all protecting groups was brought about by its treatment with liquid hydrogen fluoride–anisole (9:1), giving free peptide **143**. The disulfide bonds of the protein were formed by air oxidation of this crude material. The product was then submitted to gel-permeation chromatography, ion-exchange chromatography, tryptic digestion (the natural enzyme is trypsin resistant), and, finally, fractional precipitation. The overall yield was calculated to be 2.9%, based on the level of substitution of the initial Boc-Val resin.[440]

2.6.2 Human Immunodeficiency Virus Protease

Human immunodeficiency virus (HIV-1) protease is a 99-residue enzyme that is necessary for the replication of the virus responsible for acquired immunodeficiency syndrome (AIDS). The inhibition of this protein *in vivo* might lead to a treatment for AIDS, and there was, consequently, considerable interest in obtaining it in sufficient amounts for study. The isolation of the protease from the virus itself is difficult and hazardous so that it was necessary to explore other alternatives. Of the two main approaches investigated, one involved expression of the cloned gene in suitable host systems and the other was chemical total synthesis. The goal of producing useful amounts of HIV-1 protease became something of a race between biologists and chemists that ended when Kent,[441,442] in a magnificent achievement, carried out the chemical synthesis of the enzyme and demonstrated the crucial role that organic chemistry can play in biological studies. Very shortly afterward the Merck, Sharp and Dohme group also announced a successful chemical synthesis of a similar protein,[443,444] again by linear SPPS, further underlining its power.

The methods used in these syntheses are highly optimized and incorporate several improvements introduced over the years since the solid-phase synthesis of ribonuclease A. In that earlier synthesis, the main problems encountered were acidolytic loss both of peptide from the resin and of amino acid side chain–protecting groups, together with incomplete coupling steps and side reactions during the hydrogen fluoride treatment. In the case of the HIV-1 protease, on the other hand, the specific solutions to these problems that are now available[307] allowed much more efficient syntheses to be carried out. The procedure used by Kent is outlined in Scheme 2.45.

HIV-1 Protease Monomer

H-Pro-Gln-Ile-Thr-Leu-Trp-Gln-Arg-Pro-Leu-Val-Thr-Ile-Arg-Ile-Gly-Gly-Gln-Leu-Lys-
Glu-Ala-Leu-Leu-Asp-Thr-Gly-Ala-Asp-Asp-Thr-Val-Leu-Glu-Glu-Met-Asn-Leu-Pro-Gly-
Lys-Trp-Lys-Pro-Lys-Met-Ile-Gly-Gly-Ile-Gly-Gly-Phe-Ile-Lys-Val-Arg-Gln-Tyr-Asp-
Gln-Ile-Pro-Val-Glu-Ile-**Aba**-Gly-His-Lys-Ala-Ile-Gly-Thr-Val-Leu-Val-Gly-Pro-Thr-
Pro-Val-Asn-Ile-Ile-Gly-Arg-Asn-Leu-Leu-Thr-Gln-Ile-Gly-**Aba**-Thr-Leu-Asn-Phe-OH

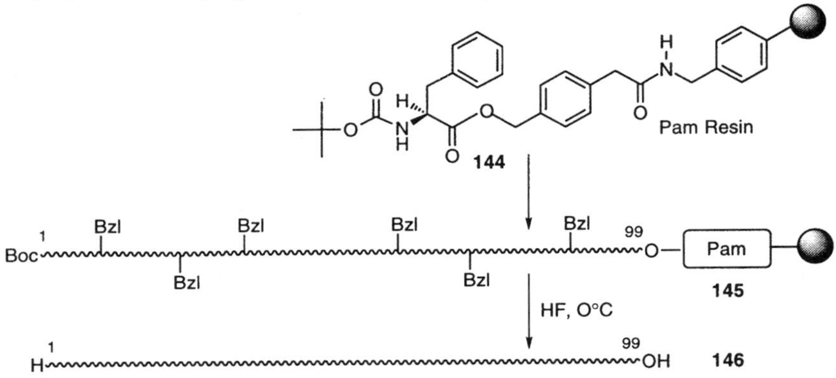

SCHEME 2.45   Synthesis of HIV-1 protease.

The solid support used was a commercially available Pam polystyrene resin **144**, incorporating the Phe C-terminal residue of the HIV-1 protease. This solid support is up to 100 times more stable to acid than a conventional polystyrene resin, and, consequently, acidolytic detachment of peptide chains on removal of the N^α-Boc group with trifluoroacetic acid is greatly reduced. With regard to side chain protection of the various amino acids, benzyl groups were used for the side chains of Ser and Thr and the BrZ group for the phenol of Tyr. Asp and Glu were protected as cyclohexyl esters, Lys was protected with the ClZ group, and Arg with the Tos group. His was introduced as the Dnp derivative, Trp protected with the For group, and Met without side chain protection. Whereas Asn was protected with the Xan group, it was not deemed necessary to protect Gln in this way, presumably since it is less prone to undergo side reactions. The choice of these side chain–protecting groups was dictated by the experience gained over the years with respect to their performance in SPPS — this combination represents an optimum for Boc/Bzl synthesis and loss of side chain–protecting groups during chain elongation is minimal.

Amino acid–coupling reactions were optimized by various modifications of the standard Boc/Bzl approach. The removal of N^α-Boc groups was carried out using neat trifluoroacetic acid, rather than a solution in dichloromethane, to reduce deprotection times and to ensure efficient swelling of the peptide–resin. The only solvent used throughout the synthesis was dimethylformamide, in order to provide maximum solvation of the peptide resin; all the amino acid derivatives used are also soluble in this solvent. Couplings were performed using PSAs and each amino acid was routinely double coupled, in order to maximize yields. These modifications allowed each of the 99 amino acids of the HIV-1 protease to be incorporated in yields greater than 99.8%. Such high coupling yields are a prerequisite if the amounts of deletion peptide impurities produced in a synthesis are to be maintained at manageable levels.

The HIV-1 protease monomer contains two Cys residues at positions 65 and 95 and, since these are not involved in disulfide bridge formation, a later synthesis[445,446] was simplified by their substitution by L-α-amino-n-butyric acid (Aba). The machine-assisted synthesis took 3.3 days to complete and peptide–resin **145** was then subjected to hydrogen fluoride-mediated cleavage. Again, advances that have taken place since Gutte and Merrifield's synthesis of ribonuclease A have made it possible to avoid or at least reduce significantly the side reactions that might occur at this stage. After purification of the crude product **146** by gel filtration and by semipreparative high-performance liquid chromatography, the protein was dissolved in a solution of 6 M guanidine hydrochloride, buffered to pH 7, and was folded by slow dialysis against decreasing concentrations of guanidine hydrochloride until a solution of the pure, folded protein was obtained. The natural, active protein is a non-covalently bound dimer of **126**. After concentration, crystalline [Aba[65,95]]-HIV-1 protease was obtained, which had full enzymatic activity.[447]

The solid-phase synthesis of HIV-1 protease has been refined even further[448] by using coupling protocols based upon the use of uronium reagents and *in situ* neutralization. The D-enzyme has even been prepared, which shows all the properties expected of the enantiomer.[449]

2.6.3   Ubiquitin

Both of the above syntheses were carried out using Merrifield's original Boc/Bzl approach to SPPS, but the newer Fmoc/tBu synthesis developed and advocated by Sheppard can also be applied to the synthesis of large peptides. The synthesis of crystalline ubiquitin, a small protein consisting of 76 residues, carried out by Ramage,[450] demonstrates again that highly pure proteins can be synthesized by linear SPPS. This synthesis is outlined in Scheme 2.46.

A Wang polystyrene resin, to which Fmoc-Gly-OH had been attached by refluxing the amino acid and resin together in the presence of thionyl chloride, was used as the solid support, **147**. Chain elongation was performed automatically and each amino acid, except Gly, was routinely coupled three times. The first coupling was with PSAs and was followed by two more, each using preformed hydroxybenzotriazolyl active esters. Both the anhydrides and the active esters were produced from the N^α-protected amino acids using DIPCDI as the activating agent. For Gly, only one coupling was performed in order to avoid the incorporation of Gly-Gly sequences into the peptide. The side chains of Asp, Glu, Ser, Thr, and Tyr were protected with *tert*-butyl groups, Asn and Gln with the Mbh group, His with the Trt group, Lys with the Boc group, and Arg with the Pmc group. Cleavage of the crude peptide from the solid support was brought about by treatment of peptide–resin **148** with trifluoroacetic acid–water (16:1). Purification was carried out by high-resolution gel-filtration and ion-exchange chromatography. Pure crystalline ubiquitin was obtained in an estimated overall yield, based upon the initial resin substitution, of 4.3%.

Ubiquitin

H-Met-Gln-Ile-Phe-Val-Lys-Thr-Leu-Thr-Gly-Lys-Thr-Ile-Thr-Leu-
Glu-Val-Glu-Pro-Ser-Asp-Thr-Ile-Glu-Asn-Val-Lys-Ala-Lys-Ile-
Gln-Asp-Lys-Glu-Gly-Ile-Pro-Pro-Asp-Gln-Gln-Arg-Leu-Ile-Phe-
Ala-Gly-Lys-Gln-Leu-Glu-Asp-Gly-Arg-Thr-Leu-Ser-Asp-Tyr-Asn-
Ile-Gln-Lys-Glu-Ser-Thr-Leu-His-Leu-Val-Leu-Arg-Leu-Arg-Gly-
Gly-OH

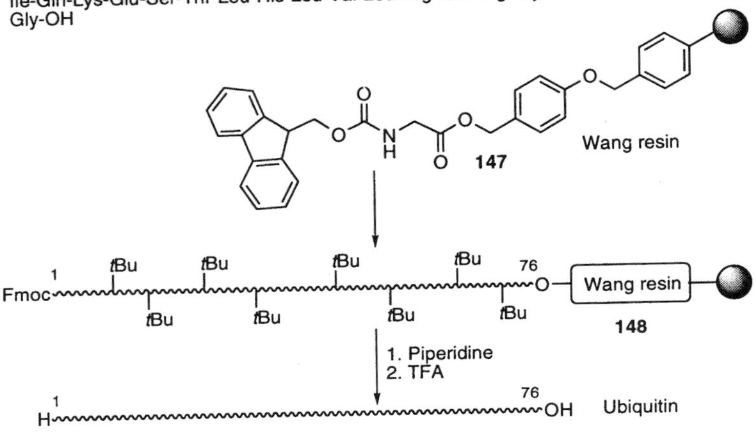

SCHEME 2.46 Synthesis of ubiquitin.

2.6.4 β-Amyloid Protein

The 43-residue βA4 protein is the main constituent of extracellular proteinaceous deposits known as amyloid plaques, present in sufferers of Alzheimer's disease. This peptide provides an illustration that synthetic difficulties are not necessarily a function of the size of the molecule. βA4 is very sparingly soluble and has an elevated tendency to aggregate on solid supports during SPPS. It therefore has the characteristics of a difficult sequence (see Section 2.4.2.2.2), and although its preparation by conventional linear SPPS has been reported,[451-453] it has been difficult to establish with certainty whether or not the product is free from deletion or termination peptides.

One successful approach to the synthesis of a very closely related peptide used a convergent synthetic strategy,[454] in order to minimize on-resin aggregation. This is discussed in more detail in Section 4.1.4.3. The synthesis of pure β-amyloid protein is also possible by modified linear SPPS using secondary structure-disrupting Hmb protection (see Section 2.4.2.2.2) for the amide bonds of selected amino acid residues.[455,456] This latter approach is outlined in Scheme 2.47.

The solid support used was a kieselguhr-supported polyamide resin (known commercially as Pepsyn KA), to which Thr(tBu) had been anchored, as in **149**. Hmb amide bond protection was introduced at Gly[25,29,33,38] and Phe[20] using the Fmoc-protected amino acid pentafluorophenyl active esters **124**. Coupling onto the Hmb-protected N-terminus of the peptide chain after Fmoc group removal was carried out using a tenfold excess of the relevant Fmoc UNCA derivative, conditions that

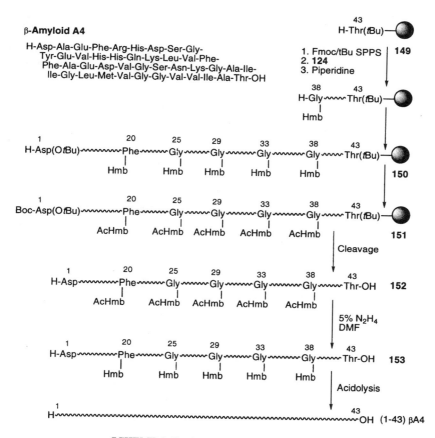

SCHEME 2.47 Synthesis of β-amyloid protein.

give the highest yields in this type of reaction.[400,401] Completion of the chain assembly and removal of the *N*-terminal Fmoc group gave the peptide–resin **150**.

Hmb amide bond protection is normally removed under the acidolytic conditions required for cleavage of the peptide from the resin. However, in the case of βA4, such removal would lead to the production of a very insoluble peptide, whose purification would be difficult. It was advisable, therefore, to maintain amide bond protection in place after cleavage, in order to maintain good solubility. This was done by simple acetylation of the Hmb groups of **150**, prior to acidolysis.[457] The AcHmb groups produced in peptide–resin **151** are appreciably less acid labile and are retained on acidolytic detachment of the peptide from the solid support. The peptide **152** retained good solubility and could be purified by standard chromatographic methods. Once pure, hydrazinolysis smoothly regenerated the Hmb groups, as in **153**, which were then removed acidolytically, with trifluoroacetic acid. Pure βA4 was obtained in an overall yield of 7.5%.

REFERENCES

1. Merrifield, R. B., *J. Am. Chem. Soc.*, 85, 2149, 1963.
2. Barany, G. and Merrifield, R. B., in *The Peptides. Analysis, Synthesis, Biology*, Vol. 2, *Special Methods in Peptide Synthesis, Part A*, Gross, E. and Meienhofer, J., Eds., Academic Press, New York, 1979, 1.
3. Fields, G. B., Tian, Z., and Barany, G., in *Synthetic Peptides. A User's Guide*, Grant, G. A., Ed., W. H. Freeman, New York, 1992, 77.
4. Merrifield, R. B., in *Peptides. Synthesis, Structures and Applications*, Gutte, B., Ed., Academic Press, New York, 1995, 93.
5. Merrifield, R. B., *Science*, 150, 178, 1965.
6. Sarin, V. K., Kent, S. B. H., and Merrifield, R. B., *J. Am. Chem. Soc.*, 102, 5463, 1980.
7. Live, D. and Kent, S. B. H., in *Elastomers and Rubber Elasticity*, Mark, J. and Lol, J., Eds., American Chemical Society, Washington, D.C., 1982, 501.
8. Feinberg, R. S. and Merrifield, R. B., *Tetrahedron*, 30, 3209, 1974.
9. Mitchell, A. R., Kent, S. B. H., Erickson, B. W., and Merrifield, R. B., *Tetrahedron Lett.*, 3795, 1976.
10. Erickson, B. W. and Merrifield, R. B., *J. Am. Chem. Soc.*, 95, 3757, 1973.
11. Wang, S. S., *J. Org. Chem.*, 40, 1235, 1975.
12. Atherton, E. and Sheppard, R. C., in *Peptides 1974. Proceedings of the 13th European Peptides Symposium*, Wolman, Y., Ed., John Wiley & Sons, New York, 1975, 123.
13. Atherton, E., Clive, D. L. J., and Sheppard, R. C., *J. Am. Chem. Soc.*, 97, 6584, 1975.
14. Atherton, E. and Sheppard, R. C., in *Peptides. Proceedings of the 5th American Peptides Symposium*, Goodman, M. and Meienhofer, J., Eds., John Wiley & Sons, New York, 1977, 503.
15. Atherton, E., Gait, M. J., Sheppard, R. C., and Williams, B. J., *Bioorg. Chem.*, 8, 351, 1979.
16. Atherton, E. and Sheppard, R. C., *Solid Phase Peptide Synthesis. A Practical Approach*, IRL Press, Oxford, 1989, 1.
17. Arshady, R., Atherton, E., Clive, D. L. J., and Sheppard, R. C., *J. Chem. Soc. Perkin Trans.*, 1, 529, 1981.
18. Atherton, E., Brown, E., and Sheppard, R., *J. Chem. Soc. Chem. Commun.*, 1151, 1981.
19. Small, P. W. and Sherrington, D. C., *J. Chem. Soc. Chem. Commun.*, 1589, 1989.
20. Mendre, C., Sarrade, V., and Calas, B., *Int. J. Pept. Protein Res.*, 39, 278, 1992.
21. Epton, R., Wellings, D. A., and Williams, A., *React. Polym.*, 6, 143, 1987.
22. Baker, P. A., Coffey, A. F., and Epton, R., in *Innovation and Perspectives in Solid Phase Synthesis and Related Technologies. Peptides, Polypeptides and Oligonucleotides. Macro-Organic Reagents and Catalysts*, Epton, R., Ed., SPCC (U.K.) Ltd, Birmingham, 1990, 435.
23. Becker, H., Lucas, H.-W., Maul, J., Pillai, V. N. R., Anzinger, H., and Mutter, M., *Makromol. Chem. Rapid Commun.*, 3, 217, 1982.
24. Hellermann, H., Lucas, H.-W., Maul, J., Pillai, V. N. R., and Mutter, M., *Makromol. Chem.*, 184, 2603, 1983.
25. Zalipsky, S., Albericio, F., and Barany, G., in *Peptides. Structure and Function. Proceedings of the 9th American Peptides Symposium*, Deber, C. M., Hruby, V. J., and Kopple, K. D., Eds., Pierce Chemical Company, Rockford, IL, 1985, 257.
26. Zalipsky, S., Chang, J. L., Albericio, F., and Barany, G., *React. Polym.*, 22, 243, 1994.
27. Bayer, E., Hemmasi, B., Albert, K., Rapp, W., and Dengler, M., in *Peptides. Structure and Function. Proceedings of the 8th American Peptides Symposium*, Hruby, V. J. and Rich, D. H., Eds., Pierce Chemical Company, Rockford, IL, 1983, 87.
28. Bayer, E., Dengler, M., and Hemmasi, B., *Int. J. Pept. Protein Res.*, 25, 178, 1985.
29. Hellstern, H. and Hemmasi, B., *Biol. Chem. Hoppe-Seyler*, 369, 289, 1988.
30. Bayer, E., *Angew. Chem. Int. Ed. Engl.*, 30, 113, 1991.
31. Bayer, E. and Rapp, W., in *Poly(Ethylene) Glycol Chemistry: Biotechnological and Biomedical Applications*, Harris, J. M., Ed., Plenum Press, New York, 1992, 325.
32. Daniels, S. B., Bernatowicz, M. S., Coull, J. M., and Köster, H., *Tetrahedron Lett.*, 30, 4345, 1989.
33. Eichler, J., Bienert, M., Stierandova, A., and Lebl, M., *Peptide Res.*, 4, 296, 1991.
34. Eichler, J., Beyermann, M., and Bienert, M., *Coll. Czech. Chem. Commun.*, 54, 1746, 1989.

35. Englebretsen, D. R. and Harding, D. R. K., *Int. J. Pept. Protein Res.,* 40, 487, 1992.
36. Meldal, M., *Tetrahedron Lett.,* 33, 3077, 1992.
37. Büttner, K., Zahn, H., and Fischer, W. H., in *Peptides. Chemistry and Biology. Proceedings of the 10th American Peptides Symposium,* Marshall, G. R., Ed., ESCOM, Leiden, 1988, 210.
38. Tregear, G. W., in *Chemistry and Biology of Peptides. Proceedings of the 3rd American Peptides Symposium,* Meienhofer, J., Ed., Ann Arbor Science Publishers, Ann Arbor, MI, 1972, 175.
39. Kent, S. B. H. and Merrifield, R. B., *Israel J. Chem.,* 17, 243, 1978.
40. Albericio, F., Ruiz-Gayo, M., Pedroso, E., and Giralt, E., *React. Polym.,* 10, 259, 1989.
41. Berg, R. H., Almdal, K., Pedersen, W. B., Holm, A., Tam, J. P., and Merrifield, R. B., *J. Am. Chem. Soc.,* 111, 8024, 1989.
42. Carpino, L. A., *J. Am. Chem. Soc.,* 79, 4427, 1957.
43. McKay, F. C. and Albertson, N. F., *J. Am. Chem. Soc.,* 79, 4686, 1957.
44. Anderson, G. W. and McGregor, A. C., *J. Am. Chem. Soc.,* 79, 6180, 1957.
45. Carpino, L. A. and Han, G. A., *J. Org. Chem.,* 37, 3404, 1972.
46. Moroder, L., Hallett, A., Wünsch, E., Keller, O., and Wersin, G., *Hoppe-Seyler's Z. Physiol. Chem.,* 357, 1651 1976.
47. Itoh, M., Hagiwara, D., and Kamiya, T., *Tetrahedron Lett.,* 4393, 1975.
48. Lapatsanis, L., Milias, G., Froussious, K., and Kolovos, M., *Synthesis,* 671, 1983.
49. Paquet, A., *Can. J. Chem.,* 54, 733, 1976.
50. Tessier, M., Albericio, F., Pedroso, E., Grandas, A., Eritja, R., Giralt, E., Granier, C., and van Rietschoten, J., *Int. J. Pept. Protein Res.,* 22, 125, 1983.
51. Bolin, D. R., Sytwu, I.-I., Humiec, F., and Meienhofer, J., *Int. J. Pept. Protein Res.,* 33, 353, 1989.
52. Tam, J. P. and Merrifield, R. B., in *The Peptides. Analysis, Synthesis, Biology,* Vol. 9, *Special Methods in Peptide Synthesis, Part C,* Udenfriend, S. and Meienhofer, J., Eds., Academic Press, New York, 1987, 185.
53. O' Ferrall, R. A. M. and Slae, S., *J.Chem. Soc. (B),* 260, 1970.
54. O' Ferrall, R. A. M., *J. Chem. Soc. (B),* 274, 1970.
55. Wade, J. D., Bedford, J., Sheppard, R. C., and Tregear, G. W., *Pept. Res.,* 4, 194, 1991.
56. Kates, S. A., Solé, N. A., Beyermann, M., Barany, G., and Albericio, F., *Pept. Res.,* 9, 106, 1996.
57. Ueki, M. and Amemiya, M., *Tetrahedron Lett.,* 28, 6617, 1987.
58. Kurtz, A. C., *J. Biol. Chem.,* 140, 705, 1941.
59. Schwyzer, R. and Rittel, W., *Helv. Chim. Acta,* 44, 159, 1961.
60. Bezas, B. and Zervas, L., *J. Am. Chem. Soc.,* 83, 719, 1961.
61. Sturm, K., Geiger, R., and Siedel, W., *Chem. Ber.,* 96, 609, 1963.
62. Weygand, F. and Csendes, E., *Angew. Chem.,* 64, 136, 1952.
63. Ohno, M., Eastlake, A., Ontjes, D., and Anfinsen, C. B., *J. Am. Chem. Soc.,* 91, 6842, 1969.
64. Ontjes, D. A. and Anfinsen, C. B., *Proc. Natl. Acad. Sci. U.S.A.,* 64, 428, 1969.
65. Matsueda, R. and Walter, R., *Int. J. Pept. Protein Res.,* 16, 392, 1980.
66. Rajagopalan, S., Heck, T., Iwamoto, T., and Tomich, J. M., *Int. J. Pept. Protein Res.,* 45, 173, 1995.
67. Bycroft, B. W., Chan, W. C., Chhabra, S. R., and Hone, N. D., *J. Chem. Soc. Chem. Commun.,* 778, 1993.
68. Lyttle, M. H. and Hudson, D., in *Peptides. Chemistry and Biology. Proceedings of the 12th American Peptides Symposium,* Smith, J. A. and Rivier, J. E., Eds., ESCOM, Leiden, 1992, 583.
69. Loffet, A. and Zhang, H. X., *Int. J. Pept. Protein Res.,* 42, 346, 1993.
70. Genet, J. P., Blart, E., Savignac, M., Lemeune, S., Lemaire-Audoire, S., and Bernard, J. M., *Synlett,* 680, 1993.
71. Dangles, O., Guibé, F., Balavoine, G., Lavielle, S., and Marquet, A., *J. Org. Chem.,* 52, 4984, 1987.
72. Kunz, H. and Unversagt, C., *Angew. Chem. Int. Ed. Engl.,* 23, 436, 1984.
73. Jones, D. A., Mikulec, R. A., and Mazur, R. H., *J. Org. Chem.,* 38, 2865, 1973.
74. Künzi, H., Manneberg, M., and Studer, R. O., *Helv. Chim. Acta,* 57, 566, 1974.
75. Yamashiro, D., Blake, J., and Li, C. H., *J. Am. Chem. Soc.,* 94, 2855, 1972.
76. Zervas, L., Winitz, M., and Greenstein, J. P., *J. Org. Chem.,* 22, 1515, 1957.
77. Jäger, G. and Geiger, R., *Chem. Ber.,* 103, 1727, 1970.
78. Verdini, A. S., Lucietto, P., Fossati, G., and Giordani, C., *Tetrahedron Lett.,* 33, 6541, 1992.
79. Wu, Y., Matsueda, G. R., and Bernatowicz, M., *Synth. Commun.,* 23, 3055, 1993.

80. Kossel, A. and Kennaway, E. L., *Hoppe-Seyler's Z. Physiol. Chem.*, 72, 486, 1911.
81. Bergmann, M., Zervas, L., and Rinke, H., *Hoppe-Seyler's Z. Physiol. Chem.*, 224, 40, 1934.
82. Turán, A., Patthy, A., and Bajusz, S., *Acta Chim. Acad. Sci. Hung.*, 85, 327, 1975.
83. Ramachandran, J. and Li, C. H., *J. Org. Chem.*, 27, 4006, 1962.
84. Fujino, M., Nishimura, O., Wakimasu, M., and Kitada, C., *J. Chem. Soc. Chem. Commun.*, 668, 1980.
85. Wakimasu, M., Kitada, C., and Fujino, M., *Bull. Chem. Soc. Jpn.*, 30, 2766, 1982.
86. Ramage, R., Green, J., and Blake, A. J., *Tetrahedron*, 47, 6353, 1991.
87. Carpino, L. A., Schroff, H., Triolo, S. A., Mansour, E.-S. M. E., Wenschuh, H., and Albericio, F., *Tetrahedron Lett.*, 34, 7829, 1993.
88. Fischer, P. M., Retson, K. V., Tyler, M. I., and Howden, M. E. H., *Int. J. Pept. Protein Res.*, 40, 19, 1992.
89. Habeeb, A. F. S. A., *Can. J. Biochem. Physiol.*, 38, 493, 1960.
90. Maryanoff, C., Stanzione, R. C., Plampin, J. N., and Mills, J. E., *J. Org. Chem.*, 51, 1882, 1986.
91. Miller, A. E. and Bischoff, J. J., *Synthesis*, 777, 1986.
92. Bernatowicz, M. S., Wu, Y., and Matsueda, G. R., *J. Org. Chem.*, 57, 2497, 1992.
93. Bodanszky, M., Ondetti, M. A., Birkhimer, C. A., and Thomas, P. L., *J. Am. Chem. Soc.*, 86, 4452, 1964.
94. Cosand, W. L. and Merrifield, R. B., *Proc. Natl. Acad. Sci. U.S.A.*, 74, 2771, 1977.
95. Granier, C., Pedroso Muller, E., and van Rietschoten, J., *Eur. J. Biochem.*, 82, 293, 1978.
96. du Vigneaud, V. and Behrens, O. K., *J. Biol. Chem.*, 117, 27, 1937.
97. Chillemi, F. and Merrifield, R. B., *Biochemistry*, 8, 4344, 1969.
98. Bell, J. R. and Jones, J. H., *J. Chem. Soc. Perkin Trans. 1*, 2336, 1974.
99. Shaltiel, S., *Biochem. Biophys. Res. Commun.*, 29, 178, 1967.
100. Shaltiel, S. and Fridkin, M., *Biochemistry*, 9, 5122, 1970.
101. Siepmann, E. and Zahn, H., *Biochim. Biophys. Acta*, 82, 412, 1964.
102. Beyerman, H. C., Hirt, J., Kranenburg, P., Syrier, J. L. M., and van Zon, A., *Recl. Trav. Chim. Pays-Bas*, 93, 256, 1974.
103. Sakakibara, S. and Fujii, T., *Bull. Chem. Soc. Jpn.*, 42, 1466, 1969.
104. Fujii, T. and Sakakibara, S., *Bull. Chem. Soc. Jpn.*, 43, 3954, 1970.
105. van der Eijk, J. M., Nolte, R. J. M., and Zwikker, J. W., *J. Org. Chem.*, 45, 547, 1980.
106. Fujii, T. and Sakakibara, S., *Bull. Chem. Soc. Jpn.*, 47, 3146, 1974.
107. Fujii, T., Kimura, T., and Sakakibara, S., *Bull. Chem. Soc. Jpn.*, 49, 1595, 1976.
108. Brown, T. and Jones, J. H., *J. Chem. Soc. Chem. Commun.*, 648, 1981.
109. Brown, T., Jones, J. H., and Richards, J. D., *J. Chem. Soc. Perkin Trans. 1*, 1553, 1982.
110. Gesquière, J.-C., Diesis, E., and Tartar, A., *J. Chem. Soc. Chem. Commun.*, 1402, 1990.
111. Mitchell, M. A., Runge, T. A., Mathews, W. R., Ichhpurani, A. K., Harn, N. K., Dobrolowski, P. J., and Eckenrode, F. M., *Int. J. Pept. Protein Res.*, 36, 350, 1990.
112. Kumagaye, K. Y., Inui, T., Nakajima, K., Kimura, T., and Sakakibara, S., *Pept. Res.*, 4, 84, 1991.
113. Gesquière, J. C., Najib, J., Diesis, E., Barbry, D., and Tartar, A., in *Peptides. Chemistry and Biology, Proceedings of the 12th American Peptides Symposium*, Smith, J. A. and Rivier, J. E., Eds., ESCOM, Leiden, 1992, 641.
114. Colombo, R., Colombo, F., and Jones, J. H., *J. Chem. Soc. Chem. Commun.*, 292, 1984.
115. Jones, J. H., *The Chemical Synthesis of Peptides*, Clarendon Press, Oxford, 1991, 1.
116. Sieber, P. and Riniker, B., *Tetrahedron Lett.*, 28, 6031, 1987.
117. Harding, S. J., Heslop, I., Jones, J. H., and Wood, M. E., in *Peptides 1994. Proceedings of the 23rd European Peptides Symposium*, Maia, H. L. S., Ed., ESCOM, Leiden, 1995, 189.
118. Fontana, A., Marchiori, F., Moroder, L., and Scoffone, E., *Tetrahedron Lett.*, 2985, 1966.
119. Jaeger, E., Thamm, P., Knof, S., Wünsch, E., Löw, M., and Kisfaludy, L., *Hoppe-Seyler's Z. Physiol. Chem.*, 359, 1617, 1978.
120. Jaeger, E., Thamm, P., Knof, S., and Wünsch, E., *Hoppe-Seyler's Z. Physiol. Chem.*, 359, 1629, 1978.
121. Löw, M., Kisfaludy, L., Jaeger, E., Thamm, P., Knof, S., and Wünsch, E., *Hoppe-Seyler's Z. Physiol. Chem.*, 359, 1637, 1978.
122. Löw, M., Kisfaludy, L., and Sohár, P., *Hoppe-Seyler's Z. Physiol. Chem.*, 359, 1643, 1978.

123. Masui, Y., Chino, N., and Sakakibara, S., *Bull. Chem. Soc. Jpn.*, 53, 464, 1980.
124. Sieber, P., *Tetrahedron Lett.*, 28, 1637, 1987.
125. Hudson, D., Kain, D., and Ng, D., in *Peptide Chemistry 1985. Proceedings of the 23rd Symposium on Peptide Chemistry*, Kiso, Y., Ed., Protein Research Foundation, Osaka, 1986, 413.
126. Yamashiro, D. and Li, C. H., *J. Org. Chem.*, 38, 2594, 1973.
127. Merrifield, R. B., Vizioli, L. D., and Boman, H. G., *Biochemistry*, 21, 5020, 1982.
128. Geiger, R. and König, W., in *The Peptides. Analysis, Synthesis, Biology*, Vol. 3, *Protection of Functional Groups in Peptide Synthesis*, Gross, E. and Meienhofer, J., Eds., Academic Press, New York, 1981, 1.
129. Franzén, H., Grehn, L., and Ragnarsson, U., *J. Chem. Soc. Chem. Commun.*, 1699, 1984.
130. White, P., in *Peptides. Chemistry and Biology. Proceedings of the 12th American Peptides Symposium*, Smith, J. A. and Rivier, J. E., Eds., ESCOM, Leiden, 1992, 537.
131. Fields, C. G. and Fields, G. B., *Tetrahedron Lett.*, 34, 6661, 1993.
132. Paul, R. and Kende, A. S., *J. Am. Chem. Soc.*, 86, 4162, 1964.
133. Kashelikar, D. V. and Ressler, C., *J. Am. Chem. Soc.*, 86, 2467, 1964.
134. Mojsov, S., Mitchell, A. R., and Merrifield, R. B., *J. Org. Chem.*, 45, 555, 1980.
135. Gausepohl, H., Kraft, M., and Frank, R. W., *Int. J. Pept. Protein Res.*, 34, 287, 1989.
136. Kisfaludy, L., Schön, I., Renyei, M., and Görög, S., *J. Am. Chem. Soc.*, 97, 5588, 1975.
137. Schnabel, E., Klostermeyer, H., Dahlmans, J., and Zahn, H., *Liebigs Ann. Chem.*, 707, 227, 1967.
138. Barany, G. and Merrifield, R. B., *Cold Spring Harbour Symp. Quant. Biol.*, 37, 121, 1973.
139. Shimonishi, Y., Sakakibara, S., and Akabori, S., *Bull. Chem. Soc. Jpn.*, 35, 1966, 1962.
140. König, W. and Geiger, R., *Chem. Ber.*, 103, 2041, 1970.
141. Funakoshi, S., Tamamura, H., Fujii, N., Yoshizawa, K., Yajima, H., Muyasaka, K., Funakoshi, A., Ohta, M., Inagaki, Y., and Carpino, L., *J. Chem. Soc. Chem. Commun.*, 1588, 1988.
142. Sieber, P. and Riniker, B., in *Innovation and Perspectives in Solid Phase Synthesis and Related Technologies. Peptides, Polypeptides and Oligonucleotides. Macro-Organic Reagents and Catalysts*, Epton, R., Ed., SPCC (U.K.) Ltd, Birmingham, 1990, 577.
143. DiMarchi, R. D., Tam, J. P., and Merrifield, R. B., *Int. J. Pept. Protein Res.*, 19, 270, 1982.
144. Tam, J. P., Riemen, M. W., and Merrifield, R. B., *Pept. Res.*, 1, 6, 1988.
145. Bodanszky, M. and Kwei, J. Z., *Int. J. Pept. Protein Res.*, 12, 69, 1978.
146. Ledger, R. and Stewart, F. H. C., *Aust. J. Chem.*, 18, 1477, 1965.
147. van Heeswijk, W. A. R., Eenink, M. J. D., and Feijen, J., *Synthesis*, 744, 1982.
148. Okada, Y., Iguchi, S., and Kawasaki, K., *J. Chem. Soc. Chem. Commun.*, 1532, 1987.
149. Okada, Y. and Iguchi, S., *J. Chem. Soc. Perkin Trans. 1*, 2129, 1988.
150. Albericio, F., Nicolás, E., Rizo, J., Ruiz-Gayo, M., Pedroso, E., and Giralt, E., *Synthesis*, 119, 1990.
151. Belshaw, P. J., Mzengeza, S., and Lajoie, G. A., *Synth. Commun.*, 20, 3157, 1990.
152. Wünsch, E. and Jentsch, J., *Chem. Ber.*, 97, 2490, 1964.
153. Mizoguchi, T., Levin, G., Woolley, D. W., and Stewart, J. M., *J. Org. Chem.*, 33, 903, 1968.
154. Hruby, V. J. and Ehler, K. W., *J. Org. Chem.*, 35, 1690, 1970.
155. Sugano, H. and Miyoshi, M., *J. Org. Chem.*, 41, 2352, 1976.
156. Chang, C.-D., Waki, M., Ahmad, M., Meienhofer, J., Lundell, E. O., and Hang, J. D., *Int. J. Pept. Protein Res.*, 15, 59, 1980.
157. Adamson, J. G., Blaskowitch, M. A., Groenvelt, H., and Lajoie, G. A., *J. Org. Chem.*, 56, 3447, 1991.
158. Wünsch, E., Fries, G., and Zwick, A., *Chem. Ber.*, 91, 542, 1958.
159. Iselin, B., *Helv. Chim. Acta*, 45, 1510, 1962.
160. Erickson, B. W. and Merrifield, R. B., *J. Am. Chem. Soc.*, 95, 3750, 1973.
161. Yamashiro, D. and Li, C. H., *J. Org. Chem.*, 38, 591, 1973.
162. Philosof-Oppenheimer, R., Pecht, I., and Fridkin, M., *Int. J. Pept. Protein Res.*, 45, 116, 1995.
163. Hofmann, K., Haas, W., Smithers, M. J., Wells, R. D., Wolman, Y., Yanaihara, N., and Zanetti, G., *J. Am. Chem. Soc.*, 87, 620, 1965.
164. Norris, K., Halstrom, J., and Brunfeldt, K., *Acta Chem. Scand.*, 25, 945, 1971.
165. Neumann, N. P., Moore, S., and Stein, W. H., *Biochemistry*, 1, 68, 1962.
166. Iselin, B., *Helv. Chim. Acta*, 44, 61, 1961.
167. Noble, R. L., Yamashiro, D., and Li, C. H., *J. Am. Chem. Soc.*, 98, 2324, 1976.

168. Guttmann, S. and Boissonnas, R. A., *Helv. Chim. Acta*, 41, 1852, 1958.
169. Guttmann, S. and Boissonnas, R. A., *Helv. Chim. Acta*, 42, 1257, 1959.
170. Houghten, R. A. and Li, C. H., *Anal. Biochem.*, 98, 36, 1979.
171. Fujii, N., Kuno, S., Otaka, A., Funakoshi, S., Takagi, K., and Yajima, H., *Chem. Pharm. Bull.*, 33, 4587, 1985.
172. Futaki, S., Yagami, T., Taike, T., Akita, T., and Kitagawa, K., *J. Chem. Soc. Perkin Trans. 1*, 653, 1990.
173. Tam, J. P., Heath, W. F., and Merrifield, R. B., *Tetrahedron Lett.*, 23, 2939, 1982.
174. Nicolás, E., Vilaseca, M., and Giralt, E., *Tetrahedron*, 51, 5701, 1995.
175. Hofmann, K., Haas, W., Smithers, M. J., and Zanetti, G., *J. Am. Chem. Soc.*, 87, 631, 1965.
176. Sifferd, R. H. and du Vigneaud, V., *J. Biol. Chem.*, 108, 753, 1935.
177. Akabori, S., Sakakibara, S., Shimonishi, Y., and Nobuhara, Y., *Bull. Chem. Soc. Jpn.*, 37, 433, 1964.
178. Munson, M. C., García-Echeverría, C., Albericio, F., and Barany, G., *J. Org. Chem.*, 57, 3013, 1992.
179. Munson, M. and Barany, G., *J. Am. Chem. Soc.*, 115, 10203, 1993.
180. Kamber, B. and Rittel, W., *Helv. Chim. Acta*, 51, 2061, 1968.
181. Kamber, B., *Helv. Chim. Acta*, 56, 1371, 1973.
182. Sieber, P., Kamber, B., Hartmann, A., Jöhl, A., Riniker, B., and Rittel, W., *Helv. Chim. Acta*, 60, 27, 1977.
183. Veber, D. F., Milkowski, J. D., Varga, S. L., Denkewalter, R. G., and Hirschmann, R., *J. Am. Chem. Soc.*, 94, 5456, 1972.
184. Bodanszky, M. and Bednarek, M. A., *Int. J. Pept. Protein Res.*, 20, 434, 1982.
185. Ruiz-Gayo, M., Albericio, F., Pedroso, E., and Giralt, E., *J. Chem. Soc. Chem. Commun.*, 1501, 1986.
186. Weber, U. and Hartter, P., *Hoppe-Seyler's Z. Physiol. Chem.*, 351, 1384, 1970.
187. Matsueda, R., Kimura, T., Kaiser, E. T., and Matsueda, G. R., *Chem. Lett.*, 737, 1981.
188. Bernatowicz, M. S., Matsueda, R., and Matsueda, G. R., *Int. J. Pept. Protein Res.*, 28, 107, 1986.
189. Wünsch, E., in *Methoden der Organischen Chemie (Houben-Weyl)*, Vol. 15/1, Müller, E., Ed., Georg Thieme Verlag, Stuttgart, 1974, 469.
190. Yajima, H., Funakoshi, S., Fujii, N., Akaji, K., and Irie, H., *Chem. Pharm. Bull.*, 27, 1060, 1979.
191. Yajima, H., Akaji, K., Funakoshi, S., Fujii, N., and Irie, H., *Chem. Pharm. Bull.*, 28, 1942, 1980.
192. Patchornik, A. and Sokolovsky, M., *J. Am. Chem. Soc.*, 86, 1206, 1964.
193. Hallinan, E. A., *Int. J. Pept. Protein Res.*, 38, 601, 1991.
194. Lukszo, J., Patterson, D., Albericio, F., and Kates, S. A., *Lett. Pept. Sci.*, 3, 157, 1996.
195. Bodanszky, M. and Bodanszky, A., *J. Chem. Soc. Chem. Commun.*, 591, 1967.
196. Kaiser, E. T., Nicholson, G. J., Kohlbau, H. J., and Voelter, W., *Tetrahedron Lett.*, 37, 1187, 1996.
197. Atherton, E., Hardy, P. M., Harris, D. E., and Matthews, B. H., in *Peptides 1990. Proceedings of the 21st European Peptides Symposium*, Giralt, E. and Andreu, D., Eds., ESCOM, Leiden, 1991, 243.
198. Fujiwara, Y., Akaji, K., and Kiso, Y., in *Peptide Chemistry 1993. Proceedings of the 31st Symposium on Peptide Chemistry*, Okada, Y., Ed., Protein Research Foundation, Osaka, 1994, 29.
199. Gisin, B. F., *Helv. Chim. Acta*, 56, 1476, 1973.
200. Gross, E., Noda, K., and Nisula, B., *Angew. Chem.*, 85, 672, 1973.
201. Noda, K., Gazis, D., and Gross, E., *Int. J. Pept. Protein Res.*, 19, 413, 1982.
202. Barany, G. and Albericio, F., *J. Am. Chem. Soc.*, 107, 4936, 1985.
203. Matsueda, G. R. and Haber, E., *Anal. Biochem.*, 104, 215, 1980.
204. Ozols, J., *Methods Enzymol.*, 182, 587, 1990.
205. Mitchell, A. R., Erickson, B. W., Ryabtsev, M. N., Hodges, R. S., and Merrifield, R. B., *J. Am. Chem. Soc.*, 98, 7357, 1976.
206. Mitchell, A. R., Kent, S. B. H., Engelhard, M., and Merrifield, R. B., *J. Org. Chem.*, 43, 2845, 1978.
207. Tam, J. P., Kent, S. B. H., Wong, T. W., and Merrifield, R. B., *Synthesis*, 955, 1979.
208. Wang, S. S., *J. Am. Chem. Soc.*, 95, 1328, 1973.
209. Sheppard, R. C. and Williams, B. J., *Int. J. Pept. Protein Res.*, 20, 451, 1982.
210. Albericio, F. and Barany, G., *Int. J. Pept. Protein Res.*, 23, 342, 1984.

211. Atherton, E., Benoiton, N. L., Brown, E., Sheppard, R. C., and Williams, B. J., *J. Chem. Soc. Chem. Commun.*, 336, 1981.
212. Mergler, M., Tanner, R., Gosteli, J., and Grogg, P., *Tetrahedron Lett.*, 29, 4005, 1988.
213. Grandas, A., Jorba, X., Giralt, E., and Pedroso, E., *Int. J. Pept. Protein Res.*, 33, 386, 1989.
214. Kirstgen, R., Sheppard, R. C., and Steglich, W., *J. Chem. Soc. Chem. Commun.*, 1870, 1987.
215. Kirstgen, R., Olbrich, A., Rehwinkel, H., and Steglich, W., *Liebigs Ann. Chem.*, 437, 1988.
216. Sieber, P., *Tetrahedron Lett.*, 28, 6147, 1987.
217. Akaji, K., Tanaka, H., Itoh, H., Imai, J., Fujiwara, Y., Kimura, T., and Kiso, Y., *Chem. Pharm. Bull.*, 38, 3471, 1990.
218. Green, J. and Bradley, K., *Tetrahedron*, 49, 4141, 1993.
219. Granitza, D., Beyermann, M., Wenschuh, H., Haber, H., Carpino, L. A., Truran, G. A., and Bienert, M., *J. Chem. Soc. Chem. Commun.*, 2223, 1995.
220. Fuller, W. D., Krotzer, N. J., Swain, P. A., Anderson, B. L., Comer, D., and Goodman, M., in *Peptides 1992. Proceedings of the 22nd European Peptides Symposium*, Schneider, C. H. and Eberle, A. N., Eds., ESCOM, Leiden, 1993, 231.
221. Colombo, R., Atherton, E., Sheppard, R. C., and Woolley, V., *Int. J. Pept. Protein Res.*, 21, 118, 1983.
222. Mergler, M., Nyfeler, R., Tanner, R., Gosteli, J., and Grogg, P., *Tetrahedron Lett.*, 29, 4009, 1988.
223. Blankemeyer-Menge, B., Nimtz, M., and Frank, R., *Tetrahedron Lett.*, 31, 1701, 1990.
224. Brown, A. R. and Ramage, R., *Tetrahedron Lett.*, 35, 789, 1994.
225. Henriksen, D. B., Breddam, K., Moller, J., and Buchardt, O., *J. Am. Chem. Soc.*, 114, 1876, 1992.
226. Pietta, P. G. and Marshall, G. R., *J. Chem. Soc. Chem. Commun.*, 650, 1970.
227. Matsueda, G. R. and Stewart, J. M., *Peptides*, 2, 45, 1981.
228. Rink, H., *Tetrahedron Lett.*, 28, 3787, 1987.
229. Stüber, W., Knolle, J., and Breipohl, G., *Int. J. Pept. Protein Res.*, 34, 215, 1989.
230. Breipohl, G., Knolle, J., and Stüber, W., *Int. J. Pept. Protein Res.*, 34, 262, 1989.
231. Breipohl, G., Knolle, J., and Stüber, W., *Int. J. Pept. Protein Res.*, 35, 281, 1990.
232. Penke, B. and Nyerges, L., in *Peptides 1988. Proceedings of the 20th European Peptides Symposium*, Jung, G. and Bayer, E., Eds., Walter de Gruyter, Berlin, 1988, 142.
233. Albericio, F. and Barany, G., *Int. J. Pept. Protein Res.*, 30, 206, 1987.
234. Albericio, F., Kneib-Cordonier, N., Biancalana, S., Gera, L., Masada, R. I., Hudson, D., and Barany, G., *J. Org. Chem.*, 55, 3730, 1990.
235. Han, Y., Bontems, S. L., Hegyes, P., Munson, M. C., Minor, C. A., Kates, S. A., Albericio, F., and Barany, G., *J. Org. Chem.*, 61, 6326, 1996.
236. Ramage, R., Irving, S. L., and McInnes, C., *Tetrahedron Lett.*, 34, 6599, 1993.
237. White, P., in *Innovation and Perspectives in Solid Phase Synthesis. Peptides, Proteins and Nucleic Acids. Biological and Biomedical Applications*, Epton, R., Ed., Mayflower Worldwide Ltd, Birmingham, 1994, 701.
238. Zhang, L., Rapp, W., Goldhammer, C., and Bayer, E., in *Innovation and Perspectives in Solid Phase Synthesis. Peptides, Proteins and Nucleic Acids. Biological and Biomedical Applications*, Epton, R., Ed., Mayflower Worldwide Ltd, Birmingham, 1994, 717.
239. Rovero, P., Quartara, L., and Fabbri, G., *Tetrahedron Lett.*, 32, 2639, 1991.
240. McMurray, J. S., *Tetrahedron Lett.*, 32, 7679, 1991.
241. Trzeciak, A. and Bannwarth, W., *Tetrahedron Lett.*, 33, 4557, 1992.
242. Tromelin, A., Fulachier, M.-H., Mourier, G., and Ménez, A., *Tetrahedron Lett.*, 33, 5197, 1992.
243. Kates, S. A., Solé, N. A., Johnson, C. R., Hudson, D., Barany, G., and Albericio, F., *Tetrahedron Lett.*, 34, 1549, 1993.
244. Kates, S. A., Solé, N. A., Albericio, F., and Barany, G., in *Peptides: Design, Synthesis and Biological Activity*, Basava, C. and Anantharamaiah, G. M., Eds., Birkhäuser, Boston, 1994, 39.
245. Alsina, J., Rabanal, F., Giralt, E., and Albericio, F., *Tetrahedron Lett.*, 35, 9633, 1994.
246. Jensen, K. J., Songster, M. F., Vágner, J., Alsina, J., Albericio, F., and Barany, G., in *Peptides. Proceedings of the 14th American Peptides Symposium*, Kaumaya, P. T. P. and Hodges, R. S., Eds., Mayflower Scientific Ltd, Kingswinford, U.K., 1996, 30.
247. Sheehan, J. C. and Hess, G. P., *J. Am. Chem. Soc.*, 77, 1067, 1955.

248. Sarantakis, D., Teichman, J., Lien, E. L., and Fenichel, R., *Biochem. Biophys. Res. Commun.*, 73, 336, 1976.
249. Khorana, H. G. and Todd, A. R., *J. Chem. Soc.*, 2257, 1953.
250. Khorana, H. G., *Chem. Ind.*, 1087, 1955.
251. Smith, M., Moffatt, J. G., and Khorana, H. G., *J. Am. Chem. Soc.*, 80, 6204, 1958.
252. DeTar, D. F. and Silverstein, R., *J. Am. Chem. Soc.*, 88, 1020, 1966.
253. Merrifield, R. B., Gisin, B. F., and Bach, A. N., *J. Org. Chem.*, 42, 1291, 1977.
254. DeTar, D. F. and Silverstein, R., *J. Am. Chem. Soc.*, 88, 1013, 1966.
255. Rebek, J. and Feitler, D., *J. Am. Chem. Soc.*, 96, 1606, 1974.
256. Wendlberger, G., in *Methoden der Organischen Chemie (Houben-Weyl)*, Vol. 15/2, Müller, E., Ed., Georg Thieme Verlag, Stuttgart, 1974, 101.
257. Rich, D. H. and Singh, J., in *The Peptides. Analysis, Synthesis, Biology*, Vol. 1, *Major Methods of Peptide Bond Formation*, Gross, E. and Meienhofer, J., Eds., Academic Press, New York, 1979, 241.
258. Bates, H. S., Jones, J. H., and Witty, M. J., *J. Chem. Soc. Chem. Commun.*, 773, 1980.
259. Bates, H. S., Jones, J. H., Ramage, W. I., and Witty, M. J., in *Peptides 1980. Proceedings of the 16th European Peptide Symposium*, Brunfeldt, K., Ed., Scriptor, Copenhagen, 1981, 185.
260. Sheehan, J. C., Goodman, M., and Hess, G. P., *J. Am. Chem. Soc.*, 78, 1367, 1956.
261. Merrifield, R. B. and Woolley, D. W., *J. Am. Chem. Soc.*, 78, 4646, 1956.
262. DeTar, D. F., Silverstein, R., and Rogers, F. F., *J. Am. Chem. Soc.*, 88, 1024, 1966.
263. Izdebski, J., Lebek, M., and Drabarek, S., *Rocz. Chem. (Ann. Soc. Chim. Polon.)*, 51, 81, 1977.
264. Hegarty, A. F. and Bruice, T. C., *J. Am. Chem. Soc.*, 92, 6561, 1970.
265. Balcom, B. J. and Petersen, N. O., *J. Org. Chem.*, 54, 1922, 1989.
266. Bergmann, M. and Zervas, L., *Biochem. Z.*, 280, 1928.
267. Benoiton, N. L., *Biopolymers (Pept. Sci.)*, 40, 245, 1996.
268. Hudson, D., *J. Org. Chem.*, 53, 617, 1988.
269. Wünsch, E. and Drees, F., *Chem. Ber.*, 99, 110, 1966.
270. Weygand, F., Hoffmann, D., and Wünsch, E., *Z. Naturforsch.*, 21b, 426, 1966.
271. König, W. and Geiger, R., *Chem. Ber.*, 103, 788, 1970.
272. König, W. and Geiger, R., in *Chemistry and Biology of Peptides. Proceedings of the 3rd American Peptides Symposium*, Meienhofer, J., Ed., Ann Arbor Science Publishers, Ann Arbor, MI, 1972, 343.
273. König, W. and Geiger, R., *Chem. Ber.*, 103, 2034, 1970.
274. McCarthy, D. G., Hegarty, A. F., and Hathaway, B. J., *J. Chem. Soc. Perkin Trans. 2*, 224, 1977.
275. Horiki, K., *Tetrahedron Lett.*, 1897, 1977.
276. Barlos, K., Papaioannou, D., and Theodoropoulos, D., *Int. J. Pept. Protein Res.*, 23, 300, 1984.
277. Barlos, K., Papaioannou, D., Sanida, C. *Liebigs Ann. Chem.*, 1308, 1984.
278. Barlos, K., Papaioannou, D., Voliotis, S., Prewo, R., and Bieri, J. H., *J. Org. Chem.*, 50, 696, 1985.
279. Horiki, K., *Tetrahedron Lett.*, 1901, 1977.
280. König, W. and Geiger, R., *Chem. Ber.*, 103, 2024, 1970.
281. Carpino, L. A., El-Faham, A., and Albericio, F., *J. Org. Chem.*, 60, 3561, 1995.
282. Atherton, E., Holder, J. L., Meldal, M., Sheppard, R. C., and Valerio, R. M., *J. Chem. Soc. Perkin Trans. 1*, 2887, 1988.
283. Carpino, L. A., *J. Am. Chem. Soc.*, 115, 4397, 1993.
284. Angell, Y. M., García-Echeverría, C., and Rich, D. H., *Tetrahedron Lett.*, 35, 5981, 1994.
285. Carpino, L. A., El-Fahan, A., Minor, C. A., and Albericio, F., *J. Chem. Soc. Chem. Commun.*, 201, 1994.
286. Izdebski, J., Pachulska, M., and Orlowska, A., *Int. J. Pept. Protein Res.*, 44, 414, 1994.
287. Gibson, F. S., Park, M. S., and Rapoport, H., *J. Org. Chem.*, 59, 7503, 1994.
288. Gawne, G., Kenner, G., and Sheppard, R. C., *J. Am. Chem. Soc.*, 91, 5669, 1969.
289. Yamada, S. and Takeuchi, Y., *Tetrahedron Lett.*, 3595, 1971.
290. Wieland, T. and Seeliger, A., *Chem. Ber.*, 104, 3992, 1971.
291. Bates, A. J., Galpin, I. J., Hallett, A., Hudson, D., Kenner, G. W., Ramage, R., and Sheppard, R. C., *Helv. Chim. Acta*, 58, 688, 1975.
292. Castro, B. and Dormoy, J.-R., *Bull. Soc. Chim.*, 3359, 1973.

293. Castro, B., Dormoy, J. R., Evin, G., and Selve, C., *Tetrahedron Lett.*, 1219, 1975.
294. Castro, B., Dormoy, J.-R., Evin, G., and Selve, C., *J. Chem. Res. (S)*, 182, 1977.
295. Coste, J., Le-Nguyen, D., and Castro, B., *Tetrahedron Lett.*, 31, 205, 1990.
296. Campagne, J.-M., Coste, J., and Jouin, P., *J. Org. Chem.*, 60, 5214, 1995.
297. Kim, M. H. and Patel, D. V., *Tetrahedron Lett.*, 35, 5603, 1994.
298. Coste, J. and Campagne, J.-M., *Tetrahedron Lett.*, 36, 4253, 1995.
299. Dourtoglou, V., Ziegler, J.-C., and Gross, B., *Tetrahedron Lett.*, 1269, 1978.
300. Dourtoglou, V., Gross, B., Lambropoulou, V., and Ziodrou, C., *Synthesis*, 572, 1984.
301. Knorr, R., Trzeciak, A., Bannwarth, W., and Gillessen, D., *Tetrahedron Lett.*, 30, 1927, 1989.
302. Carpino, L., El-Faham, A., and Albericio, F., *Tetrahedron Lett.*, 35, 2279, 1994.
303. Abdelmoty, I., Albericio, F., Carpino, L. A., Foxman, B. M., and Kates, S. A., *Lett. Pept. Sci.*, 1, 57, 1994.
304. Story, S. C. and Aldrich, J. V., *Int. J. Pept. Protein Res.*, 43, 292, 1994.
305. Coste, J., Dufour, M. N., Pantaloni, A., and Castro, B., *Tetrahedron Lett.*, 31, 669, 1990.
306. Frérot, E., Coste, J., Pantaloni, A., Dufour, M.-N., and Jouin, P., *Tetrahedron*, 47, 259, 1991.
307. Schnölzer, M., Alewood, P., Jones, A., Alewood, D., and Kent, S. B. H., *Int. J. Pept. Protein Res.*, 40, 180, 1992.
308. Wijkmans, J. C. H. M., Blok, F. A. A., van der Marel, G. A., van Boom, J. H., and Bloemhoff, W., *Tetrahedron Lett.*, 36, 4643, 1995.
309. Bodanszky, M., in *The Peptides. Analysis, Synthesis, Biology*, Vol. 1, *Major Methods of Peptide Bond Formation*, Gross, E. and Meienhofer, J., Eds., Academic Press, New York, 1979, 105.
310. Menger, F. M. and Smith, J. H., *J. Am. Chem. Soc.*, 94, 3824, 1972.
311. Elliott, D. F. and Russell, D. W., *Biochem. J.*, 66, 49P, 1957.
312. Rothe, M. and Kunitz, F.-W., *Liebigs Ann. Chem.*, 609, 88, 1957.
313. Bodanszky, M. and du Vigneaud, V., *Nature (London)*, 183, 1324, 1959.
314. Jakobsen, M. H., Buchardt, O., Engdahl, T., and Holm, A., *Tetrahedron Lett.*, 32, 6199, 1991.
315. Atherton, E., Cameron, L., Meldal, M., and Sheppard, R. C., *J. Chem. Soc. Chem. Commun.*, 1763, 1986.
316. Atherton, E., Cameron, L. R., and Sheppard, R. C., *Tetrahedron*, 44, 843, 1988.
317. Hudson, D., *Pept. Res.*, 3, 51, 1990.
318. Fields, C. G., Fields, G. B., Noble, R. L., and Cross, T. A., *Int. J. Pept. Protein Res.*, 33, 298, 1989.
319. Otvos, L., Elekes, I., and Lee, V. M.-Y., *Int. J. Pept. Protein Res.*, 34, 129, 1989.
320. Yamashiro, D., *Int. J. Pept. Protein Res.*, 30, 9, 1987.
321. Wallace, C. J. A., Mascagni, P., Chait, B. T., Collawn, J. F., Paterson, Y., Proudfoot, A. E. I., and Kent, S. B. H., *J. Biol. Chem.*, 264, 15199, 1989.
322. Merrifield, R. B., Mitchell, A. R., and Clarke, J. E., *J. Org. Chem.*, 39, 660, 1974.
323. Merrifield, R. B., Singer, J., and Chait, B. T., *Anal. Biochem.*, 174, 399, 1988.
324. Benoiton, N. L. and Chen, F. M. F., in *Peptides 1986. Proceedings of the 19th European Peptides Symposium*, Theodoropoulos, D., Ed., Walter de Gruyter, Berlin, 1987, 127.
325. Harrison, J. L., Petrie, G. M., Noble, R. L., Beilan, H. S., and McCurdy, S. N., in *Techniques in Protein Chemistry*, Vol. 1, Hugli, T. E., Ed., Academic Press, San Diego, 1989, 506.
326. Fuller, W. D., Goodman, M., Naider, F., and Zhu, Y.-F., *Biopolymers (Pept. Sci.)*, 40, 183, 1996.
327. Halstrom, J., Brunfeldt, K., and Kovács, K., *Acta Chem. Scand.*, 33B, 685, 1979.
328. Halstrom, J., Qasim, M. A., Brunfeldt, K., and Nebelin, E., *Hoppe-Seyler's Z. Physiol. Chem.*, 362, 593, 1981.
329. Mobashery, S. and Johnston, M., *J. Am. Chem. Soc.*, 107, 2200, 1985.
330. Loffet, A. and Zhang, H. X., in *Innovation and Perspectives in Solid Phase Synthesis. Peptides, Proteins and Nucleic Acids. Biological and Biomedical Applications*, Epton, R., Ed., Mayflower Worldwide Ltd, Birmingham, 301, 1994.
331. Fuller, W. D., Krotzer, N. J., Naider, F. R., Xue, C.-B., and Goodman, M., in *Peptides 1992. Proceedings of the 22nd European Peptides Symposium*, Schneider, C. H. and Eberle, A. N., Eds., ESCOM, Leiden, 1993, 229.
332. Fuller, W. D., Cohen, M. P., Shabankareh, M., Blair, R. K., Goodman, M., and Naider, F. R., *J. Am. Chem. Soc.*, 112, 7414, 1990.
333. Swain, P. A., Anderson, B. L., Goodman, M., and Fuller, W. D., *Pept. Res.*, 6, 147, 1993.

334. Spencer, J. R., Antonenko, V. V., Delaet, N. G. J., and Goodman, M., *Int. J. Pept. Protein Res.,* 40, 282, 1992.

335. Xue, C.-B. and Naider, F., *J. Org. Chem.,* 58, 350, 1993.

336. Carpino, L. A., Cohen, B. J., Stephens, K. E., Sadat-Aalee, S. Y., Tien, J.-H., and Langridge, D. C., *J. Org. Chem.,* 51, 3734, 1986.

337. Carpino, L. A., Chao, H. G., Beyermann, M., and Bienert, M., *J. Org. Chem.,* 56, 2635, 1991.

338. Carpino, L. A., Sadat-Aalee, D., Chao, H. G., and DeSelms, R., *J. Am. Chem. Soc.,* 112, 9651, 1990.

339. Carpino, L. A., Mansour, E.-S. M. E., and Sadat-Aalee, D., *J. Org. Chem.,* 56, 2611, 1991.

340. Bertho, J.-N., Loffet, A., Pinel, C., Reuther, F., and Sennyey, G., *Tetrahedron Lett.,* 32, 1303, 1991.

341. Kaduk, C., Wenschuh, H., Beyerman, H. C., Forner, K., Carpino, L. A., and Bienert, M., *Lett. Pept. Sci.,* 2, 285, 1995.

342. Carpino, L. A. and El-Faham, A., *J. Am. Chem. Soc.,* 117, 5401, 1995.

343. Wenschuh, H., Beyermann, M., El-Faham, A., Ghassemi, S., Carpino, L. A., and Bienert, M., *J. Chem. Soc. Chem. Commun.,* 669, 1995.

344. Wenschuh, H., Beyermann, M., Krause, E., Carpino, L. A., and Bienert, M., *Tetrahedron Lett.,* 34, 3733, 1993.

345. Wenschuh, H., Beyermann, M., Krause, E., Carpino, L. A., and Bienert, M., in *Innovation and Perspectives in Solid Phase Synthesis. Peptides, Proteins and Nucleic Acids. Biological and Biomedical Applications,* Epton, R., Ed., Mayflower Worldwide Ltd, Birmingham, 1994, 697.

346. Wenschuh, H., Rothemund, S., Beyermann, M., Carpino, L. A., and Bienert, M., in *Peptides 1994. Proceedings of the 23rd European Peptides Symposium,* Maia, H. L. S., Ed., ESCOM, Leiden, 1995, 187.

347. Wenschuh, H., Beyermann, M., Rothemund, S., Carpino, L. A., and Bienert, M., *Tetrahedron Lett.,* 36, 1247, 1995.

348. Gisin, B. F. and Merrifield, R. B., *J. Am. Chem. Soc.,* 94, 3102, 1972.

349. Lunkenheimer, W. and Zahn, H., *Liebigs Ann. Chem.,* 740, 1, 1970.

350. Barlos, K., Gatos, D., Kapolos, S., Papaphotiu, G., Schäfer, W., and Wenqing, Y., *Tetrahedron Lett.,* 30, 3947, 1989.

351. Barlos, K., Gatos, D., Kallitsis, J., Papaphotiu, G., Sotiriu, P., Wenqing, Y., and Schäfer, W., *Tetrahedron Lett.,* 30, 3943, 1989.

352. Akaji, K., Kiso, Y., and Carpino, L. A., *J. Chem. Soc. Chem. Commun.,* 584, 1990.

353. Khosla, M. C., Smeby, R. R., and Bumpus, F. M., *J. Am. Chem. Soc.,* 94, 4721, 1972.

354. Rothe, M. and Mazánek, J., *Angew. Chem. Int. Ed. Engl.,* 11, 293, 1972.

355. Meienhofer, J., *J. Am. Chem. Soc.,* 92, 3771, 1970.

356. Albericio, F. and Barany, G., *Int. J. Pept. Protein Res.,* 26, 92, 1985.

357. Pedroso, E., Grandas, A., de las Heras, X., Eritja, R., and Giralt, E., *Tetrahedron Lett.,* 27, 743, 1986.

358. Gairí, M., Lloyd-Williams, P., Albericio, F., and Giralt, E., *Tetrahedron Lett.,* 31, 7363, 1990.

359. Capasso, S., Sica, F., Mazzarella, L., Balboni, G., Guerrini, R., and Salvadori, S., *Int. J. Pept. Protein Res.,* 45, 567, 1995.

360. Rothe, M. and Mazánek, J., *Liebigs Ann. Chem.,* 439, 1974.

361. Giralt, E., Eritja, R., and Pedroso, E., *Tetrahedron Lett.,* 22, 3779, 1981.

362. Suzuki, K., Nitta, K., and Endo, N., *Chem. Pharm. Bull.,* 23, 222, 1975.

363. Alsina, J., Giralt, E., and Albericio, F., *Tetrahedron Lett.,* 37, 4195, 1996.

364. Giralt, E., Celma, C., Ludevid, M. D., and Pedroso, E., *Int. J. Pept. Protein Res.,* 29, 647, 1987.

365. Celma, C., Albericio, F., Pedroso, E., and Giralt, E., *Pept. Res.,* 5, 62, 1992.

366. Dalcol, I., Rabanal, F., Ludevid, M.-D., Albericio, F., and Giralt, E., *J. Org. Chem.,* 60, 7575, 1995.

367. Yang, Y., Sweeney, W. V., Schneider, K., Thörnqvist, S., Chait, B. T., and Tam, J. P., *Tetrahedron Lett.,* 35, 9689, 1994.

368. Dölling, R., Beyermann, M., Haednal, J., Kernchen, F., Krause, E., Franke, P., Brudel, M., and Bienert, M., *J. Chem. Soc. Chem. Commun.,* 853, 1994.

369. L
iefländer, M., *Hoppe-Seyler's Z. Physiol. Chem.,* 320, 35, 1960.

370. Kates, S. A. and Albericio, F., *Lett. Pept. Sci.,* 1, 213, 1994.

371. Tam, J. P., Wong, T.-W., Riemen, M. W., Tjoeng, F.-S., and Merrifield, R. B., *Tetrahedron Lett.*, 4033, 1979.
372. Lauer, J. L., Fields, C. G., and Fields, G. B., *Lett. Pept. Sci.*, 1, 197, 1994.
373. Quibell, M., Owen, D., Packman, L. C., and Johnson, T., *J. Chem. Soc. Chem. Commun.*, 2343, 1994.
374. DiMarchi, R., Tam, J. P., Kent, S. B. H., and Merrifield, R. B., *Int. J. Pept. Protein Res.*, 19, 88, 1982.
375. Orlowska, A., Witkowska, E., and Izdebski, J., *Int. J. Pept. Protein Res.*, 30, 141, 1987.
376. Kent, S. B. H., Mitchell, A. R., Engelhard, M., and Merrifield, R. B., *Proc. Natl. Acad. Sci. U.S.A.*, 76, 2180, 1979.
377. Kent, S. B. H., *Annu. Rev. Biochem.*, 57, 957, 1988.
378. Meister, S. M. and Kent, S. B. H., in *Peptides. Structure and Function. Proceedings of the 8th American Peptides Symposium*, Hruby, V. J. and Rich, D. H., Eds., Pierce Chemical Company, Rockford, IL, 1983, 103.
379. Atherton, E. and Sheppard, R. C., in *Peptides. Structure and Function. Proceedings of the 9th American Peptides Symposium*, Deber, C. M., Hruby, V. J., and Kopple, K. D., Eds., Pierce Chemical Company, Rockford, IL, 1985, 415.
380. Kent, S. B. H., in *Peptides. Structure and Function. Proceedings of the 9th American Peptides Symposium*, Deber, C. M., Hruby, V. J., and Kopple, K. D., Eds., Pierce Chemical Company, Rockford, IL, 1985, 407.
381. Bedford, J., Johnson, T., Jun, W., and Sheppard, R. C., in *Innovation and Perspectives in Solid Phase Synthesis. Peptides, Polypeptides and Oligonucleotides*, Epton, R., Ed., Intercept Ltd, Andover, 1992, 213.
382. Atherton, E., Woolley, V., and Sheppard, R. C., *J. Chem. Soc. Chem. Commun.*, 970, 1980.
383. Epton, R., Goddard, P., and Ivin, K. J., *Polymer*, 21, 1367, 1980.
384. Giralt, E., Rizo, J., and Pedroso, E., *Tetrahedron*, 40, 4141, 1984.
385. Live, D. H. and Kent, S. B. H., in *Peptides. Structure and Function. Proceedings of the 8th American Peptides Symposium*, Hruby, V. J. and Rich, D. H., Eds., Pierce Chemical Company, Rockford, IL, 1983, 65.
386. Ludwick, A. G., Jelinski, L. W., Live, D., Kintaner, A., and Dumais, J. J., *J. Am. Chem. Soc.*, 108, 6493, 1986.
387. Westall, F. C. and Robinson, A. B., *J. Org. Chem.*, 35, 2842, 1970.
388. Yamashiro, D., Blake, J., and Li, C. H., *Tetrahedron Lett.*, 1469, 1976.
389. Milton, S. C. F. and Milton, R. C. d. L., *Int. J. Pept. Protein Res.*, 36, 193, 1990.
390. Stewart, J. M. and Klis, W. A., in *Innovation and Perspectives in Solid Phase Synthesis and Related Technologies. Peptides, Polypeptides and Oligonucleotides. Macro-Organic Reagents and Catalysts*, Epton, R., Ed., SPCC (U.K.) Ltd, Birmingham, 1990, 1.
391. Hendrix, J. C., Halverson, K. J., Jarrett, J. T., and Lansbury, P. T., *J. Org. Chem.*, 55, 4517, 1990.
392. Zhang, L., Goldhammer, C., Henkel, B., Zühl, F., Panhaus, G., Jung, G., and Bayer, E., in *Innovation and Perspectives in Solid Phase Synthesis. Peptides, Proteins and Nucleic Acids. Biological and Biomedical Applications*, Epton, R., Ed., Mayflower Worldwide Ltd, Birmingham, 1994, 711.
393. Haack, T. and Mutter, M., *Tetrahedron Lett.*, 33, 1589, 1992.
394. Haack, T., Zier, A., Nefzi, A., and Mutter, M., in *Peptides 1992. Proceedings of the 22nd European Peptides Symposium*, Schneider, C. H. and Eberle, A. N., Eds., ESCOM, Leiden, 1993, 595.
395. Haack, T., Nefzi, A., Dhanapal, D., and Mutter, M., in *Innovation and Perspectives in Solid Phase Synthesis. Peptides, Proteins and Nucleic Acids. Biological and Biomedical Applications*, Epton, R., Ed., Mayflower Worldwide Ltd, Birmingham, 1994, 521.
396. Narita, M., Ishikawa, K., Chen, J.-Y., and Kim, Y., *Int. J. Pept. Protein Res.*, 24, 580, 1984.
397. Narita, M., Ishikawa, K., Nakano, H., and Isokawa, S., *Int. J. Pept. Protein Res.*, 24, 14, 1984.
398. Weygand, F., Steglich, W., Bjarnason, J., Akhtar, R., and Khan, N. M., *Tetrahedron Lett.*, 3483, 1966.
399. Bedford, J., Hyde, C., Johnson, T., Jun, W., Owen, D., Quibell, M., and Sheppard, R. C., *Int. J. Pept. Protein Res.*, 40, 300, 1992.

400. Johnson, T., Quibell, M., Owen, D., and Sheppard, R. C., *J. Chem. Soc. Chem. Commun.,* 369, 1993.
401. Hyde, C., Johnson, T., Owen, D., Quibell, M., and Sheppard, R. C., *Int. J. Pept. Protein Res.,* 43, 431, 1994.
402. Hancock, W. S., Prescott, D. J., Vagelos, P. R., and Marshall, G. R., *J. Org. Chem.,* 38, 774, 1973.
403. Kent, S. B. H. and Merrifield, R. B., in *Peptides 1980. Proceedings of the 16th European Peptides Symposium,* Brunfeldt, K., Ed., Scriptor, Copenhagen, 1981, 328.
404. Stewart, J. M. and Young, J. D., *Solid Phase Peptide Synthesis,* Pierce Chemical Company, Rockford, IL, 1984, 1.
405. Cameron, L., Meldal, M., and Sheppard, R. C., *J. Chem. Soc. Chem. Commun.,* 270, 1987.
406. Sheppard, R. C., *Chem. Br.,* 557, 1988.
407. Kaiser, E., Colescott, R. L., Bossinger, C. D., and Cook, P. I., *Anal. Biochem.,* 34, 595, 1970.
408. Sarin, V. K., Kent, S. B. H., Tam, J. P., and Merrifield, R. B., *Anal. Biochem.,* 117, 147, 1981.
409. Edman, P., *Acta Chem. Scand.,* 4, 277, 1950.
410. Edman, P., *Acta Chem. Scand.,* 4, 283, 1950.
411. Edman, P. and Begg, G., *Eur. J. Biochem.,* 1, 80, 1967.
412. Laursen, R. A., *Eur. J. Biochem.,* 20, 89, 1971.
413. Sakakibara, S., Shimonishi, Y., Kishida, Y., Okada, M., and Sugihara, H., *Bull. Chem. Soc. Jpn.,* 40, 2164, 1967.
414. Sakakibara, S., Kishida, Y., Nishizawa, R., and Shimonishi, Y., *Bull. Chem. Soc. Jpn.,* 41, 438, 1968.
415. Brenner, M. and Curtius, H. C., *Helv. Chim. Acta,* 46, 2126, 1963.
416. Yajima, H., Fujii, N., Ogawa, H., and Kawatani, H., *J. Chem. Soc. Chem. Commun.,* 107, 1974.
417. Lenard, J. and Robinson, A. B., *J. Am. Chem. Soc.,* 89, 181, 1967.
418. Sakakibara, S. and Shimonishi, Y., *Bull. Chem. Soc. Jpn.,* 38, 1412, 1965.
419. Scotchler, J., Lozier, R., and Robinson, A. B., *J. Org. Chem.,* 35, 3151, 1970.
420. Feinberg, R. S. and Merrifield, R. B., *J. Am. Chem. Soc.,* 97, 3485, 1975.
421. Tam, J. P., Heath, W. F., and Merrifield, R. B., *J. Am. Chem. Soc.,* 105, 6442, 1983.
422. King, D. S., Fields, C. G., and Fields, G. B., *J. Chem. Soc. Perkin Trans.,* 1, 255, 1990.
423. Solé, N. A. and Barany, G., *J. Org. Chem.,* 57, 5399, 1992.
424. Ponsati, B., Giralt, E., and Andreu, D., in *Peptides. Chemistry, Structure and Biology. Proceedings of the 11th American Peptides Symposium,* Rivier, J. E. and Marshall, G. R., Eds., ESCOM, Leiden, 1990, 960.
425. Riniker, B. and Hartmann, A., in *Peptides. Chemistry, Structure and Biology. Proceedings of the 11th American Peptides Symposium,* Rivier, J. E. and Marshall, G. R., Eds., ESCOM, Leiden, 1990, 950.
426. Riniker, B. and Kamber, B., in *Peptides 1988. Proceedings of the 20th European Peptides Symposium,* Jung, G. and Bayer, E., Eds., ESCOM, Leiden, 1989, 115.
427. Gesellchen, P. D., Rothenberger, R. B., Dorman, D. E., Paschal, J. W., Elzey, T. K., and Campbell, C. S., in *Peptides. Chemistry, Structure and Biology. Proceedings of the 11th American Peptides Symposium,* Rivier, J. E. and Marshall, G. R., Eds., ESCOM, Leiden, 1990, 957.
428. Rovero, P., Pegoraro, S., Viganò, S., Bonelli, F., and Triolo, A., *Lett. Pept. Sci.,* 1, 149, 1994.
429. Pearson, D. A., Blanchette, M., Baker, M. L., and Guindon, C. A., *Tetrahedron Lett.,* 30, 2739, 1989.
430. Alewood, P. F. and Meutermans, W. D. F., in *Peptides 1994. Proceedings of the 23rd European Peptides Symposium,* Maia, H. L. S., Ed., ESCOM, Leiden, 1995, 242.
431. Gairí, M., Lloyd-Williams, P., Albericio, F., and Giralt, E., *Tetrahedron Lett.,* 35, 175, 1994.
432. Nicolás, E., Pedroso, E., Giralt, E. *Tetrahedron Lett.,* 30, 497, 1989.
433. Lenard, J., Scally, A. V., and Hess, G. P., *Biochem. Biophys. Res. Commun.,* 14, 498, 1964.
434. Partridge, S. M. and Davis, H. F., *Nature (London),* 165, 62, 1950.
435. Wu, C.-R., Wade, J. D., and Tregear, G. W., *Int. J. Pept. Protein Res.,* 31, 47, 1988.
436. Sakakibara, S., Shin, K. H., and Hess, G. P., *J. Am. Chem. Soc.,* 84, 4921, 1962.
437. Lenard, J. and Hess, G. P., *J. Biol. Chem.,* 239, 3275, 1964.
438. Anteunis, M. J. O. and van der Auwera, C., *Int. J. Pept. Protein Res.,* 31, 301, 1988.
439. Gutte, B. and Merrifield, R. B., *J. Am. Chem. Soc.,* 91, 501, 1969.
440. Gutte, B. and Merrifield, R. B., *J. Biol. Chem.,* 246, 1922, 1971.

441. Kent, S. B. H., Parker, K. F., Schiller, D. L., Woo, D. D. L., Clark-Lewis, I., and Chait, B. T., in *Peptides. Chemistry and Biology. Proceedings of the 10th American Peptides Symposium*, Marshall, G. R., Ed., ESCOM, Leiden, 1988, 173.
442. Schneider, J. and Kent, S. B. H., *Cell*, 54, 363, 1988.
443. Nutt, R. F., Brady, S. F., Darke, P. L., Ciccarone, T. M., Colton, C. D., Nutt, E. M., Rodkey, J. A., Bennett, C. D., Waxman, L. H., Sigal, I. S., Anderson, P. S., and Veber, D. F., *Proc. Natl. Acad. Sci. U.S.A.*, 85, 7129, 1988.
444. Darke, P. L., Nutt, R. F., Brady, S. F., Garsky, V. M., Ciccarone, T. M., Leu, C.-T., Lumma, P. K., Freidinger, R. M., Veber, D. F., and Sigal, I. S., *Biochem. Biophys. Res. Commun.*, 156, 297, 1988.
445. Wlodawer, A., Miller, M., Jaskólski, M., Sathyanarayana, B. K., Baldwin, E., Weber, I. T., Selk, L. M., Clawson, L., Schneider, J., and Kent, S. B. H., *Science*, 245, 616, 1989.
446. Miller, M., Schneider, J., Sathyanarayana, B. K., Toth, M. V., Marshall, G. R., Clawson, L., Selk, L., Kent, S. B. H., and Wlodawer, A., *Science*, 245, 1149, 1989.
447. Bergman, D. A., Alewood, D., Alewood, P. F., Andrews, J. L., Brinkworth, R. I., Englebretsen, D. R., and Kent, S. B. H., *Lett. Pept. Sci.*, 2, 99, 1995.
448. Kent, S. B. H., Alewood, D., Alewood, P., Baca, M., Jones, A., and Schnölzer, M., in *Innovations and Perspectives in Solid-Phase Synthesis. Peptides, Polypeptides and Oligonucleotides*, Epton, R., Ed., Intercept Ltd, Andover, 1992, 1.
449. Milton, R. C. D. L., Milton, S. C. F., and Kent, S. B. H., *Science*, 256, 1445, 1992.
450. Ramage, R., Green, J., Muir, T. W., Ogunjobi, O. M., Love, S., and Shaw, K., *Biochem. J.*, 299, 151, 1994.
451. Barrow, C. J. and Zagorski, M. G., *Science*, 253, 179, 1991.
452. Hilbich, C., Kisters-Woike, B., Reed, J., Masters, C. L., and Beyreuther, K., *J. Mol. Biol.*, 218, 149, 1991.
453. Burdick, D., Soreghan, B.; Kwon, M., Kosmoski, J., Knauer, M., Henschen, A., Yates, J., Cotman, C., and Glabe, C., *J. Biol. Chem.*, 267, 546, 1992.
454. Hendrix, J. C., Halverson, K. J., and Lansbury, P. T., *J. Am. Chem. Soc.*, 114, 7930, 1992.
455. Quibell, M., Turnell, W. G., and Johnson, T., in *Innovation and Perspectives in Solid Phase Synthesis. Peptides, Proteins and Nucleic Acids. Biological and Biomedical Applications*, Epton, R., Ed., Mayflower Worldwide Ltd, Birmingham, 1994, 653.
456. Quibell, M., Turnell, W. G., and Johnson, T., *J. Chem. Soc. Perkin Trans. 1*, 2019, 1995.
457. Quibell, M., Turnell, W. G., and Johnson, T., *Tetrahedron Lett.*, 35, 2237, 1994.

Chapter 3

Peptide Synthesis in Solution

Despite the dominance of solid-phase methods, the synthesis of peptides in solution remains one of the major chemical approaches to these molecules.[1,2] The principal advantage of synthesis in solution is that intermediates can be isolated and characterized at every step, if necessary, so that chemists can always know which molecular species they are dealing with at any point. Problems that arise can be identified and dealt with there and then, unlike in SPPS where one might discover the presence of appreciable amounts of unwanted side products only right at the end, after the peptide has been cleaved from the resin. However, this advantage is bought at a price — classical peptide synthesis in solution is much slower than SPPS and the problems that the insolubility of some of the protected peptide intermediates can cause may, in many cases, be insurmountable. Even so, there are still some areas, such as the large-scale synthesis of peptides, the synthesis of peptides composed of unusual or uncommon amino acids, the synthesis of cyclic peptides, and of cyclic or of linear depsipeptides, and the semisynthesis of protein molecules, where synthesis may often be advantageously carried out in solution.

3.1 STRATEGIC CONSIDERATIONS

Peptide molecules can be synthesized by following either of two basic strategies: linear or convergent. The linear (stepwise) coupling of amino acids in the $C{\rightarrow}N$ direction is the most successful strategy in solid-phase methods. Although it can also be used in peptide synthesis in solution, an alternative is to use convergent strategies based on the condensation of peptide segments. The choice of which to adopt depends upon several factors, such as the complexity of the target molecule, the protection scheme chosen, and considerations of economics and logistics.

Convergent strategies in which peptide segments are coupled together to give the desired target molecule appear, at least at first sight, to be the most promising for the synthesis of longer peptides, for several reasons. The condensation of peptide segments should lead to fewer problems in the isolation and purification of intermediates. The difference between the desired condensation product and the segments themselves, in terms of molecular size and chemical nature, should be sufficiently pronounced so as to permit their separation relatively easily. In linear syntheses, on the other hand, the addition of residues one at a time leads to only small differences in chemical character between the various intermediate peptides of a long synthesis, in this way generating a more difficult separation problem. In the synthesis of large peptides by convergent strategies, teams of workers can be assigned to different segments, which can be prepared simultaneously. A linear synthesis must, of necessity, be carried out in a sequential manner and probably by one person or at most by one team. A further advantage of convergent strategies is that workers are always closer to the starting materials, so that it is easier to go back to the beginning and to repeat a synthesis if difficulties are encountered, if material is lost, or if it is realized that a change in the protection scheme is necessary. If an advanced intermediate in a linear synthesis is lost, on the other hand, then the whole molecule must be synthesized again from scratch. In addition, if a range of analogues differing in one region of the peptide chain are to be produced, the convergent strategy lends itself to this with greater facility.

However, in spite of these seeming advantages, there are other, perhaps less obvious, factors that militate against convergent syntheses. First, and above all for larger segments, the low molar concentration of the components to be coupled means that reaction kinetics are slow and that yields for the coupling reaction can often be low. The use of an excess of one of the components can, of course, alleviate this to some extent, but the use of excesses of a large, valuable, and complex advanced intermediate is something that must be considered carefully. Second, in order for the condensation of peptide segments to take place in an unambiguous manner, they must normally be protected at all reactive functionalities that are not involved in the coupling reaction. (See Section 4.3 for methods that allow unprotected or partially protected peptides to be coupled.) This requirement leads to problems of solubility, again above all for large segments, which are often only poorly soluble in the commonly used solvents. In severe cases of insolubility the chemical manipulation of protected peptide segments may be rendered effectively impossible. (See Section 4.1.2.1 for some modern approaches to improving the solubility of protected peptide segments.)

Nevertheless, in general, convergent strategies have tended to be those adopted for the synthesis of complex peptides in solution, as many of the most important syntheses demonstrate (see Section 3.6 for selected examples). The protected peptide segments used in a convergent strategy are, however, usually synthesized in a stepwise manner although larger segments, of course, might be made by the condensation of smaller ones. For smaller peptide target molecules, linear strategies have been used more often, although not necessarily to the exclusion of convergent ones.

The protected peptide segments used in convergent synthesis strategies can be synthesized by conventional solution methods or by SPPS using appropriate modifications of the chemistry involved. By the same token, segment condensations can be carried out either in solution or on solid supports. In this chapter only methods for synthesis in solution are considered. The use of solid supports, either for the synthesis of protected peptide segments or for their condensation, is treated more fully in Chapter 4.

3.2 PROTECTION SCHEMES

The two major protection schemes used in SPPS, namely, the Boc/Bzl and the Fmoc/tBu approaches, are not necessarily the most suitable for synthesis in solution, although variations on them may be used. Generally speaking, a wider range of protecting groups is used, because the *C*-terminus must be protected in a way that is compatible with the overall protection scheme. In addition, other N^α-protecting groups apart from the Boc and Fmoc groups are useful in synthesis in solution, even though they have not found application in SPPS.

3.2.1 N^α Protection

Urethanes are by far the most commonly applied type of N^α protection when peptides are synthesized in solution. The most useful are the benzyloxycarbonyl[3] group **1**, known as the Z group in honor of its originator, Leonidas Zervas, and the Boc[4-6] group **2**, which is also used in SPPS (see Section 2.2.1). These two are the mainstays of peptide synthesis in solution, having been used extensively in the elaboration of complex target molecules.

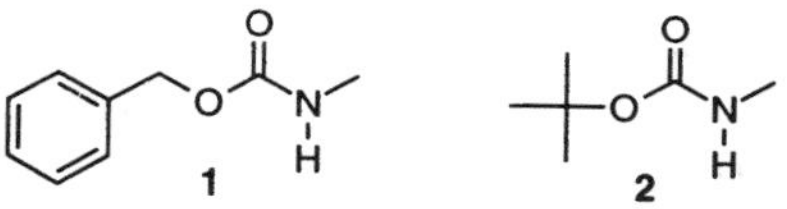

Structures are drawn to include the N^α atom of the amino acid.

Introduction and removal of the Boc group has already been discussed in Section 2.2.1. Protection of the N^α-terminus of an amino acid or peptide with the Z group is usually carried out using benzyl chloroformate[7] although occasionally, if unacceptable amounts of dipeptide formation are observed, less-reactive acylating agents may be preferred.[8] The Z group can be removed under mild conditions by catalytic hydrogenolysis as shown in Scheme 3.1 and the side products, carbon dioxide and toluene, are especially innocuous.

Hydrogenolysis, however, is not always high yielding and its use for the removal of the Z group from some complex peptides can be problematic and may sometimes require forcing conditions. Additional difficulties are sometimes caused by the presence of methionine or cysteine that can cause catalyst poisoning. Although the use of liquid ammonia as solvent[9,10] or of large excesses of catalyst[11] have been reported in such cases, it is not clear that they represent general solutions to the problem.

An alternative method for removing the Z group is acidolysis, but strong acids such as hydrogen bromide in acetic acid[12,13] are required. The stability of side chain- and *C*-terminal-protecting groups must be carefully checked if they are not to be removed in the process. The acid lability of the Z group is increased by the incorporation of electron-donating groups into the aromatic ring and the 4-methoxy-benzyloxycarbonyl (Moz) group **3** can be removed by trifluoroacetic acid

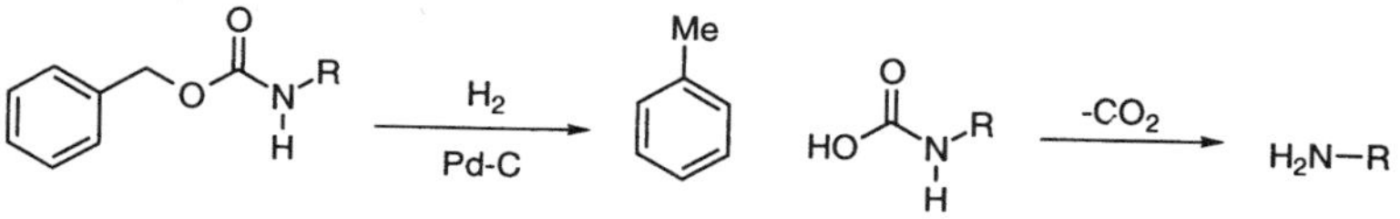

SCHEME 3.1 Hydrogenolytic removal of the Z group.

treatment[5,14-17] as well as by hydrogenolysis. Even more acid labile are the α,α-dimethyl 3, 5-dimethoxybenzyloxy[18] (Ddz) **4**, and the 2-(4-biphenylyl) isopropoxy-carbonyl[19] (Bpoc) **5** groups. These can be removed under much milder conditions than those necessary for removal of the Z group, or even of the Boc group, and solutions of acetic acid or of chloroacetic acid in dichloromethane are normally sufficient. The versatility of these N^α-protecting groups is increased by it also being possible to effect their removal by hydrogenolysis and, in the case of the Ddz group, by photolysis.

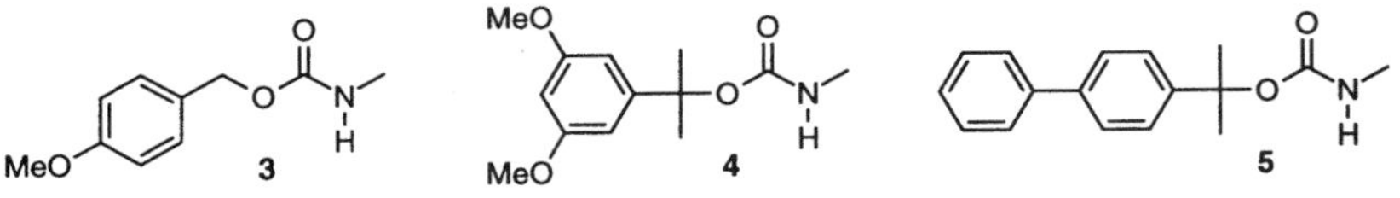

Structures are drawn to include the N^α-atom of the amino acid.

Effective N^α protection is also provided by the 2,2,2-trichloroethoxycarbonyl[20-23] (Troc) group **6**. This presents the useful property of being stable to both acid and base, but it is removed under mild conditions on treatment with zinc dust in acetic acid, following the mechanism outlined below in Scheme 3.2. The stability of the Troc group to catalytic hydrogenolysis is somewhat questionable; the hydrogenolytic removal of benzyl groups in its presence has been reported,[24] but the group is probably not to be recommended in syntheses where hydrogenolysis has to be carried out repetitively or under forcing conditions.

Many other urethane N^α-protecting groups have been used in solution synthesis, such as the *tert*-amyloxycarbonyl[25-28] **7**, adamantyloxycarbonyl[29] **8**, and allyloxycarbonyl[30] **9** groups, to name a few, although they have not, by and large, been extensively applied. The Fmoc[31] group **10**, widely used in SPPS (see Section 2.2.1), has also been applied in synthesis in solution.[32] However, its premature loss has been observed[33] and the dibenzofulvene-piperidine adduct formed on deprotection can be difficult to remove from the reaction mixture in solution and may complicate synthetic operations.

Nonurethane N^α-protecting groups have been applied far less frequently in peptide synthesis. This is mainly because many do not provide adequate protection against amino acid racemization when the carboxyl group is activated (see Section 3.3.2.1). Nevertheless, a number of nonurethane groups do provide sufficiently secure protection against racemization and perhaps the most important is the *o*-nitrophenylsulfenyl (Nps) group **11**, which has been used in complex peptide

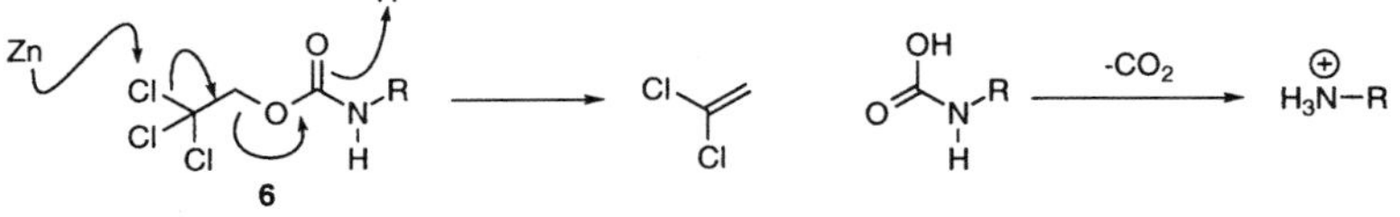

SCHEME 3.2   Zinc-mediated removal of the Troc group.

Structures are drawn to include the N^{α} atom of the amino acid.

syntheses.[34,35] It can be removed under mild conditions, which do not affect *tert*-butyl- or benzyl-based protecting groups, on treatment with 2-mercaptopyridine in acetic acid.[36] The N^{α}-trityl group[37-39] **12** is another nonurethane N^{α}-protecting group that has also been used in peptide synthesis in solution. It can be useful because its extreme acid lability allows its selective removal even in the presence of the Bpoc group.[40-43]

3.2.2 Side Chain Protection of Amino Acids

Side chain protection for the proteinogenic amino acids is treated more fully in Section 2.2.2. There are no important differences between the side chain–protecting groups used for the various amino acids in SPPS and in synthesis in solution. Side chain protection must, of course, be compatible with that of the *C*-terminus of the peptide and with the N^{α}-protecting group used. Global protection of the side chain functional groups has been more frequently applied in synthesis in solution, as it has in SPPS. However, the milder conditions that obtain in synthesis in solution when active esters or azides are used for coupling allow the use of minimal protection strategies,[44] which have been applied in several complex syntheses. Minimal protection involves leaving the side chain functional groups of various amino acids without protecting groups, and only those whose protection is obligatory are blocked. The chemistry involved is delicate, but there can be much gain in the solubility of the intermediates (see Section 4.3).

3.2.3 Protection of the *C*-Terminus

Protection of the *C*-terminus is the major differentiating factor between SPPS and peptide synthesis in solution. In the former, the insoluble polymeric support upon which synthesis is carried out may be considered to be the *C*-terminus-protecting group, whereas in peptide synthesis in solution a range of different, more conventional protecting groups is used. Protection of the *C*-terminus is not always necessary, but the unambiguous *N*-acylation of peptides having free *C*-terminal carboxylic acids requires that separately preactivated carboxyl components be used. If the carboxyl component is to be activated in the presence of the amino component, then protection of the C-terminus of the latter is obligatory. Peptide C-terminal amides do not normally need to be protected, although their formation is sometimes brought about by treating a C-terminal carboxylic alkyl ester with ammonia at some convenient point.

The C-terminus is most commonly protected as an alkyl or aryl ester. An alternative is provided by hydrazides or protected hydrazides. These latter were important in classical peptide synthesis in solution because they are easily converted into peptide acyl azides, for segment-coupling reactions.

3.2.3.1 Alkyl and Aryl Esters

Many different esters have been evaluated for protection of the C-terminus in peptide synthesis in solution.[45] The most useful for general application are the benzyl (OBzl) **13**, phenacyl (OPac) **14**, and *tert*-butyl (OtBu) **15** esters. Other alternatives include methyl (OMe) **16**, 2,2,2-trichloroethyl (OTce) **17**, carboxamidomethyl (OCam) **18**, N-benzhydryl-glycolamide (OBg) **19**, 2-trimethylsilylethyl (Tmse) **20**, and phenyl (OPh) **21** esters.

3.2.3.1.1 Benzyl and Substituted Benzyl Esters

OBzl esters **13** of amino acids are easy to form in high yields under mild conditions by acid-catalyzed esterification with benzyl alcohol,[46,47] or by reaction of the amino acid cesium carboxylate with benzyl bromide.[48] Their use as C-terminus-protecting groups parallels SPPS where the peptide–resin anchorage is usually a polymeric benzyl ester (see Section 2.3). In SPPS cleavage of the peptide–resin benzyl ester is brought about by acidolysis with strong acid and, while this possibility is also available in synthesis in solution, a more frequently used milder alternative is catalytic hydrogenolysis which, as in the case of the Z group discussed in Section 3.2.1, may not always be high yielding. Benzyl esters may also be removed by saponification or can be converted into hydrazides by treatment with hydrazine.[49-51] This latter procedure was important in classical peptide synthesis in solution since the hydrazides can then be converted into the acyl azides for peptide coupling. This effectively allows the transformation of a benzyl ester C-terminus-protecting group into an active derivative for amide bond formation.

When the benzyl ester is used for C-terminus protection a range of N^α-protecting groups may be used. This includes the Boc **2**, Moz **3**, Ddz **4**, Bpoc **5**, and Troc **6** groups, although the repetitive acidolyses necessary for Boc group removal may lead to some loss of the C-terminal benzyl ester in longer syntheses. The benzyl ester is incompatible with hydrogenolytic removal of the Z group and, although it has been used as carboxyl protection when the Z group is removed acidolytically, it is not fully stable to such conditions. The use of the appreciably more-acid-stable p-nitrobenzyl ester is a safer combination.[52,53]

3.2.3.1.2 Phenacyl Esters

Amino acid OPac esters **14** are produced in high yield upon treatment of N^α-protected amino acid carboxylates with bromoacetophenone.[54] Formally, the phenacyl ester may be regarded as a methyl ester substituted with a benzoyl group. It is, therefore, appreciably more reactive toward nucleophiles than the unsubstituted methyl ester. Deprotection of C-terminal phenacyl esters can be brought about under mild conditions by treatment of the peptide with a range of nucleophiles, of which the most commonly-used is sodium thiophenoxide.[55] Indeed, the phenacyl group is a little too sensitive to nucleophiles so that racemization of amino acid phenacyl esters **14** can result during coupling, owing to the reversible cyclization shown in Scheme 3.3. For a more detailed discussion on racemization and epimerization in peptide synthesis see Section 3.3.2.1.

This mechanism leads to racemization of the amino component in coupling reactions. It is particularly severe in the case of Pro[1,56,57] (normally the most resistant of the DNA-encoded amino acids to racemization) possibly because the rigid five-membered ring favors the conformation necessary for cyclization. Racemization by this mechanism is catalyzed by HOBt and can sometimes be suppressed by carrying out couplings in the absence of this additive. A surer but more intricate alternative is to use a different C-terminus-protecting group, such as the benzyl ester, for the formation of dipeptides and then to replace this with the phenacyl group for further chain elongation. Although glycine phenacyl esters cannot racemize, cyclization can

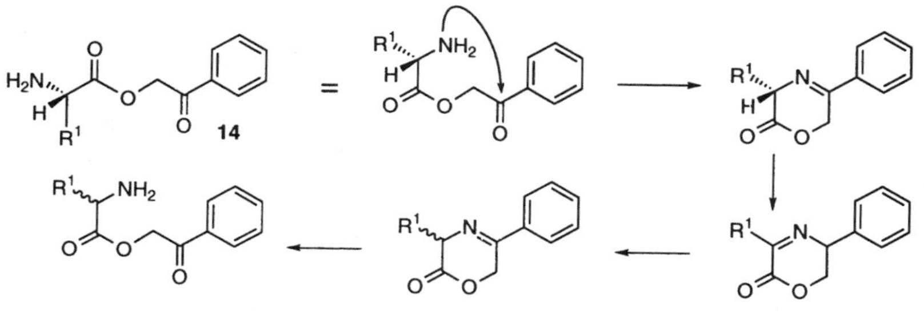

SCHEME 3.3 Mechanism of racemization of amino acid OPac esters.

occur, often in appreciable amounts, leading to poor coupling yields for the second amino acid of the sequence.[58-60]

The phenacyl group can also be removed by treatment with zinc in acetic acid[54] but is degraded, and only partially removed, by catalytic hydrogenation because of the competing reduction of the ketone function.[61] This renders it incompatible with hydrogenolytic removal of the Z group **1**. The phenacyl group is stable to acids, even to liquid hydrogen fluoride, a property that makes it very suitable for use with the Boc group **2** as N^α protection. Its use has been advocated by Sakakibara,[1] as part of a general approach to the synthesis of large peptides and proteins.

3.2.3.1.3 tert-Butyl Esters

The formation of O*t*Bu esters **15** may be brought about by treatment of the N^α-protected amino acid with isobutene in the presence of sulfuric acid,[62] by transesterification of the amino acid with *tert*-butyl acetate,[63,64] or by using *tert*-butyl 2,2,2-trichlororacetimidate.[65-67] *tert*-Butyl esters are stable to base-catalyzed hydrolysis and to nucleophiles in general; the formation of diketopiperazines in dipeptide *tert*-butyl esters is not usually a problem (see Section 3.3.3.1). However, they may not be stable under certain basic conditions, if an intramolecular nucleophile can intervene such as in the formation of aspartimides or of hydantoins (see Section 2.4.2.1.2, and Section 3.3.3.2, respectively). In general, *tert*-butyl esters are useful *C*-terminus-protecting groups and may be removed by acid hydrolysis with moderately strong acids, such as trifluoroacetic acid or solutions of hydrogen chloride in organic solvents. They are, however, sufficiently stable to weak acids to allow the washing of organic solutions of the peptides with dilute aqueous acids, in standard workup procedures.

Use of the *tert*-butyl ester-protecting group is particularly effective when the Bpoc group **5** is used for N^α protection and trityl group-based protection for the side chains, and this combination has been employed in major syntheses (see Section 5.4.2.2). Another useful combination is the use of *tert*-butyl esters with the Z group **1** as N^α protection.

3.2.3.1.4 Methyl and Substituted Methyl Esters

OMe esters **16** can be prepared simply and in high yield using a variety of procedures, one of the most popular being treatment of the relevant amino acid with a methanolic solution of hydrogen chloride.[68,69] This group is a reasonable choice when the goal is the synthesis of peptide *C*-terminal amides, since treatment of the ester with ammonia at some strategically convenient point leads to their formation in good yield. However, the methyl ester is more problematic for the synthesis of peptide *C*-terminal acids, since its removal by saponification, even under optimum conditions, can lead to unacceptable amounts of epimerization. Furthermore, certain residues, in particular Ser, Cys, and Thr, can be degraded.[70-73] In spite of these drawbacks methyl esters were used extensively in classical peptide synthesis in solution. They were important because, as with benzyl esters, their hydrazinolysis leads to peptide hydrazides that can then be converted into the corresponding acyl azides for segment coupling. A further advantage is that they are compatible with a wide variety of N^α-protecting groups and almost all of those discussed above in Section 3.2.1 can be used in combination with them.

The OTce ester[74,75] **17**, which may formally be considered to be a substituted methyl ester, can also be used for C-terminus protection. As with the Troc group **6** described above in Section 3.2.1, trichloroethyl esters can be hydrolyzed under mild conditions[76] that are compatible with acid-labile N^α protection. Again, by analogy with the Troc group, their stability to hydrogenolysis is less clear.

The OCam **18** and N-benzhydryl-glycolamide **19** esters are two other substituted methyl esters that can usefully be applied in peptide synthesis in solution.[77-79] The former has been used for C-terminus protection in enzymatic peptide synthesis.[80] Both are removed by saponification, although the latter can also be hydrazinolyzed efficiently.

3.2.3.1.5 Ethyl and Substituted Ethyl Esters

The considerations that apply to peptide methyl esters also apply to peptide ethyl (OEt) esters, with the difference that they are somewhat more resistant to saponification. This means that the various side reactions that accompany the basic hydrolysis of methyl esters are correspondingly more troublesome in the case of ethyl esters. Substituted ethyl esters can be easier to hydrolyze, and several types were proposed for peptide synthesis. The only ones that became adopted in practice to any significant extent were the OTmse esters **20**, which present the useful property of being labile to fluoride ion, [81,82] as shown in Scheme 3.4.

Their formation is achieved by esterification of the N^α-protected amino acid with 2-trimethylsilylethanol. The preferred method for their elimination is treatment of the peptide ester with a quaternary ammonium fluoride in dimethylformamide. Since they are stable to hydrogenolysis, they are compatible with the use of the Z group **1** as N^α protection. They can also be used in combination with the Boc group **2** since, although they are removed on treatment with moderately strong acids such as anhydrous trifluoroacetic acid, they are sufficiently stable to solutions of hydrogen chloride in organic solvents to allow the Boc group to be removed selectively.

3.2.3.1.6 Phenyl Esters

Amino acid OPh esters **21** are easily prepared by treatment of suitably protected derivatives with phenol in the presence of a carbodiimide[72] and are usually crystalline

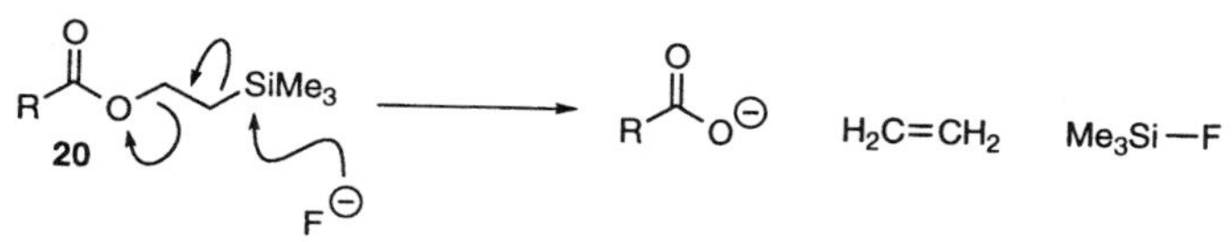

SCHEME 3.4 Fluoride ion-mediated removal of the Tmse group.

solids that are stable to acidolysis and to catalytic hydrogenolysis. They can be removed either by saponification, at rates faster than those of methyl esters, or more rapidly, under milder conditions, by the action of peroxide ion and alkali.[83,84] This latter procedure must be adopted with a certain amount of caution if the peptide contains residues sensitive to oxidation such as Met, Cys, or Trp, although it has been reported that the addition of dimethyl sulfide eliminates the problem.[72] Phenyl esters have been used as C-terminus-protecting groups for the synthesis of complex peptides, although they have not been widely adopted, probably because of concern over their stability to aminolysis. The widespread use of phenyl esters substituted by electron-withdrawing groups, such as the *p*-chlorophenyl or *p*-nitrophenyl esters, as activated derivatives (see Section 3.3.1.3) for coupling reactions, illustrates that the dividing line between a protecting group and an activating group can be a fine one.

Substitution of the aromatic ring with electron-donating groups has been suggested for reducing their susceptibility to aminolysis, but few such esters have been adopted into general practice. An interesting example is the *o*-benzyloxphenyl[85,86] ester **22**, which is a member of the so-called "safety-catch"[87] class of protecting groups. These are stable in their original forms but can be converted, by simple chemical reactions, into groups that are much more reactive.

The benzyloxyphenyl ester **22** is generally unreactive toward nucleophiles but upon hydrogenolysis is converted into the 2-hydroxyphenyl ester **23**, as shown in Scheme 3.5. Ester **23** undergoes aminolysis much more rapidly. A similar state of affairs obtains when the 2-phenacyloxyphenyl ester is used, but in this case conversion to the 2-hydroxyphenyl ester **23** is achieved by treatment with zinc dust in acetic acid.[88] Such chemistry has been used in the elaboration of several polypeptides.[89-91]

3.2.3.2 Hydrazides

Peptide hydrazides **24** are formed simply by treating the peptide methyl, ethyl, or benzyl ester with hydrazine.[49-51] The hydrazide can serve as a protecting group for the carboxyl function and may be converted into the acyl azide by treatment with nitrous acid,[92] or by treatment with organic nitrites.[93] The disadvantage of the hydrazide as a protecting group for the carboxyl terminal is that it is not by any means inert to the acylating species used for coupling. The second amino group of the hydrazide is sufficiently reactive to undergo acylation by activated amino acids or peptides. For protection of the C-terminus as the hydrazide to be useful, only

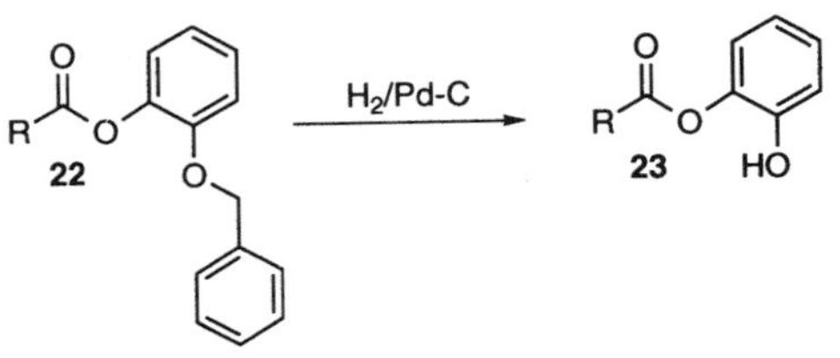

SCHEME 3.5 The safety-catch principle. The stable benzyloxyphenyl ester is converted into the reactive 2-hydroxyphenyl ester by hydrogenolysis.

moderately active acylating derivatives, such as active esters, can be used for chain elongation.[94]

Unwanted acylation at the amino group of peptide C-terminal hydrazides can be avoided by using N-protected hydrazides. The most useful of this class of compounds are the benzyloxycarbonyl **25**, *tert*-butoxycarbonyl **26**, and 2,2,2-trichloroethoxycarbonyl **27** protected derivatives.[95-102] These C-terminus protecting groups can be prepared by reaction of the appropriately protected hydrazine derivative with an N^α-protected amino acid, using standard coupling procedures.[103] With the C-terminus protected, the peptide is synthesized by chain elongation at the N-terminus until, at the appropriate time, the hydrazide protecting group is removed, releasing the hydrazide that is then converted into the azide for coupling at the C-terminus. This method of C-terminus protection is very flexible and found extensive use in the synthesis of complex peptides by segment coupling strategies. The disadvantages are that coupling must be performed using peptide acyl azides as activated derivatives with the attendant slow coupling kinetics that this often involves. Furthermore, two mutually compatible amine-protecting groups are required, making the overall protection scheme correspondingly more involved.

3.3 CHAIN ELONGATION

As has been discussed in Section 3.1 above, a combination of linear and convergent strategies may be used when peptide molecules are synthesized in solution, so that chain elongation can take place both by the coupling of single amino acids and of peptide segments. Although the chemical methods used in both of these types of couplings are broadly similar, there may be important differences in reaction kinetics and solubility. However, above and beyond such considerations there is the danger of epimerization of the C-terminal amino acid of the carboxyl component when peptide segments are coupled. The minimization of such epimerization is vital if successful syntheses of biologically active peptides are to be achieved. The problem of racemization and epimerization in peptide synthesis is considered in more detail below in Section 3.3.2.1.

3.3.1 Amino Acid Coupling

All of the major methods for effecting amide bond formation between amino acids for the synthesis of peptides were originally investigated for peptide synthesis in solution. Several of those methods that were important in their time have now been superseded by more modern procedures, often as a result of experience gained in solid-phase work. In SPPS the use of coupling reagents, such as carbodiimides or phosphonium and uronium salts (see Section 2.4.1.1), is the most popular way of making peptide bonds. Although this option may also be applied in peptide synthesis in solution, other alternatives are also available. The most important methods for forming amide bonds in solution are considered below.

3.3.1.1 Acyl Azides

Acyl azides may be formed from the corresponding acyl hydrazides **24** by treatment with nitrous acid or organic nitrites.[92,93] These activated derivatives are not isolated, but rather are generated *in situ,* so that addition of the amino component suffices to bring about coupling. The use of acid azides in peptide synthesis was first proposed and demonstrated by Theodor Curtius[104] in 1902. However, it became obsolete as a method for the coupling of amino acids, rather than peptide segments, when more rapid and cleaner procedures were developed. Its use in the coupling of protected peptide segments in convergent synthesis strategies was much longer lived, however, and is discussed in more detail in Section 3.3.2.2.2.

3.3.1.2 Anhydrides

Several classes of anhydride are used as activated derivatives in peptide chemistry. In general, reactions tend to be fast even at lower temperatures, and the side products are usually innocuous so that coupling reactions are clean and workup is straightforward.

3.3.1.2.1 *Mixed Carboxylic Acid Anhydrides*

There are two problems inherent in the use of mixed anhydrides formed from protected amino acids or peptides and carboxylic acids. First of all, since the anhydride has two carbonyl groups, there is a regiochemical problem. Attack of the amino component must take place at the correct site if the desired product is to be obtained. The success of the method depends upon it being possible to reduce attack at the "wrong" carbonyl group completely, since it leads to the formation of *N*-acylated (capped) amino derivatives that can take no further part in peptide synthesis. This can be achieved by using carboxylic acids, or their derivatives, which present severe steric hindrance. Mixed anhydrides formed with isovaleroyl chloride **28** or pivaloyl chloride **30** are the most successful of this class of coupling reagent.[105-108] Steric hindrance at the wrong carbonyl of anhydrides **29** or **31** is now so severe that acylation occurs more or less exclusively at the other (see Scheme 3.6).

A second problem associated with the use of carboxylic mixed anhydrides is that they can disproportionate, giving mixtures of the two possible symmetrical anhydrides, which again leads to capping of the amino component. Disproportionation can be avoided by preparing the anhydride immediately prior to coupling, in a separate step, by treatment of the N^{α}-protected amino acid carboxylate with the requisite acid chloride.[107] The resulting solution is then allowed to react immediately with the amino component.

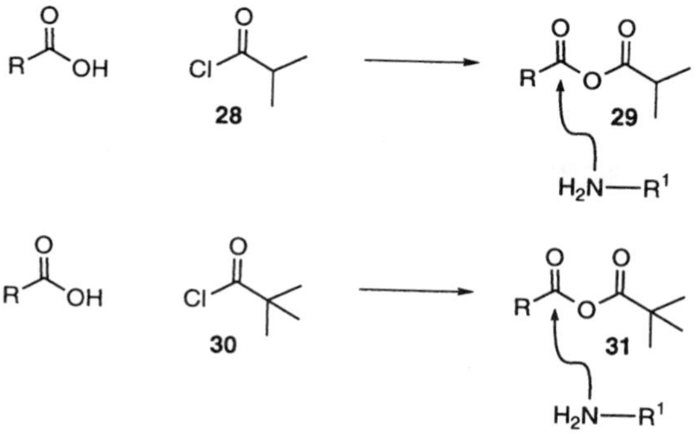

SCHEME 3.6 Formation of mixed anhydrides from carboxylic acid chlorides.

3.3.1.2.2 Mixed Carbonic Acid Anhydrides

A more popular type of mixed anhydride is that formed with carbonic acids instead of carboxylic acids. Typically, ethyl chloroformate **32**, isobutyl chloroformate **34**, or isopropenyl chloroformate **36** are used.[109-112] Isobutyl chloroformate became one of the most widely used reagents in peptide chemistry and still finds application today.[113] Anhydrides such as **33**, **35**, and **37** show an elevated tendency to react at the carbonyl group of the original amino acid or peptide component because the other is flanked by two electron-releasing oxygen atoms that decrease its electro-philicity (see Scheme 3.7).

For this class of reagent, capping of the amino component is not usually a problem except when there is severe steric hindrance to coupling. As with mixed carboxylic anhydrides, they are usually generated immediately prior to coupling and are normally not isolated as such. An advantage of the method is that the side

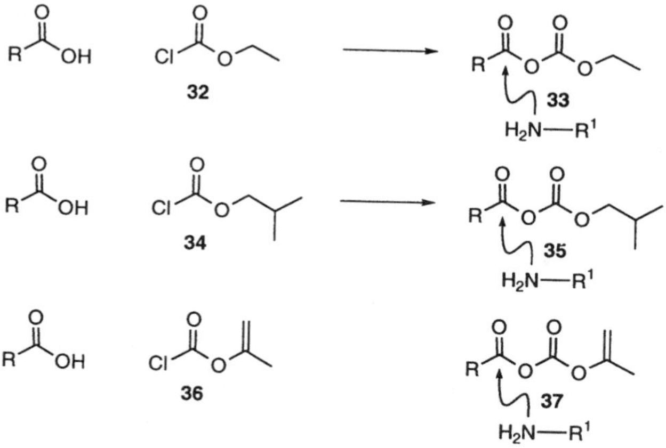

SCHEME 3.7 Formation of mixed anhydrides from chloroformates.

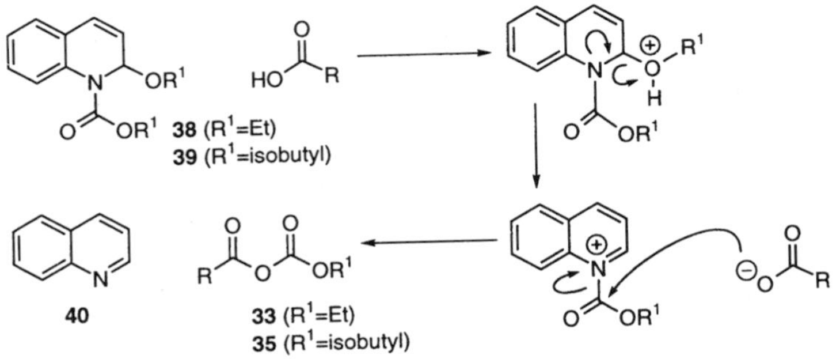

SCHEME 3.8 Mechanism of EEDQ- and IIDQ-mediated coupling.

products generated, carbon dioxide and the corresponding alcohol (acetone in the case of isopropenyl chloroformate), are innocuous and are easy to remove. The reaction conditions are mild, usually being carried out at low temperatures, in this way reducing the risks of other side reactions that might compete with amide bond formation. These advantages led to mixed carbonic anhydrides being proposed for use in the so-called repetitive mixed anhydride approach (REMA) to peptide synthesis.[114,115] This technique is discussed in more detail in Section 3.5.2.1.

Mixed carbonic acid anhydrides are also generated *in situ* when the coupling reagents 1-ethoxycarbonyl-2-ethoxy-1,2-dihydroquinoline (EEDQ) **38**, (R^1 = Et) and 1-isobutyloxycarbonyl-2-isobutoxy-1,2-dihydroquinoline (IIDQ) **39**, (R^1 = isobutyl) are used.[116,117] These reagents react only slowly with amines but rapidly with carboxylic acids, generating the corresponding mixed carbonic anhydride **33** or **35**, as shown below in Scheme 3.8.

An advantage of using EEDQ or IIDQ for the generation of mixed carbonic anhydrides is that the activated carboxyl component can undergo aminolysis immediately so that there is less likelihood of troublesome side reactions intervening. As might be expected, IIDQ leads to less acylation at the wrong carbonyl group than does EEDQ. The by-products of the reaction are carbon dioxide, quinoline **40**, and ethanol or isobutanol, so that coupling is clean and workup straightforward.

3.3.1.2.3 N-Carboxyanhydrides and Urethane N-Protected Carboxyanhydrides

N-Carboxyanhydrides (NCAs) **41** are cyclic amino acid anhydrides, also known as Leuchs anhydrides.[118,119] They can be prepared by the action of phosgene on α-amino acids[120] or by thermal decomposition of benzyloxycarbonyl-α-amino acid chlorides.[121,122] When NCAs are attacked by nucleophiles, the carbonyl group of the original amino acid is the one that reacts preferentially. Acylation gives rise to a carbamic acid that decarboxylates, generating an amino group that can then be further acylated. The sole side product of the reaction is carbon dioxide, so that the coupling reaction is very clean, and it is this factor that has been the main impulse behind efforts to apply these compounds more widely in peptide synthesis. Unfortunately, NCAs present some serious problems in use, and these have severely limited their application. They are sensitive to water, which can provoke their polymerization, and they can also be acylated at the wrong carbonyl group. Additionally, premature decarboxylation of the carbamic acid produced during the coupling reaction liberates an amino group that is then free to react with any unreacted NCA that may be present, resulting in uncontrolled reaction. Although these problems can be minimized, peptide synthesis with NCAs requires rigorously optimized reaction conditions for each specific case. Consequently, they have not been generally adopted for the synthesis of complex peptides.[122-127] They are, however, very useful for the preparation of homopolymeric α-amino acids, constituting the method of choice for the preparation of such peptides.[128-130]

Many of the problems associated with the use of NCAs can be solved by protection of the N^α-amino group as a urethane, forming urethane-protected N-carboxyanhydrides (UNCAs), such as the Boc derivative **42** or the Z derivative **43**. The use of these in SPPS has already been discussed in Section 2.4.1.4 and they are much more promising reagents for complex peptide synthesis.

In such derivatives, acylation at the unwanted carbonyl is reduced even further. Since they are protected at the N^α group, the premature decarboxylation of the carbamic acid that is produced when NCAs are used cannot occur, so that uncontrolled coupling is prevented. UNCAs have been used for the construction of several shorter peptides, usually without isolation of synthetic intermediates. They are also useful in difficult coupling reactions involving α,α-dialkylated amino acids.[131-136] The rapid reactions and the lack of troublesome side products make UNCAs very attractive reagents for peptide synthesis in solution.

3.3.1.2.4 *Mixed Anhydrides with Acids Derived from Phosphorus*

Several coupling procedures are based upon the use of phosphorus reagents. Those that proceed via acyloxyphosphonium salt intermediates are treated separately in Section 3.3.1.5 (see also Section 2.4.1.2). The main type of reactive intermediate formed by the reagents considered here is thought to be a mixed anhydride of the amino acid or peptide, on the one hand, and a phosphorus-derived acid, on the other.

The most widely used of this type of phosphorus reagent are diphenylphosphoroazidate[137-139] (DPPA) **44**, diethylphosphorocyanidate[139] (DEPC) **46**, diphenylphosphinyl chloride[140,141] **48**, and N,N'-bis(2-oxo-3-oxazolinidyl) phosphinic chloride[142,143] (BOP-Cl) **50**. These compounds are thought to generate the mixed anhydrides **45**, **47**, **49**, and **51**, respectively. In the cases of **44** and **46**, which may

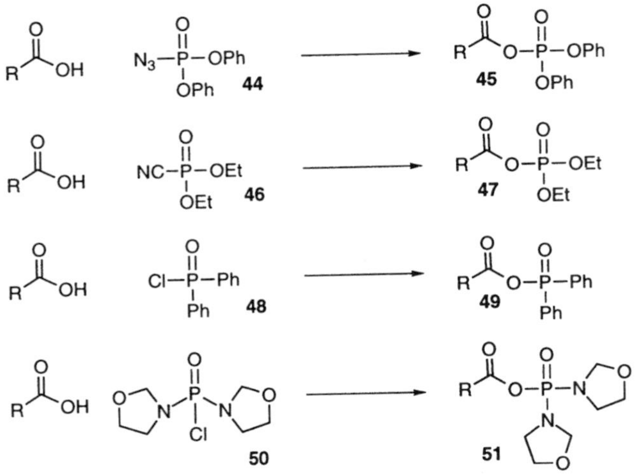

SCHEME 3.9 Formation of mixed anhydrides from phosphorus-based acid derivatives.

be used as direct coupling reagents, the amino acid or peptide acyl azides or acyl cyanides may also be involved (see Scheme 3.9). The advantages of the use of such phosphorus-based acids are that disproportionation of the mixed anhydride is substantially reduced and that the acylation is, to all intents and purposes, regiospecific.

3.3.1.2.5 N$^\alpha$-Protected Amino Acid Symmetrical Anhydrides

Regioselective acylation using anhydrides may be achieved by using symmetrical anhydrides of the N$^\alpha$-protected amino acids themselves. Such preformed symmetrical anhydrides can be prepared from the amino acid using a variety of reagents such as, for example, dicyclohexylcarbodiimide. Symmetrical anhydrides are discussed in more detail in Section 2.4.1.4. The advantage of using them is that, as with other anhydrides, they are highly activated, but acylation is now unambiguous. The disadvantages are that they are relatively wasteful in terms of the initial amino acid (although the ready commercial availability of Boc, Fmoc and Z-protected amino acids makes this consideration one of secondary importance) and that some amino acids can undergo side reactions when activated in this way. Side reactions may occur, however, when any of the other anhydrides discussed in this section are used for activation. Symmetrical anhydrides are also produced as transient intermediates in amino acid couplings when carbodiimides are used and, consequently, are of considerable importance in modern peptide chemistry.

3.3.1.3 Active Esters

Active esters have already been discussed in Section 2.4.1.3. They have been used in many syntheses in solution, and their use has been especially advocated by Bodanszky.[144] Since active esters are at a lower level of activation than some of the active species produced in other coupling methods, synthetic peptide intermediates can be less completely protected than is necessary when other methods for chain elongation are used. Such "semiglobal" protection strategies[44] can have favorable effects with respect to the solubility of the intermediates, especially in solvents such as alcohols. Residues that can normally be left without side chain–protecting groups in this type of strategy include Asn, Gln, Ser, Thr, Tyr, and, less commonly, Asp and Glu.

Although other coupling methods can be used, active esters tend to give the best results when the amino component in a coupling reaction has a free carboxylic acid.[145] They have also been widely used for the stepwise elongation of peptides in solution without isolation of the synthetic intermediates since the only side product produced in the various coupling reactions is the alcohol or phenol component of the active esters themselves, and this accumulates only slowly throughout the synthesis.[144]

Some active esters can also be used for temporary protection of the *C*-terminus, in so-called "backing-off" procedures. These are used to prepare protected peptide active esters when, for whatever reason, their direct preparation from the protected peptide carboxylic acid and the coupling reagent and substituted phenol is undesirable.[146] This might be because unacceptable amounts of side products (such as diketopiperazines) are formed or because extensive racemization occurs. A typical backing-off maneuver,[147-149] used to prepare dipeptide active ester **54**, is shown below in Scheme 3.10.

The operation involves removal of the N^α-protecting group from amino acid active ester **52**, followed by its coupling to an activated derivative of another amino acid **53**. It is important that the carboxyl *C*-terminus of **53** be strongly activated in order to promote rapid acylation and to avoid self-condensation of the active ester: mixed anhydrides are often used to ensure this. Once **54** has been formed, it can then be used to acylate a suitably protected amino component **55** in a subsequent coupling reaction.

3.3.1.4 Carbodiimides

The use of carbodiimides as coupling reagents for peptide synthesis has been discussed in more detail in Section 2.4.1.1. Carbodiimides are also very useful coupling

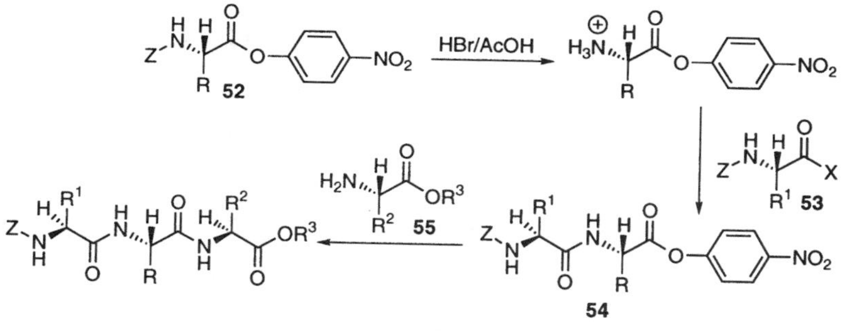

SCHEME 3.10 The "backing-off" maneuver.

reagents for the synthesis of peptides in solution. As in SPPS, the most popular member of this class of compounds is dicyclohexylcarbodiimide (DCC), although the urea by-product can be difficult to remove completely because it may have solubility properties that are similar to those of the peptide itself. The protected peptide intermediates produced in solution synthesis often require the use of dipolar aprotic solvents, such as dimethylformamide, in order to achieve acceptably high substrate concentrations, and, unfortunately, this can lead to slower activation and to the formation of more N-acylurea than is the case in dichloromethane or acetonitrile.[150-152]

The water-soluble carbodiimide (WSCDI) **56** can sometimes be advantageously used in solution synthesis since the corresponding urea can be removed by washing a solution of the peptide with aqueous acid. The mechanism of action of this carbodiimide is analogous to that for DCC as shown in Scheme 2.27, Section 2.4.1.1.

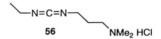

56

The combination of WSCDI, particularly as the free base, either with or without added 1-hydroxybenzotriazole has been advocated as a general method for the synthesis of large peptides in solution.[1,153] The problems caused by the poor solubility of the protected peptide intermediates can be overcome to a large extent by using solvent mixtures containing perfluorinated alcohols.[154] These solvents are appreciably acidic and can cause protonation of N^α-amino groups, reducing coupling efficiency. The basic dimethylamino group of the WSCDI, when used as the free base, serves as added tertiary amine, deprotonating any N^α-ammonium salt formed and promoting efficient amide bond formation.

3.3.1.5 Phosphonium and Uronium Reagents

The phosphonium and uronium reagents discussed in Section 2.4.1.2 are equally effective as coupling reagents in the synthesis in solution of peptides.[79,155,156] The so-called oxidation–reduction condensation, in which acyloxyphosphonium species **58** are generated as intermediates[157-159] has been applied in synthesis in solution, as well as in SPPS.[160-163] It involves treatment of a solution of the carboxyl component, whether an N^α-protected amino acid or a protected peptide, with triphenylphosphine and 2,2'-dipyridyldisulfide. The mechanism is probably that shown below in Scheme 3.11.

Initial attack of triphenylphosphine on 2,2'-dipyridyldisulfide gives the phosphonium salt **57**, which reacts with the carboxylate of the amino acid or peptide, generating acyloxyphosphonium salt **58**, the acylating species formed in this procedure. The by-products are triphenylphosphine oxide and thiopyridione. The oxidation–reduction condensation has been reported to proceed without side reactions, such as racemization or dehydration of Asn or Gln, but has not been extensively adopted in modern peptide synthesis, presumably because better-established methods are available that give similar results.

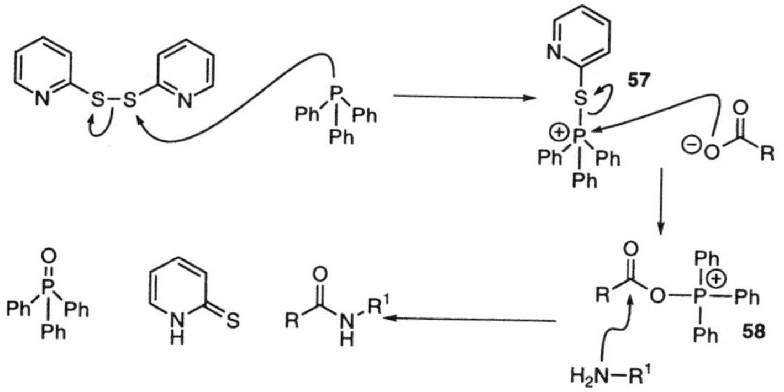

SCHEME 3.11 Mechanism of the oxidation–reduction coupling procedure.

3.3.1.6 Acyl Halides

Amino acid chlorides were used as activated derivatives in the very infancy of peptide chemistry.[68,164] However, the drastic conditions necessary to generate them, such as treatment with thionyl chloride or phosphorus pentachloride, were incompatible with the presence of sensitive amino acid residues and were a barrier to the method gaining any significant popularity, especially when numerous milder alternatives became available. The preparation of amino acid chlorides can be carried out under much less vigorous conditions nowadays,[165,166] but, even so, protecting groups based on the *tert*-butyl or trityl groups may not survive their formation except perhaps at low temperatures.[167] Boc-N^α-protected amino acid chlorides are not, consequently, practical activated derivatives for peptide synthesis.

Although the acid chlorides of Z-N^α-protected amino acids **59** can be formed, they are unstable, either cyclizing to give NCAs **41** upon warming, this being one of the standard methods for forming such anhydrides,[121,125] or giving oxazolones **60** on treatment with base (see Scheme 3.12). Simple acylamino acid chlorides also cyclize spontaneously giving oxazolones, leading to racemization.

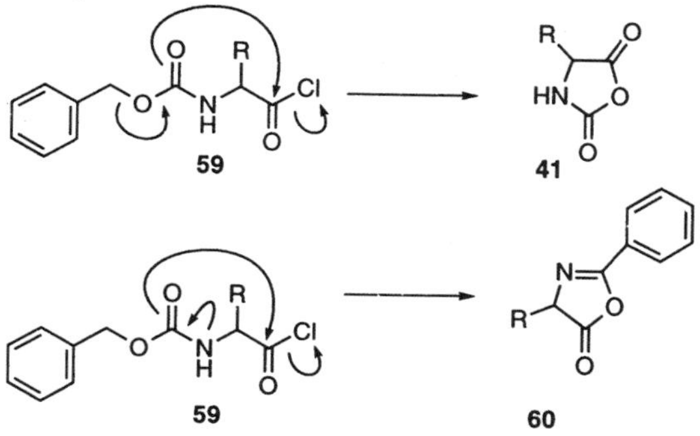

SCHEME 3.12 Side reactions suffered by benzyloxycarbonylamino acid chlorides.

Fmoc-N^α-protected amino acids on the other hand are stable, isolable interme-diates.[168] Although the conditions required for their formation are again incompatible with the presence of *tert*-butyl-based protecting groups, they can be used for peptide synthesis if the amino acids involved either do not require side chain protection or if it is benzyl or cyclohexyl based. The high level of activation of the carbonyl group leads to rapid coupling, and amino acid chlorides have been used in certain difficult coupling reactions where severe steric hindrance is present.[169]

The related amino acid fluorides are more promising as generally applicable activated derivatives since they are compatible with the presence of the Boc group for N^α protection and *tert*-butyl-based side chain protection.[167,170-172] These acid halide derivatives have, however, been more extensively applied in SPPS than in synthesis in solution (see Section 2.4.1.5).

3.3.2 Peptide-Segment Coupling

The most demanding requirement for any coupling method applied to the formation of an amide bond between two protected peptide segments is that of reducing epimerization of the *C*-terminal amino acid of the carboxyl component to a mini-mum. This can be made all the more difficult by the low molar quantities of reactants and the often poor solubility of the intermediates. Together, these tend to lead to slow coupling rates, allowing side reactions such as epimerization to compete effec-tively with peptide bond formation (see Section 3.1). Nevertheless, efficient methods for the coupling of peptide segments are available and can usually be made to proceed cleanly and in good yield, even when larger segments are involved.

3.3.2.1 Racemization and Epimerization in Peptide Synthesis

The DNA-encoded amino acids, with the exception of glycine, are chiral. Peptides are, therefore, assembled by amide bond formation between optically active mono-mers. This being so, the possibility of loss of chiral integrity (also known as stere-omutation) must be considered. If biologically important peptides are to be produced efficiently, such loss must be reduced to minimal levels.

When a molecule containing one stereogenic center suffers stereomutation, then the term most proper to the process is *enantiomerization*. It is, however, far more commonly referred to as *racemization,* a term that, strictly speaking, implies the conversion of one enantiomer into a 1:1 mixture of enantiomers. In peptide synthesis, therefore, if an amino acid having only one stereogenic center undergoes a loss of stereochemical integrity during any synthetic operation, it can be said to have racemized. On the other hand, when a molecule containing two or more stereogenic carbon atoms, such as a peptide or amino acids like Ile or Thr, undergoes a similar loss of chiral integrity at one of these atoms, then the process is best described as epimerization. However, again, such terminology is rather loosely used in peptide chemistry, and, despite pleas for greater precision,[173] racemization is commonly used to mean the loss of any degree of chiral integrity in amino acids or peptides.

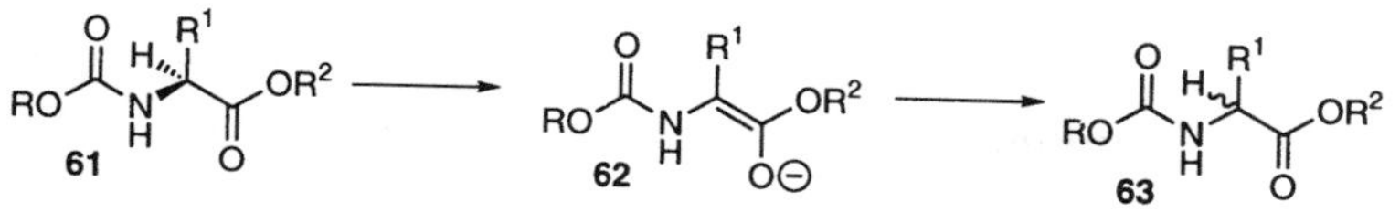

SCHEME 3.13 Direct enolization of amino acid and peptide derivatives.

Racemization or epimerization can occur, in principle, during any of the steps involved in making a peptide. In practical terms, however, the risk is usually negligible except during activation and coupling, where two main mechanisms, both base catalyzed, can operate.

3.3.2.1.1 Direct Enolization

Abstraction of the α-proton of an amino acid derivative **61**, generates a carbanion **62** that can then reprotonate on either side leading to racemized products **63**, as shown in Scheme 3.13. If the carboxyl terminal of the amino acid or peptide derivative is in the form of the free carboxylic acid (**61**, $R^2 = H$), then racemization under the normal conditions of peptide synthesis does not occur by this mechanism. This is because the formation of the carboxylate anion, in effect, protects the α-stereogenic center by making the abstraction of its proton more difficult. Racemization can occur, however, either when the carboxyl terminus is protected as an ester or when it is in the form of an activated derivative for coupling. The amount observed will depend upon the base, the protecting groups present, and the solvent used. Nevertheless, this pathway alone is thought to account for relatively little of the racemization observed in peptide synthesis.[174] It may operate to some extent when couplings are slow and the C-terminal activated derivative is present for appreciable lengths of time in a basic medium. On the other hand, it is known to be the dominant process in some specific cases where abstraction of the α-proton in question is facilitated, either because the anion that results is stabilized in some way or because the reaction is carried out in strongly basic media.

Aryl glycines. Aryl glycines are nonproteinogenic amino acids often found in microbial peptides. They are easily racemized under basic conditions because the α-proton, being benzylic, is easily abstracted and the anion generated is stabilized by the aromatic ring.[175] This facile racemization can cause quite serious problems when peptides containing phenylglycine or its derivatives are synthesized.[176-179] Careful attention must be paid to the reaction conditions at all times, but especially on activation and coupling.

S-Protected cysteine derivatives. The facile racemization of S-protected cysteine derivatives,[180] on the other hand, is somewhat more puzzling. In the case of S-benzyl protected Cys, it was initially suggested that it occurred via base-catalyzed elimination of benzylthiol,[175] but subsequent work appeared to rule this out.[181-183] It is possible that racemization of Cys is facilitated by stabilization of the carbanion formed on α-proton abstraction, by interaction with the d-orbitals of the sulfur atom.[184,185] That Met does not racemize to any appreciable extent under similar conditions might be a consequence of such overlap being less favored in its case.[186] Ser, however, in which such interaction is impossible, also undergoes similar racemization, albeit to a lesser extent.[187,188] The stabilization of the enol form in structures such as **64** might be a contributing factor here; similar stabilization is also possible in **65**, but is much less likely in **66**.

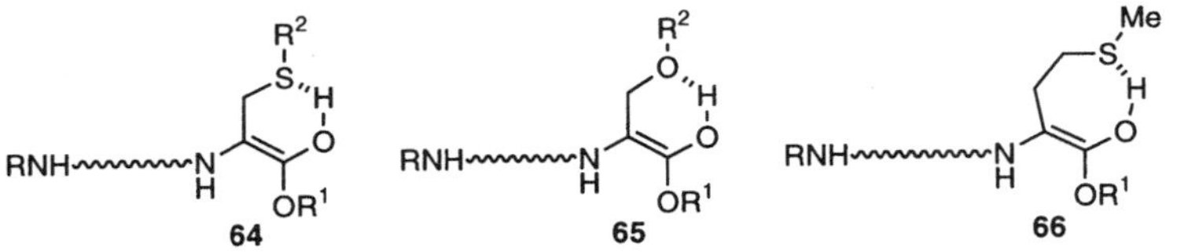

Saponification of amino acid or peptide esters. Despite it being quite a widely used synthetic operation,[45] unacceptable amounts of racemization may be observed on saponification of amino acid (even those having N^α-urethane protection) or peptide esters, if conditions are not carefully controlled. The strong bases used are powerful enough to abstract α-protons: their use in peptide synthesis should be avoided where possible. Similar reservations apply to the formation of C-terminal peptide amides by ammonolysis of peptide esters.

3.3.2.1.2 Formation and Epimerization of 5(4H)-Oxazolones

The main base-catalyzed mechanistic pathway for racemization in peptide synthesis proceeds via the formation of 5(4H)-oxazolones.[174,189-191] These intermediates can be formed rapidly on base treatment of activated amino acid or peptide derivatives **67**, and the process may be subject to either general or specific base catalysis. The former, in which abstraction of the amide proton is the rate-determining step[192] and is simultaneous with ring closure to form the oxazolone **68**, is illustrated in Scheme 3.14.

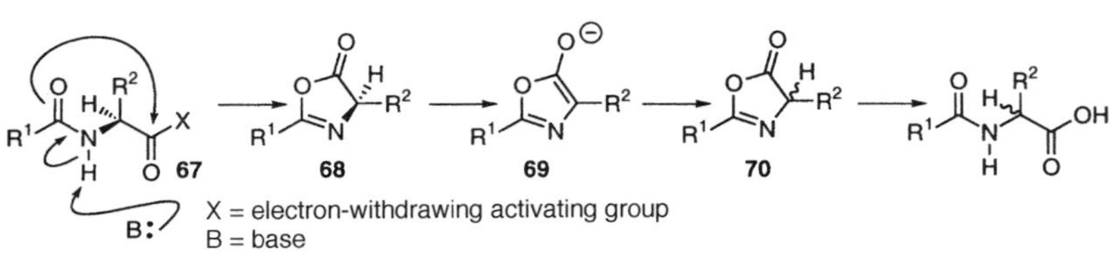

SCHEME 3.14 Racemization of amino acid and peptide derivatives via oxazolone formation.

Abstraction of the amino acid α-proton from **68** then gives rise to a resonance-stabilized carbanion **69**. This can reprotonate on either side, leading to the formation of a racemic oxazolone **70**, which on opening gives the racemized amino acid or peptide derivative. In the case of specific base catalysis,[193] abstraction of the amide proton gives rise to an equilibrium concentration of the amide oxyanion **71**, which then cyclizes to give **68**, as shown in Scheme 3.15.

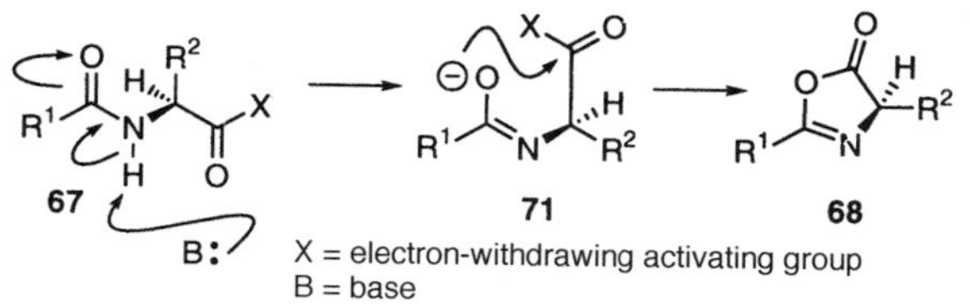
67
71
68
X = electron-withdrawing activating group
B = base

SCHEME 3.15 Specific base catalysis in oxazolone formation.

The rate of formation of the oxazolone depends quite critically upon the nature of the substituent present at the N^{α}-amino group. If this is a simple acyl group, as in an acetyl or benzoyl N^{α}-substituted amino acid, or if it is an amide bond to another amino acid in a peptide, then formation of the amino acid or peptide oxazolone can be rapid. Racemization will then occur by proton abstraction. Consequently, when peptide segments are to be coupled, methods for minimizing these processes must be used and the most important are discussed below in Section 3.3.2.2.

On the other hand, single α-amino acids, protected with urethane groups at their *N*-termini can be activated and coupled with minimal racemization.[194] The consequence of this is that both in linear SPPS and linear peptide synthesis in solution, racemization on coupling is not normally an issue that has to be addressed, except under exceptional circumstances. This fact is central to modern peptide chemistry, but the reasons it should be so are not entirely clear. It may be that activated amino acids having urethane N^{α} protection form oxazolones less readily[195,196] and that, if formed, the urethane group destabilizes the carbanion that would be generated by proton abstraction. Urethane N^{α}-protected amino acids do, nevertheless, racemize to a greater or lesser extent in certain circumstances. The esterification of an amino acid to a solid support using excess dimethylaminopyridine as catalyst is one such case and prolonged contact between the activated derivative and a basic medium, another. These conditions can, however, normally be avoided.

Oxazolones can also form in the absence of base catalysis,[197] as shown in Scheme 3.16, where the oxazolonium cation **72** is generated on cyclization. The racemization of *N*-acyl-amino acids on treatment with acetic anhydride at reflux[198-202] occurs by this mechanism. These conditions are, however, too harsh for it to be relevant in normal peptide synthesis. A more important example of the formation of oxazolonium ions in the absence of base catalysis is provided by the relatively facile racemization of N^{α}-methylated amino acids **73**. These can racemize, even when they are urethane-N^{α}-protected, as shown in Scheme 3.17.

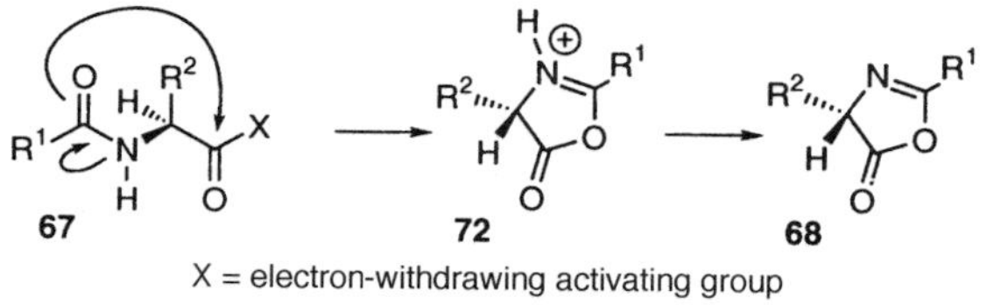
67
72
68
X = electron-withdrawing activating group

SCHEME 3.16 Oxazolone formation in the absence of base catalysis.

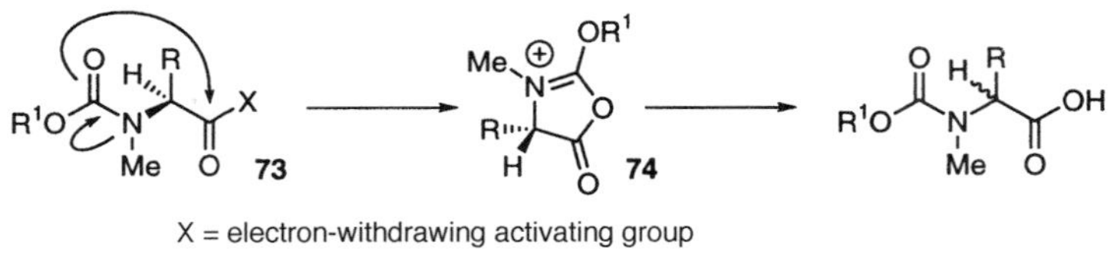

X = electron-withdrawing activating group

Scheme 3.17 Racemization of *N*-methylated amino acids by oxazolonium ion formation.

Here formation of the oxazolonium salt[203-207] **74** is facilitated by the increased nucleophilicity of the amide oxygen, consequence of the electron-donating effect of the *N*-methyl group in **73**. Since racemization of such salts can occur even under normal peptide synthesis conditions, special care must be taken in the synthesis of peptides containing *N*-methylated amino acids. Although proline may be considered to be an *N*-alkylated amino acid, formation of the oxazolonium salt does not occur except under forcing conditions, probably because the rigid five-membered ring disfavors cyclization. The result is that proline is very resistant to racemization under most conditions, although its phenacyl ester, H-Pro-OPac, can racemize easily (see Section 3.2.3.1.2).

The formation of oxazolones at the *C*-terminus of peptide segments can also adversely affect the chiral integrity of the penultimate amino acid residue. Here the initially formed oxazolone **75** can give rise to the stabilized anion **76** on loss of a proton, leading to racemization at this residue, as shown in Scheme 3.18. Although examples have been reported,[208] it is not normally a serious problem in peptide synthesis.

In principle, all of the DNA-encoded amino acids can suffer racemization by the oxazolone mechanisms discussed above. However, not all racemize to the same extent under a given set of conditions, and, furthermore, the order of the sensitivity of the various amino acids depends upon the solvent and coupling method used. In polar solvents such as dimethylformamide, the hindered residues Ile and Val racemize more than Ala, Leu, or Phe, but in dichloromethane the order is inverted, with the latter amino acids racemizing more.[209-211] These factors apart, certain amino acids

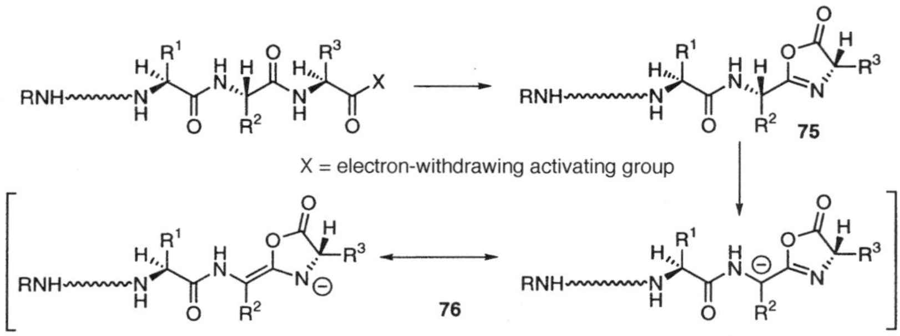

X = electron-withdrawing activating group

SCHEME 3.18 Racemization of the penultimate amino acid residue of a peptide as a consequence of oxazolone formation.

are notoriously prone to racemize, the most troublesome being His and Cys but in these cases other mechanistic factors are also at work (see Sections 2.2.2.3 and 3.3.2.1.1, respectively).

3.3.2.1.3 Acid-Catalyzed Racemization

The racemization of amino acids and peptides can also occur by acid-catalyzed mechanisms. One possibility is protonation of the amide oxygen atom followed by enolization, but this is likely to occur only in strongly acidic media.[212] Nevertheless, peptides are subjected to strongly acidic conditions, such as liquid hydrogen fluoride or hydrogen bromide, in the final deprotection reaction. Although the large amount of work carried out using these reagents indicates that racemization is not provoked to any appreciable extent, it is prudent not to use temperatures any higher than those necessary to ensure complete deprotection.

The hydrolysis of peptides for amino acid analysis is carried out in strong acid medium at elevated temperatures, conditions that can lead to appreciable amounts of racemization.[213] Methods are available for detecting and quantifying the amounts suffered by the various individual amino acids in the hydrolysis itself.[214] This allows it to be deducted from the total racemization observed, so that the amount observed in other processes, such as coupling or deprotection, can be quantified, if necessary. When all that is required from the amino acid analysis is an estimate of the amount of each amino acid present in a peptide, racemization is not important because modern amino acid analyzers do not differentiate between enantiomers under routine operating conditions.

3.3.2.2 Suppression of Racemization in the Coupling of Peptide Segments

There are two general approaches to the minimization of racemization in the coupling of peptide segments. The first is to select only those segments having either Gly or Pro as the C-terminal amino acid. In the former, racemization cannot occur, and in the latter, it does not occur under normal synthesis conditions. An alternative to this strategy is to use coupling methods that reduce racemization to a minimum when coupling peptide segments that do not have Pro or Gly at the C-terminus.

3.3.2.2.1 Glycine and Proline as C-Terminal Amino Acids

This is the surest way of avoiding racemization of the C-terminal amino acid in the coupling of peptide segments. The problem with it is that it depends upon there being a reasonably even and consistent distribution of these two residues in the peptide to be synthesized. If the target peptide contains no Gly or Pro then, obviously, it cannot be applied. Somewhat more common in practice, however, is that the distribution of these residues tends to be uneven. This means that, if the strategy is to be applied rigorously, some quite long protected peptides have to be synthesized and coupled, leading inevitably to problems of solubility and of low coupling yields.

Furthermore, it is possible that the placing of Pro at the *C*-terminus of a peptide segment is inadvisable since several studies suggest[215-220] that solubility is improved by incorporating tertiary amide bonds at more central positions within a segment (see Section 4.1.2.1.1).

3.3.2.2.2 Coupling Methods that Proceed with Minimal Racemization

A more flexible but at the same time less secure approach is to couple segments using methods that are known to give rise to little or no racemization. Several such methods have been used time and again in the condensation of peptide segments, and the most important are treated below.

Acid azides. For a long time it was believed that the coupling of peptide acyl azides proceeded without racemization, and this belief led to the method becoming one of the most widely used for the coupling of peptide segments in solution. However, it is now known that racemization can occur to an appreciable extent when peptide acyl azides are coupled, although the amount is usually relatively low and, if certain precautions are taken, can be maintained at very low levels indeed.[221] Why so little racemization should occur with this method when compared with others, such as anhydrides, carbodiimides, or even active esters, has never been convincingly explained. It has been speculated that the relatively low level of activation provided by the acyl azide may be a contributing factor. Another theory, initially proposed by Young,[222] is that the aminolysis of peptide acid azides **77** is governed by intramolecular general base catalysis, as shown in Scheme 3.19. Such catalysis would not operate for oxazolone formation since the nucleophile does not have a hydrogen atom (compare transition state **90** in Scheme 2.32). The overall result is to favor aminolysis over racemization. The azide method is, nonetheless, little used in contemporary peptide chemistry. The low level of activation of the carboxyl group leads to slow coupling reactions and the competing Curtius rearrangement[223] can give products that closely resemble the desired peptide molecule and that are difficult to remove. Newer, more rapid coupling procedures are available that proceed with similar low amounts of racemization. There are now few occasions when the use of a peptide acyl azide provides sufficient advantages to justify its use.

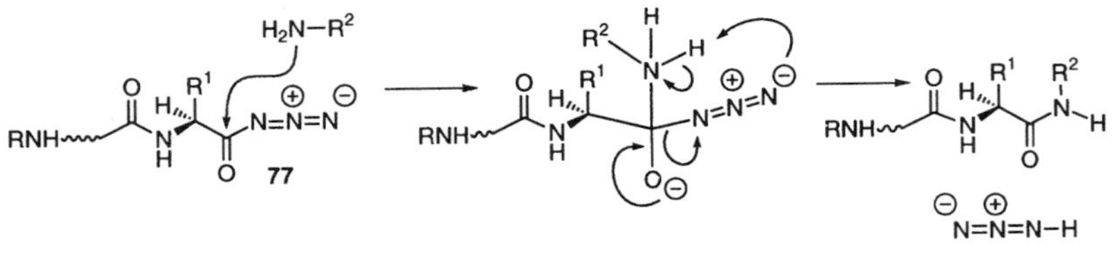

SCHEME 3.19 The aminolysis of peptide acyl azides.

Carbodiimides in the presence of additives. The coupling of protected peptide segments using carbodiimides alone proceeds with extensive racemization of the C-terminal amino acid of the segment.[224] However, such racemization can be drastically reduced by using additives such as HOSu **78**, HOBt **79**, HODhbt **80**, or HOAt **81**. These compounds are discussed in more detail in Section 2.4.1.1. They are thought to suppress racemization by intercepting highly activated derivatives, such as O-acylisoureas or symmetrical anhydrides, converting them into the corresponding active esters that are much less prone to form oxazolones. A secondary function of additives is that the hydroxyl group provides a proton source to "mop up" any excess base in the reaction medium. This prevents abstraction of the amino acid α-proton, in this way suppressing racemization by enolization.

The combination of WSCDI **56** and HOBt **79**, or HODhbt **80** has been used extensively for the coupling of peptide segments in solution. Racemization in such couplings can be maintained at very low levels if these reagents are used in a mixed solvent system consisting of perfluorinated alcohols and dichloromethane.[1,154] In segment-coupling model studies, even lower levels of racemization have been reported[225] when carbodiimide **56** is used in conjunction with additive **81**.

Phosphonium or uronium reagents in the presence of additives. The growing popularity of phosphonium and uronium salts as coupling reagents in peptide synthesis (see Section 2.4.1.2) has led to some of them being tested in peptide segment couplings in solution. Significant racemization has been observed when the BOP reagent is used alone, but the addition of 1-hydroxybenzotriazole reduces it to acceptable levels.[226] Uronium reagents have not as yet been extensively applied to peptide segment coupling in solution. Model studies with several uronium reagents[225] indicate that relatively low levels of racemization are produced. These can be reduced significantly by the addition of additives, especially **80** or **81**.

3.3.3   Side Reactions during Chain Elongation

Side reactions during chain assembly in synthesis in solution can occur in much the same way as in SPPS, although the milder coupling conditions obtaining when acyl azides or active esters are used in solution synthesis can mean that fewer secondary reactions are observed. The types of side reactions that occur are basically the same, with the formation of aspartimides or of pyroglutamic acid being among the most common undesired processes. There are, however, aspects that are peculiar to synthesis in solution.

3.3.3.1 Diketopiperazine Formation

The formation of diketopiperazines in the synthesis in solution of peptides parallels that in SPPS (see Section 2.4.2.1.1). Care must be taken to minimize this side reaction during the coupling of dipeptide esters, especially those containing amino acids that can easily adopt the *cis*-amide configuration. Diketopiperazine formation with dipeptide *tert*-butyl esters is normally not severe but for other esters significant amounts of diketopiperazines may be formed if precautions are not taken to avoid them.[227]

Diketopiperazine formation usually occurs when the N^α-amino group is liberated, but in certain cases even the urethane-protected N^α-amino group is a good enough nucleophile. This is often the case in dipeptide active esters and N^α-protected Gly–Pro dipeptide active esters **82** have an elevated tendency to cyclize under basic conditions[204] giving acyl diketopiperazines **83**, as shown in Scheme 3.20.

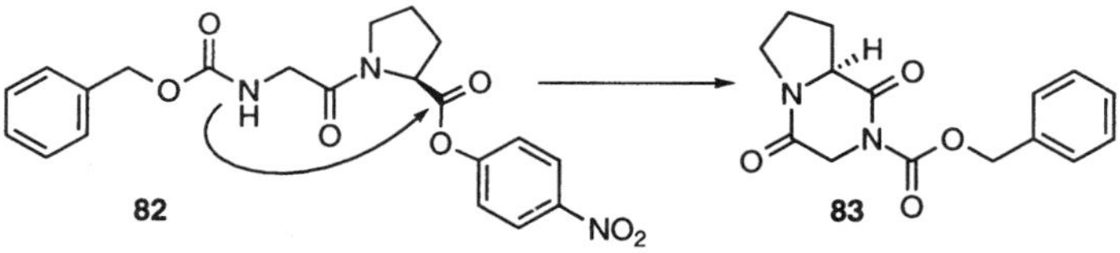

SCHEME 3.20 Diketopiperazine formation in Pro–Gly dipeptides.

Nevertheless, diketopiperazine formation tends to be more of a problem in SPPS than in synthesis in solution. Whereas in SPPS chain elongation always takes place in the $C{\rightarrow}N$ direction, in solution the tripeptide in question can be synthesized in more than one way, depending upon how severe the formation of diketopiperazines is at the dipeptide stage. As in SPPS, methods relying on the *in situ* neutralization of the N^α-amino group of the dipeptide ester can also be useful for avoiding dipeptide cyclization.

3.3.3.2 Hydantoin Formation

Peptides, such as **84**, which have benzyloxycarbonyl N^α-amino protection and Gly as the second residue, can undergo base-catalyzed ring closure on ammonolysis or saponification, yielding hydantoins[228-230] **85** or **86**, respectively. In the latter case, the hydantoin can undergo ring opening to give compounds of the type **87**. These processes are shown in Scheme 3.21. Since the synthetic operations that cause their formation can usually be avoided, hydantoins do not usually present a serious problem.

3.4   FINAL REMOVAL OF PROTECTING GROUPS

After a peptide chain has been constructed in solution, the question of the final removal of all the protecting groups remains to be considered. By far the most common reaction for the final deprotection of peptide molecules, whether synthesized in solution or by SPPS, is acidolysis. For peptides synthesized in solution the

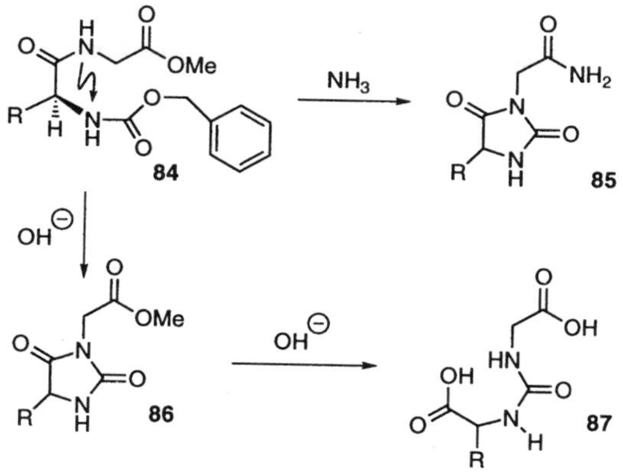

SCHEME 3.21 Hydantoin formation.

techniques used and the problems encountered in this final step are essentially the same as for the corresponding stage in SPPS. The main difference is that in SPPS the peptide molecule must be detached from the polymeric solid support. In practice, this difference is relatively insignificant, such that no in-depth treatment of this stage is given here and the reader is referred to Section 2.5.

3.5 SPECIAL APPROACHES TO THE SYNTHESIS OF PEPTIDES IN SOLUTION

Although historically peptide synthesis was carried out in solution for several decades before the invention of SPPS, such is the importance of this latter technique that it has influenced peptide synthesis in solution to a considerable degree. There have been several ingenious attempts to make modern peptide synthesis in solution less time-consuming than the conventional "classical" approaches, while at the same time preserving its inherent advantage of allowing the chemist to monitor the synthesis at any given stage. In general, the main methods for achieving this have been to carry out stepwise syntheses from the *C*-terminal amino acid and, to a large extent, to dispense with the purification and characterization of synthetic intermediates, since this is the most time-consuming aspect of classical synthesis in solution. Some of the most important developments in the field are considered below.

3.5.1 Use of Picolyl Ester-Protecting Groups

The use of polar picolyl esters as *C*-terminal protection **88** was developed[231-235] in an attempt to facilitate the isolation of the desired peptide intermediate from the reaction medium, something that is not always straightforward.

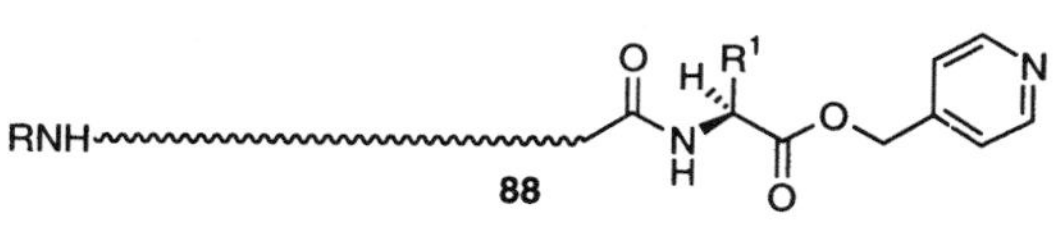

The basic picolyl group provides a "handle" on the peptide that can be used to facilitate the workup procedure. In principle, the peptide can be extracted into dilute aqueous acid by washing a solution of the crude reaction product in, say, dichloromethane with dilute aqueous citric acid. The protonated picolyl handle increases solubility of the peptide in the aqueous medium while the lower-molecular-weight side products remain in the organic phase. Another possibility is to pass the crude mixture through an appropriate ion-exchange resin so that the peptide is retained but the by-products are washed through. The purified peptide can then be recovered by washing the ion-exchange resin with a suitable buffer. The idea is simple and practical but was almost immediately superseded by SPPS, so that it was never extensively adopted in practice.

3.5.2 Rapid Synthesis in Solution

The most time-consuming aspect of peptide synthesis in solution is the isolation and characterization of each and every synthetic intermediate. This classical procedure is the most rigorous way of synthesizing peptides and allows the chemist to detect problems when they occur. For large peptides, however, the time and effort involved become prohibitive, especially so when large peptides can be synthesized by SPPS. Although chemical reactions can always go awry, the majority of deprotection and coupling reactions do proceed as expected and in high yields because of the optimization of the chemistry that has taken place over the years. This being the case, the isolation, characterization, and purification of every intermediate in a synthesis is a redundant activity. The process may be speeded up simply by not isolating *every* intermediate but just the occasional one, to check that things are going according to plan. The most extreme form of this philosophy is not to isolate *any* of the intermediates but to go right to the end of the synthesis, as in SPPS, and then to characterize the final products.

There have been several attempts to put this into practice and successful syntheses of medium-sized peptides have been carried out in this way. Such methods depend upon using deprotection schemes and coupling reactions in which the side products formed are innocuous and build up only slowly throughout the synthesis.

3.5.2.1 Repetitive Mixed Anhydride Approach

The REMA approach to peptide synthesis is based upon the activation of N^α-protected amino acids as mixed carbonic anhydrides and their coupling in a linear fashion from the C- to the N-terminus of the peptide chain.[114,115,236] When a coupling reaction is complete the excess mixed anhydride is hydrolyzed by the addition of aqueous potassium hydrogen carbonate; the protected peptide normally precipitates and the by-products are usually water soluble, providing a simple separation procedure. The intermediate peptides are isolated by filtration and can be characterized if desired. Alternatively, the synthesis can be completed and purification carried out at the end. The method is useful for the synthesis of peptides having 10 to 15 amino acids.[237-239] Longer peptides have also been synthesized, in particular, the hormone secretin, which has 27 residues.[240]

3.5.2.2 Bodanszky's Reactor

Another technique for the facilitation of stepwise peptide synthesis in solution is the use of a reactor that permits all of the mechanical and chemical operations necessary for peptide synthesis to be carried out in it, without having to transfer the contents to any other vessel at any time. This has also been called *in situ* peptide synthesis and was first proposed by Bodanszky.[241] The reaction vessel is essentially a modified centrifuge tube in which deprotection and coupling reactions (usually using aryl active ester) are carried out. The solvent is then evaporated, and a "non-solvent" such as ethyl acetate is added to the residue in order to precipitate the peptide intermediate and to dissolve the by-products. A problem that arises is that many shorter peptides are appreciably soluble in the nonsolvents. This means that *in situ* synthesis can only be carried out when the peptide intermediates have reached a certain length, normally about four but sometimes up to eight residues. The shorter peptides must therefore be synthesized by other methods. As with REMA, medium-sized peptides can be synthesized in this way and the synthesis of longer peptides has been reported.

3.5.2.3 Rapid Continuous Peptide Synthesis

This technique, developed by Carpino[168,172,242,243] relies on the use of Fmoc N^α-protected amino acid chlorides or fluorides as activated derivatives. Coupling reactions are carried out in a two-phase system consisting of chloroform and aqueous sodium bicarbonate. Once coupling is complete, the mixture is treated with 4-aminomethylpiperidine or, alternatively, tris(2-aminomethyl)amine. These react with any excess acid chloride present, converting it to the corresponding amide, and, in addition, deprotect the peptide N^α-amino group and trap the resulting dibenzo-fulvene. All subproducts are then removed by simple extraction with a phosphate buffer at pH 5.5. Coupling cycles are repeated until the desired peptide has been produced. The method is simple and quick and is useful for the synthesis of multi-gram quantities of short- and medium-sized peptides.

3.5.3 Liquid-Phase Peptide Synthesis

In liquid-phase peptide synthesis, as in SPPS, the growing peptide chain remains attached to a polymer throughout the synthesis.[244-246] The difference between the two techniques resides in the nature of the polymeric support. As the name implies, in liquid-phase peptide synthesis the support is a polyethylene glycol that is soluble in the majority of organic solvents. Deprotection and acylation reactions take place in homogeneous solution but, in order to purify intermediates, the peptide–polymer is precipitated by the addition of ether.[247] This solid is then washed extensively, as in SPPS, in order to remove low-molecular-weight impurities. Monitoring of the synthetic intermediates by spectroscopy, particularly by NMR techniques, is also easier than in SPPS since the peaks due to the polymer do not usually obscure those belonging to the peptide products.

The method is versatile and allows peptides to be constructed either by stepwise or convergent strategies. It has never become particularly popular as a general method for peptide synthesis because for the majority of cases other alternatives are available. Perhaps the most important applications of soluble polyethylene glycol polymers in peptide chemistry have been in the synthesis and conformational analysis of peptides that would normally be too insoluble to work with. A major advantage in such cases is that the solubilizing polymer permits analysis by techniques such as infrared and NMR spectroscopy and by circular dichroism; such studies indicate that the polymeric C-terminus has little effect on the conformation adopted by the peptide.[248-253]

3.6 EXAMPLES OF PEPTIDE SYNTHESIS IN SOLUTION

3.6.1 Porcine Gastrin I

The gastrins are peptide hormones that, as the name suggests, stimulate gastric secretion in the stomach in addition to having other physiological functions. There are various different types, derived from a common precursor, and they may or may not be sulfated at the Tyr residue. The structural elucidation and first synthesis of one of these hormones were carried out by Kenner and his group[254-257] in 1966. Their approach was influential in determining the strategy adopted in subsequent syntheses. The preparation of porcine gastrin I, a 17-residue peptide amide is outlined in Scheme 3.22.

The strategy adopted revolved around constructing the peptide chain from three segments of roughly equal size. The protected derivatives **89**, **90**, and **91** were synthesized from the constituent amino acids using DCC, 2,4,5-trichlorophenyl active esters or acyl azides, as appropriate. Segment **89** had pyroglutamic acid at the N-terminus and did not require N^α protection. This amino acid was incorporated

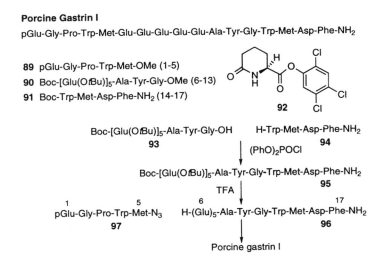

SCHEME 3.22 Synthesis of porcine gastrin I.

by reaction of its trichlorophenyl active ester **92** with the tetrapeptide methyl ester H-Gly-Pro-Trp-Met-OMe, itself synthesized by the coupling of H-Trp-Met-OMe with Boc-Gly-Pro-OH, followed by removal of the Boc group. Segment **90** was synthesized in a stepwise manner starting from H-Gly-OMe by the coupling of Z-N^α-protected amino acids, using DCC for activation in the cases of Tyr and Ala and trichlorophenyl active esters for each of the five Glu residues; the final Glu residue was, however, incorporated with Boc N^α protection. Methyl ester **90** was then selectively saponified and activation of the carboxyl group as a phosphoric acid mixed anhydride, on treatment with diphenylphosphoric chloride, allowed coupling of **93** with **94**, itself produced by removal of the Boc group from **91**. The resulting protected dodecapeptide **95** was treated with trifluoroacetic acid to remove all *tert*-butyl-based protecting groups and the free peptide **96** was purified, prior to its coupling with the acid azide **97**, derived from segment **89**. Purification of the resulting peptide by affinity chromatography and gel filtration yielded the synthetic hormone with full biological activity. This same synthetic approach was subsequently used to prepare human gastrin I, which differs from the porcine hormone in having Leu rather than Met at position 5.

3.6.2   Human Parathyroid Hormone

Human parathyroid hormone is a protein consisting of 84 amino acid residues, and the first synthesis of the molecule was carried out in solution by Sakakibara [153,258,259] using the convergent segment-coupling strategy advocated by him for the synthesis of large peptides and proteins. The overall protection scheme is that of the Boc/Bzl approach to SPPS, except that OPac esters are used as *C*-terminus protection (see Section 3.2.3.1.2 for a discussion of the use of OPac esters as *C*-terminus protection). Protected peptide segments having the *N*-terminus protected with the Boc group and the *C*-terminus protected as the OPac ester can be used either as amino components or as carboxyl components in segment-coupling reactions, since the Boc group and the OPac ester are orthogonal. This lends the strategy a useful degree of flexibility. The synthesis is outlined in Scheme 3.23.

The peptide segments themselves were all synthesized by stepwise coupling of the constituent amino acids using WSCDI **56** and HOBt **79**. The protected peptide segments were deprotected either at the *N*-terminus by treatment with trifluoroacetic acid or at the *C*-terminus by treatment with zinc dust in acetic acid, as appropriate. Segment coupling was carried out again using the WSCDI–HOBt combination of reagents. Once the protected peptide chain had been constructed, all protecting groups were removed by acidolysis with liquid hydrogen fluoride in the presence of anisole, methionine, dimethylsulfide, and ethanedithiol as scavengers. The crude peptide was purified by gel filtration, and the purified hormone had biological and chemical properties comparable with those of the native material.

3.6.3   Bovine Pancreatic Ribonuclease A

Since the primary structure of bovine ribonuclease A became known at the beginning of the 1960s, the total synthesis of this enzyme became the goal of a number of

Human Parathyroid Hormone

H-Ser-Val-Ser-Glu-Ile-Gln-Leu-Met-His-Asn-Leu-Gly-Lys-His-Leu-
Asn-Ser-Met-Glu-Arg-Val-Glu-Trp-Leu-Arg-Lys-Lys-leu-Gln-Asp-
Val-His-Asn-Phe-Val-Ala-Leu-Gly-Ala-Pro-Leu-Ala-Pro-Arg-Asp-
Ala-Gly-Ser-Gln-Arg-Pro-Arg-Lys-Lys-Glu-Asp-Asn-Val-Leu-Val-
Glu-Ser-His-Glu-Lys-Ser-Leu-Gly-Glu-Ala-Asp-Lys-Ala-Asp-Val-
Asp-Val-Leu-Thr-Lys-Ala-Lys-Ser-Gln-OH

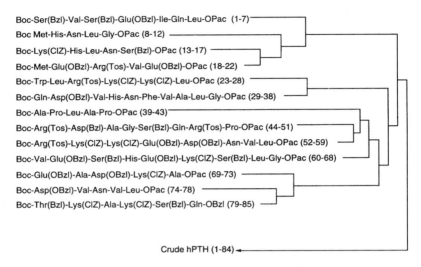

Boc-Ser(Bzl)-Val-Ser(Bzl)-Glu(OBzl)-Ile-Gln-Leu-OPac (1-7)

Boc Met-His-Asn-Leu-Gly-OPac (8-12)

Boc-Lys(ClZ)-His-Leu-Asn-Ser(Bzl)-OPac (13-17)

Boc-Met-Glu(OBzl)-Arg(Tos)-Val-Glu(OBzl)-OPac (18-22)

Boc-Trp-Leu-Arg(Tos)-Lys(ClZ)-Lys(ClZ)-Leu-OPac (23-28)

Boc-Gln-Asp(OBzl)-Val-His-Asn-Phe-Val-Ala-Leu-Gly-OPac (29-38)

Boc-Ala-Pro-Leu-Ala-Pro-OPac (39-43)

Boc-Arg(Tos)-Asp(Bzl)-Ala-Gly-Ser(Bzl)-Gln-Arg(Tos)-Pro-OPac (44-51)

Boc-Arg(Tos)-Lys(ClZ)-Lys(ClZ)-Glu(OBzl)-Asp(OBzl)-Asn-Val-Leu-OPac (52-59)

Boc-Val-Glu(OBzl)-Ser(Bzl)-His-Glu(OBzl)-Lys(ClZ)-Ser(Bzl)-Leu-Gly-OPac (60-68)

Boc-Glu(OBzl)-Ala-Asp(OBzl)-Lys(ClZ)-Ala-OPac (69-73)

Boc-Asp(OBzl)-Val-Asn-Val-Leu-OPac (74-78)

Boc-Thr(Bzl)-Lys(ClZ)-Ala-Lys(ClZ)-Ser(Bzl)-Gln-OBzl (79-85)

Crude hPTH (1-84)

SCHEME 3.23 Synthesis of human parathyroid hormone.

research groups. The chemical synthesis of material with ribonuclease activity was reported simultaneously by Gutte and Merrifield[260,261] (see Section 2.6.1) and by Denkewalter and co-workers[51,262-265] in 1969. The former was a solid-phase synthesis (see Section 2.6.1), while the latter was carried out in solution. However, since in both cases only tiny amounts of the final target molecule were produced, it was not possible to fully characterize it. The chemical preparation of pure, crystalline ribonuclease A with full enzymatic activity was not achieved until 1981 when Yajima and Fujii,[102,266-273] in a modification of their earlier solution synthesis, were able to produce crystalline material in an overall yield of 6.6%. It was a milestone in synthetic organic chemistry and marked the end of an era — the culmination of classical peptide synthesis in solution. By the time it had been accomplished, solid-phase methods had begun to dominate peptide chemistry and the advantages of the newer technique had become apparent to all but a few determined skeptics.

Yajima and Fujii carried out the synthesis of ribonuclease A by coupling together 30 relatively short, protected peptide segments, starting with the 118–124 C-terminal segment and proceeding from this in a C to N direction by condensation of the others. An advantage of this strategy was that, since the protected peptide segments were short, they were all reasonably easy to synthesize and were also reasonably soluble. This being so, as the length of the peptide chain increased, larger excesses of the relevant segments could be used to drive coupling reactions to completion. The synthesis is outlined in Scheme 3.24.

Ribonuclease A

H-Lys-Glu-Thr-Ala-Ala-Ala-Lys-Phe-Glu-Arg-Gln-His-Met-Asp-Ser-Ser-
Thr-Ser-Ala-Ala-Ser-Ser-Ser-Asn-Tyr-Cys-Asn-Gln-Met-Met-Leys-Ser-
Arg-Asn-Leu-Thr-Lys-Asp-Arg-Cys-Lys-Pro-Val-Asn-Thr-Phe-Val-His-
Glu-Ser-Leu-Ala-Asp-Val-Gln-Ala-Val-Cys-Ser-Gln-Lys-Asn-Val-Ala-Cys-
Lys-Asn-Gly-Gln-Thr-Asn-Cys-Tyr-Gln-Ser-Tyr-Ser-Thr-Met-Ser-Ile-Thr-Asp-
Cys-Arg-Glu-Thr-Gly-Ser-Ser-Lys-Tyr-Pro-Asn-Cys-Ala-Tyr-Lys-Thr-Thr-
Gln-Ala-Asn-Lys-His-Ile-Ile-Val-Ala-Cys-Glu-Gly-Asn-Pro-Tyr-Val-Pro-Val-
His-Phe-Asp-Ala-Ser-Val-OH

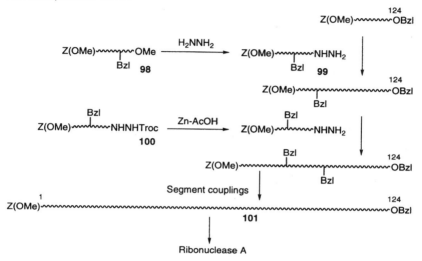

SCHEME 3.24 Synthesis of ribonuclease A.

A semiglobal side chain-protecting scheme was adopted and the side chains of Asn, Gln, His, Ser, Thr, and Tyr were systematically left without protection. Those of Asp and Glu were protected as benzyl esters (except in one case where Asp was protected as the *tert*-butyl ester in order to avoid aspartimide formation on hydrazinolysis). The side chain of Lys was protected with the Z group, that of Arg with the *p*-methoxybenzenesulfonyl group, and Met was protected as the sulfoxide. The Cys residues were all incorporated with *p*-methoxybenzyl protection, which was removed in the final acidolytic deprotection step. This represents the most basic approach to Cys residue management and was possible only because, as was known at the time, reduced ribonuclease A can be refolded and oxidized to give the native three-dimensional structure with the correct arrangement of the four disulfide bridges. The success of the synthesis probably hinged upon this fact. The *p*-methoxybenzyloxycarbonyl [Z(MeO)] group, labile to treatment with solutions of trifluoroacetic acid was chosen as the temporary amino acid N^α-protecting group.

The protected peptide segments were built up from their constituent amino acids usually by DCC-mediated coupling although *p*-nitrophenyl active esters and occasionally mixed anhydrides were also used. Except in the cases of some segments having *C*-terminal Pro or Gly, where pentafluorophenol active ester coupling proceeded without risk of racemization, the segments themselves were coupled by the acid azide procedure. Acyl hydrazides **99** were formed either by the treatment of the corresponding protected peptide methyl esters **98** with hydrazine, or, alternatively,

N-Troc protected hydrazides **100** of the segments were prepared (see Section 3.2.3.2). Usually a two- to threefold excess of azide was sufficient to ensure complete acylation of the growing peptide chain, although as its length increased acylation reactions became increasingly sluggish and larger excesses of azide became necessary. Problems with the solubility of the *N*-terminal components also became more and more apparent with increasing chain length and required the use of a mixed solvent system consisting of dimethylformamide, dimethylsulfoxide, and *N*-methyl pyrrolidinone. Attempts to couple the acid azide of the 1–20 section of the protein (known as the S peptide) to the segment corresponding to residues 21–124 (known as the S protein) could not be achieved in acceptable yield, and the synthesis was successfully completed by the condensation of two smaller peptides spanning the S-peptide region of the molecule. The condensation of these two segments required the use of 30 equivalents of acylating agent.

When the fully protected 124-residue peptide chain **101** had been constructed, it was treated with thiophenol to reduce the Met(O) residues and any sulfoxides of Cys(MeOBzl) that might have been formed during synthetic manipulation. This reduced, protected ribonuclease A was subjected to acidolytic treatment with trifluoromethanesulfonic acid containing thioanisole as a scavenger, in trifluoroacetic acid, to remove all protecting groups. After acidolysis, the crude material was treated with 2-mercaptoethanol and dithiothreitol to reduce any remaining sulfoxides and disulfide bridges that might have formed. Extensive purification by affinity chromatography and gel filtration followed by air oxidation, to form the four disulfide bridges, then gave ribonuclease A. Crystallization gave pure synthetic material with full enzymatic activity.

REFERENCES

1. Sakakibara, S., *Biopolymers (Pept. Sci.)*, 37, 17, 1995.
2. Kiso, H. and Yajima, H., in *Peptides. Synthesis, Structures and Applications*, Gutte, B., Ed., Academic Press, New York, 1995, 40.
3. Bergmann, M. and Zervas, L., *Ber. Dtsch. Chem. Ges.*, 65, 1192, 1932.
4. Carpino, L. A., *J. Am. Chem. Soc.*, 79, 4427, 1957.
5. McKay, F. C. and Albertson, N. F., *J. Am. Chem. Soc.*, 79, 4686, 1957.
6. Anderson, G. W. and McGregor, A. C., *J. Am. Chem. Soc.*, 79, 6180, 1957.
7. Bodanszky, M. and Bodanszky, A., *The Practice of Peptide Synthesis*, Springer-Verlag, Berlin, 1984, 1.
8. Wünsch, E., Graf, W., Keller, O., and Wersin, G., *Synthesis*, 958, 1986.
9. Kuromizu, K. and Meienhofer, J., *J. Am. Chem. Soc.*, 96, 4978, 1974.
10. Felix, A. M., Jimenez, M. H., Mowles, T., and Meienhofer, J., *Int. J. Pept. Protein Res.*, 11, 329, 1978.
11. Fukase, K., Kitazawa, M., Sano, A., Shimbo, K., Fujita, H., Horimoto, S., Wakamiya, T., and Shiba, T., *Tetrahedron Lett.*, 29, 795, 1988.
12. Homer, R. B., Moodie, R. B., and Rydon, H. N., *J. Chem. Soc.*, 4403, 1965.
13. Bláha, K. and Rudinger, J., *Coll. Czech. Chem. Commun.*, 30, 585, 1965.
14. Weygand, F. and Hunger, K., *Chem. Ber.*, 95, 1, 1962.
15. Weygand, F. and Hunger, K., *Chem. Ber.*, 95, 7, 1962.
16. Ohno, M., Kuromizu, K., Ogawa, H., and Izumiya, N., *J. Am. Chem. Soc.*, 93, 5251, 1971.
17. Wang, S. S., Chen, S. T., Wang, K. T., and Merrifield, R. B., *Int. J. Pept. Protein Res.*, 30, 662, 1987.

18. Birr, C., Lochinger, W., Stahlnke, G., and Land, P., *Liebigs Ann. Chem.*, 763, 162, 1972.
19. Sieber, P. and Iselin, B., *Helv. Chim. Acta*, 51, 622, 1968.
20. Woodward, R. B., Heuser, K., Gosteli, J., Naegeli, P., Oppolzer, W., Ramage, R., Ranganathan, S., and Vorbrüggen, H., *J. Am. Chem. Soc.*, 88, 852, 1966.
21. Windholz, T. B. and Johnston, D. B., *Tetrahedron Lett.*, 2555, 1967.
22. Lapatsanis, L., Milias, G., Froussios, K., and Kolovos, M., *Synthesis*, 671, 1983.
23. Dong, Q., Anderson, C. E., and Ciufolini, M. A., *Tetrahedron Lett.*, 36, 5681, 1995.
24. Jou, G., González, I., Albericio, F., Lloyd-Williams, P., and Giralt, E., *J. Org. Chem.*, 354, 62, 1996.
25. Sakakibara, S. and Fujino, M., *Bull. Chem. Soc. Jpn.*, 39, 947, 1966.
26. Sakakibara, S. and Itoh, M., *Bull. Chem. Soc. Jpn.*, 40, 646, 1967.
27. Honda, I., Shimonishi, Y., and Sakakibara, S., *Bull. Chem. Soc. Jpn.*, 40, 2415, 1967.
28. Sakakibara, S., Honda, I., Takada, K., Miyoshi, M., Ohnishi, T., and Okumura, K., *Bull. Chem. Soc. Jpn.*, 42, 809, 1969.
29. Haas, W. L., Krumkalns, E. V., and Gerzon, K., *J. Am. Chem. Soc.*, 88, 1988, 1966.
30. Kunz, H. and Unversagt, C., *Angew. Chem. Int. Ed. Engl.*, 23, 436, 1984.
31. Carpino, L. A. and Han, G. A., *J. Org. Chem.*, 37, 3404, 1972.
32. Bodanszky, M., Tolle, J. C., Gardner, J. D., Walker, M. D., and Mutt, V., *Int. J. Pept. Protein Res.*, 16, 402, 1980.
33. Bodanszky, M., Deshmane, S. S., and Martinez, J., *J. Org. Chem.*, 44, 1622, 1979.
34. Wünsch, E. and Spangenberg, R., *Chem. Ber.*, 105, 740, 1972.
35. Zervas, L., Borovas, D., and Gazis, E., *J. Am. Chem. Soc.*, 85, 3660, 1963.
36. Tun-Kyi, A., *Helv. Chim. Acta*, 61, 1086, 1978.
37. Helferich, B., Moog, L., and Jünger, A., *Ber. Dtsch. Chem. Ges.*, 58, 872, 1925.
38. Zervas, L. and Theodoropoulos, D. M., *J. Am. Chem. Soc.*, 78, 1359, 1956.
39. Barlos, K., Papaiannou, D., Patrianakou, S., and Tsegenidis, T., *Liebigs Ann. Chem.*, 1950, 1986.
40. Kamber, B., *Helv. Chim. Acta*, 54, 398, 1971.
41. Hiskey, R. G., Beacham, L. M., and Matl, V. G., *J. Org. Chem.*, 37, 2472, 1972.
42. Hiskey, R. G., Wolters, E. T., Ülkü, G., and Rao, V. R., *J. Org. Chem.*, 37, 2478, 1972.
43. Riniker, B., Kamber, B., and Sieber, P., *Helv. Chim. Acta*, 58, 1086, 1975.
44. Inman, J. K., in *The Peptides. Analysis, Synthesis, Biology*, Vol. 3, *Protection of Functional Groups in Peptide Synthesis*, Gross, E. and Meienhofer, J., Eds., Academic Press, New York, 1981, 253.
45. Roeske, R. W., in *The Peptides. Analysis, Synthesis, Biology*, Vol. 3, *Protection of Functional Groups in Peptide Synthesis*, Gross, E. and Meienhofer, J., Eds., Academic Press, New York, 1981, 101.
46. Sachs, H. and Brand, E., *J. Am. Chem. Soc.*, 75, 4610, 1953.
47. Zervas, L., Winitz, M., and Greenstein, J. P., *J. Org. Chem.*, 22, 1515, 1957.
48. Wang, S. S., Gisin, B. F., Winter, D. P., Makofske, R., Kulesha, I. D., Tzougraki, C., and Meienhofer, J., *J. Org. Chem.*, 42, 1286, 1977.
49. Gillessen, D., Schnabel, E., and Meienhofer, J., *Liebigs Ann. Chem.*, 667, 164, 1963.
50. Maclaren, J. A., *Aust. J. Chem.*, 11, 345, 1958.
51. Strachan, R. G., Paleveda, W. J., Nutt, R. F., Vitali, R. A., Veber, D. F., Dickinson, M. J., Garsby, V., Deak, J. E., Walton, E., Jenkins, S. R., Holly, F. W., and Hirschmann, R., *J. Am. Chem. Soc.*, 91, 503, 1969.
52. Schwyzer, R. and Sieber, P., *Helv. Chim. Acta*, 42, 972, 1959.
53. Schwarz, H. and Arakawa, K., *J. Am. Chem. Soc.*, 81, 5691, 1959.
54. Hendrickson, J. B. and Kandall, C., *Tetrahedron Lett.*, 343, 1970.
55. Stelakatos, G. C., Paganou, A., and Zervas, L., *J. Chem. Soc. (C)*, 1191, 1966.
56. Kubo, S., Kuroda, H., Chino, N., Watanabe, T. X., Kimura, T., and Sakakibara, S., in *Peptide Chemistry 1989. Proceedings of the 27th Symposium on Peptide Chemistry*, Yanaihara, N., Ed., Protein Research Foundation, Osaka, 1990, 257.
57. Kuroda, H., Kubo, S., Chino, N., Kimura, T., and Sakakibara, S., *Int. J. Pept. Protein Res.*, 40, 114, 1992.
58. Tjoeng, F. S., Tam, J. P., and Merrifield, R. B., *Int. J. Pept. Protein Res.*, 14, 262, 1979.
59. Birr, C., Wengert-Muller, M., and Buku, A., in *Peptides. Proceedings of the 5th American Peptides Symposium*, Goodman, M. and Meienhofer, J., Eds., John Wiley & Sons, New York, 1977, 510.

60. Birr, C. and Wengert-Muller, M., *Angew. Chem. Int. Ed. Engl.*, 18, 147, 1979.
61. Taylor-Papadimitriou, J., Yovanidis, C., Paganou, A., and Zervas, L., *J. Chem. Soc. (C)*, 1830, 1967.
62. Roeske, R. W., *Chem. Ind.*, 1121, 1959.
63. Taschner, E., Wasielewski, C., Sokolowska, T., and Biernat, J. F., *Liebigs Ann. Chem.*, 646, 127, 1961.
64. Taschner, E., Chimiak, A., Bator, B., and Sokolowska, T., *Liebigs Ann. Chem.*, 646, 134, 1961.
65. Armstrong, A., Brackenridge, I., Jackson, R. F. W., and Kirk, J. M., *Tetrahedron Lett.*, 29, 2483, 1988.
66. Riniker, B., Flörsheimer, A., Fretz, H., Sieber, P., and Kamber, B., *Tetrahedron*, 49, 9307, 1993.
67. Riniker, B., Flörsheimer, A., Fretz, H., and Kamber, B., in *Peptides 1992. Proceedings of the 22nd European Peptides Symposium*, Schneider, C. H. and Eberle, A. N., Eds., ESCOM, Leiden, 1993.
68. Curtius, T., *Ber. Dtsch. Chem. Ges.*, 16, 753, 1883.
69. Bodanszky, M., *Int. J. Pept. Protein Res.*, 23, 111, 1984.
70. Guttmann, S. and Boissonnas, R. A., *Helv. Chim. Acta*, 41, 1852, 1958.
71. Maclaren, J. A., *Aust. J. Chem.*, 11, 360, 1958.
72. Galpin, I. J., Hardy, P. M., Kenner, G. W., McDermott, J. R., Ramage, R., Seely, J. H., and Tyson, R. G., *Tetrahedron*, 35, 2577, 1979.
73. McDermott, J. R. and Benoiton, N. L., *Can. J. Chem.*, 51, 2555, 1973.
74. Eckstein, F., *Angew. Chem.*, 77, 912, 1965.
75. Woodward, R. B., *Angew. Chem.*, 78, 557, 1966.
76. Just, G. and Grozinger, K., *Synthesis*, 457, 1976.
77. Martinez, J., Laur, J., and Castro, B., *Tetrahedron Lett.*, 24, 5219, 1983.
78. Martinez, J., Laur, J., and Castro, B., *Tetrahedron*, 41, 739, 1985.
79. Amblard, M., Rodriguez, M., and Martinez, J., *Tetrahedron*, 44, 5101, 1988.
80. Kuhl, P., Zacharias, U., Burckhardt, H., and Jakubke, H.-D., *Monatsh. Chem.*, 117, 1195, 1986.
81. Sieber, P., Andreatta, R. H., Eisler, K., Kamber, B., Riniker, B., and Rink, H., in *Peptides. Proceedings of the 5th American Peptides Symposium*, Goodman, M. and Meienhofer, J., Eds., John WIley & Sons, New York, 1977, 543.
82. Sieber, P., *Helv. Chim. Acta*, 60, 2711, 1977.
83. Kenner, G. W., *Angew. Chem.*, 71, 741, 1959.
84. Kenner, G. W. and Seely, J. H., *J. Am. Chem. Soc.*, 94, 3259, 1972.
85. Jones, J. H. and Young, G. T., *J. Chem. Soc. (C)*, 436, 1968.
86. Jones, J. H., *J. Chem. Soc. Chem. Commun.*, 1436, 1969.
87. Rudinger, J., *Pure Appl. Chem.*, 7, 335, 1967.
88. Trudelle, Y., *J. Chem. Soc. Chem. Commun.*, 639, 1971.
89. Jones, J. H. and Walker, J., *J. Chem. Soc. Perkin Trans. 1*, 2923, 1972.
90. Bell, J. R., Jones, J. H., Regester, D. M., and Webb, T. C., *J. Chem. Soc. Perkin Trans. 1*, 1961, 1974.
91. Trudelle, Y., *J. Chem. Soc. Perkin Trans. 1*, 1001, 1973.
92. Stelzel, P., in *Methoden der Organische Chemie (Houben-Weyl)*, Vol. 15/2, Müller, E., Ed., Georg Thieme Verlag, Stuttgart, 1974, 296.
93. Honzl, J. and Rudinger, J., *Coll. Czech. Chem. Commun.*, 26, 2333, 1961.
94. Cheung, H. T. and Blout, E. R., *J. Org. Chem.*, 30, 315, 1965.
95. Hofmann, K., Magee, M. Z., and Lindenmann, A., *J. Am. Chem. Soc.*, 72, 2814, 1950.
96. Boissonnas, R. A., Guttmann, S., and Jaquenoud, P.-A., *Helv. Chim. Acta*, 43, 1349, 1960.
97. Yajima, H. and Kiso, Y., *Chem. Pharm. Bull.*, 19, 420, 1971.
98. Camble, R., Dupuis, G., Kawasaki, K., Romovacek, H., Yanaihara, N., and Hofmann, K., *J. Am. Chem. Soc.*, 94, 2091, 1972.
99. Storey, H. T., Beacham, J., Cernosek, S. F., Finn, F. M., Yanaihara, C., and Hofmann, K., *J. Am. Chem. Soc.*, 94, 6170, 1972.
100. Moroder, L., Borin, G., Marchiori, F., and Scoffone, E., *Biopolymers*, 11, 2191, 1972.
101. Romovacek, H., Drabarek, S., Kawasaki, K., Dowd, S. R., Obermeier, R., and Hofmann, K., *Int. J. Pept. Protein Res.*, 6, 435, 1974.
102. Yajima, H. and Fujii, N., *J. Chem. Soc. Chem. Commun.*, 115, 1980.

103. Wang, S. S., Kulesha, I. D., Winter, D. P., Makofske, R., Kutney, R., and Meienhofer, J., *Int. J. Pept. Protein Res.*, 11, 297, 1978.
104. Curtius, T., *Ber. Dtsch. Chem. Ges.*, 35, 3226, 1902.
105. Vaughn, J. R. and Osato, R. L., *J. Am. Chem. Soc.*, 73, 5553, 1951.
106. du Vigneaud, V., Ressler, C., Swain, J. M., Roberts, C. W., Katsoyannis, P. G., and Gordon, S., *J. Am. Chem. Soc.*, 75, 4879, 1953.
107. Zaoral, M., *Coll. Czech. Chem. Commun.*, 27, 1273, 1962.
108. Galpin, I. J., Handa, B. K., Hudson, B. K., Jackson, A. G., Kenner, G. W., Ohlsen, S. R., Ramage, R., Singh, B., and Tyson, R. G., in *Peptides 1976. Proceedings of the 14th European Peptides Symposium*, Loffet, A., Ed., Editions de l'Université de Bruxelles, Bruxelles, 1976, 247.
109. Wieland, T. and Bernhardt, H., *Liebigs Ann. Chem.*, 572, 190, 1951.
110. Boissonnas, R. A., *Helv. Chim. Acta*, 34, 874, 1951.
111. Vaughn, J. R., *J. Am. Chem. Soc.*, 73, 3547, 1951.
112. Jaouadi, M., Selve, C., Dormoy, J. R., and Castro, B., *Bull. Soc. Chim.*, II, 409, 1984.
113. Stavropoulos, G., Karagiannis, K., Anagnostides, S., Ministrouski, I., Selinger, Z., and Chorev, M., *Int. J. Pept. Protein Res.*, 45, 508, 1995.
114. Tilak, M. A., *Tetrahedron Lett.*, 849, 1970.
115. Beyerman, H. C., in *Chemistry and Biology of Peptides. Proceedings of the 3rd American Peptides Symposium*, Meienhofer, J., Ed., Ann Arbor Science Publishers, Ann Arbor, MI, 1972, 351.
116. Belleau, B. and Malek, G., *J. Am. Chem. Soc.*, 90, 1651, 1968.
117. Kiso, Y., Kai, Y., and Yajima, H., *Chem. Pharm. Bull.*, 21, 2507, 1973.
118. Leuchs, H., *Ber. Dtsch. Chem. Ges.*, 39, 857, 1906.
119. Leuchs, H. and Geiger, W., *Ber. Dtsch. Chem. Ges.*, 41, 1721, 1908.
120. Blacklock, T. J., Shuman, R. F., Butcher, J. W., Shearin, W. E., Budavari, J., and Grenda, V. J., *J. Org. Chem.*, 53, 836, 1988.
121. Hunt, M. and du Vigneaud, V., *J. Biol. Chem.*, 124, 699, 1938.
122. Ben-Ishai, D. and Katchalski, E., *J. Am. Chem. Soc.*, 74, 3688, 1952.
123. Bailey, J. L., *Nature (London)*, 164, 889, 1949.
124. Bailey, J. L., *J. Chem. Soc.*, 3461, 1950.
125. Bartlett, P. D. and Jones, R. H., *J. Am. Chem. Soc.*, 79, 2153, 1957.
126. Bartlett, P. D. and Dittmer, D. C., *J. Am. Chem. Soc.*, 79, 2159, 1957.
127. Denkewalter, R. G., Schwam, H., Strachan, R. G., Beesley, T. E., Veber, D. F., Schoenewaldt, E. F., Barbemeyer, H., Paleveda, W. J., Jacob, T. A., and Hirschmann, R., *J. Am. Chem. Soc.*, 88, 3163, 1966.
128. Farthing, A. C., *J. Chem. Soc.*, 3213, 1950.
129. Coleman, D. and Farthing, A. C., *J. Chem. Soc.*, 3218, 1950.
130. Coleman, D., *J. Chem. Soc.*, 3222, 1950.
131. Auvin-Guette, C., Frérot, E., Coste, J., Rebuffat, S., Jouin, P., and Bodo, B., *Tetrahedron Lett.*, 34, 2481, 1993.
132. Rodriguez, M., Califano, J. C., Loffet, A. and Martinez, J., in *Peptides 1992. Proceedings of the 22nd European Peptides Symposium*, Schneider, C. H. and Eberle, A. N., Eds., ESCOM, Leiden, 1993, 233.
133. Swain, P. A., Anderson, B. L., Goodman, M., and Fuller, W. D., *Pept. Res.*, 6, 147, 1993.
134. Goodman, M., Spencer, J. R., Swain, P. A., Antonenko, V. V., Delaet, N. G. J., Naider, F. R., Xue, C.-B., and Fuller, W. D., in *Peptides 1992. Proceedings of the 22nd European Peptides Symposium*, Schneider, C. H. and Eberle, A. N., Eds., ESCOM, Leiden, 1993, 29.
135. Llinares, M., Bourdel, E., Califano, J. C., Rodriguez, M., Loffet, A., and Martinez, J., in *Peptides. Chemistry, Structure and Biology. Proceedings of the 13th American Peptides Symposium*, Hodges, R. S. and Smith, J. A., Eds., ESCOM, Leiden, 1994, 56.
136. Fehrentz, J.-A., Genu-Dellac, C., Amblard, H., Winternitz, F., Loffet, A., and Martinez, J., *J. Pept. Sci.*, 1, 124, 1995.
137. Shioiri, T., Ninomiya, K., and Yamada, S., *J. Am. Chem. Soc.*, 94, 6203, 1972.
138. Shioiri, T. and Yamada, S., *Chem. Pharm. Bull.*, 22, 855, 1974.
139. Hamada, Y., Rishi, S., Shioiri, T., and Yamada, S.-I., *Chem. Pharm. Bull.*, 25, 224, 1977.

140. Jackson, A. G., Kenner, G. W., Moore, G. A., Ramage, R., and Thorpe, W. D., *Tetrahedron Lett.*, 3627, 1976.
141. Galpin, I. J. and Robinson, A. E., *Tetrahedron*, 40, 627, 1984.
142. Cabré-Castellvi, J. and Palomo-Coll, A.L., *Tetrahedron Lett.*, 21, 4179, 1980.
143. Diago-Mesequet, J., Palomo-Coll, A.L., Fernandez-Lizarbe, J.R., and Zugaza-Bilbao, A., *Synthesis*, 547, 1980.
144. Bodanszky, M., *Principles of Peptide Synthesis,* 2nd ed., Springer-Verlag, Berlin, 1993, 1.
145. Bell, J. R., Jones, J. H., and Webb, T. C., *Int. J. Pept. Protein Res.*, 7, 235, 1975.
146. Fairweather, R. and Jones, J. H., *J. Chem. Soc. Perkin Trans.,* 1, 2475, 1972.
147. Wieland, T. and Heinke, B., *Liebigs Ann. Chem.*, 615, 184, 1958.
148. Goodman, M. and Stueben, K. C., *J. Am. Chem. Soc.*, 81, 3980, 1959.
149. Iselin, B. and Schwyzer, R., *Helv. Chim. Acta*, 43, 1760, 1960.
150. Wendlberger, G., in *Methoden der Organische Chemie (Houben-Weyl)*, Vol. 15/2, Müller, E., Ed., Georg Thieme Verlag, Stuttgart, 1974, 101.
151. Bates, H. S., Jones, J. H., and Witty, M. J., *J. Chem. Soc. Chem. Commun.*, 773, 1980.
152. Bates, H. S., Jones, J. H., Ramage, W. I., and Witty, M. J., in *Peptides 1980. Proceedings of the 16th European Peptide Symposium*, Brunfeldt, K., Ed., Scriptor, Copenhagen, 1981, 185.
153. Kimura, T., Takai, M., Masui, Y., Morikawa, T., and Sakakibara, S., *Biopolymers*, 20, 1823, 1981.
154. Kuroda, H., Chen, Y.-N., Kimura, T., and Sakakibara, S., *Int. J. Pept. Protein Res.*, 40, 294, 1992.
155. Ehrlich, A., Brundel, M., Beyermann, M., Winter, R., Carpino, L. A., and Bienert, M., in *Peptides 1994. Proceedings of the 23rd European Peptides Symposium*, Maia, H. L. S., Ed., ESCOM, Leiden, 1994, 167.
156. Ruzza, P., Calderan, A., Filippi, B., Biondi, B., Deana, A. D., Cesaro, L., Pinna, L. A., and Borin, G., *Int. J. Pept. Protein Res.*, 45, 529, 1995.
157. Mukaiyama, T., Ueki, M., Maruyama, H., and Matsueda, R., *J. Am. Chem. Soc.*, 90, 4490, 1968.
158. Mukaiyama, T., Matsueda, R., Maruyama, H., and Ueki, M., *J. Am. Chem. Soc.*, 91, 1554, 1969.
159. Mukaiyama, T., Matsueda, R., and Suzuki, M., *Tetrahedron Lett.*, 1901, 1970.
160. Matsueda, R., Maruyama, H., Kitazawa, E., Takahagi, H., and Mukaiyama, T., *Bull. Chem. Soc. Jpn.*, 46, 3240, 1973.
161. Matsueda, R., Maruyama, H., Kitazawa, E., and Takahagi, H., in *Peptides. Chemistry, Structure, Biology. Proceedings of the 4th American Peptides Symposium*, Walter, R. and Meienhofer, J., Eds., Ann Arbor Science Publishers, Ann Arbor, MI, 1975, 403.
162. Matsueda, R., Maruyama, H., Kitazawa, E., Takahagi, H., and Mukaiyama, T., *J. Am. Chem. Soc.*, 97, 2573, 1975.
163. Maruyama, H., Matsueda, R., Kitasaura, E., Takahagi, H., and Mukaiyama, T., *Bull. Chem. Soc. Jpn.*, 49, 2259, 1976.
164. Fischer, E., *Ber. Dtsch. Chem. Ges.*, 39, 530, 1906.
165. Devos, A., Remion, J., Frisque-Hesbain, A. M., Colens, A., and Ghosez, L., *J. Chem. Soc. Chem. Commun.*, 1180, 1979.
166. Schmidt, U., Kroner, M., and Beutler, U., *Synthesis*, 475, 1988.
167. Carpino, L. A., Mansour, E.-S. M. E., and Sadat-Aalee, D., *J. Org. Chem.*, 56, 2611, 1991.
168. Carpino, L. A., Cohen, B. J., Stephens, K. E., Sadat-Aalee, S. Y., Tien, J.-H., and Langridge, D. C., *J. Org. Chem.*, 51, 3734, 1986.
169. Perlow, D. S., Erb, J. M., Gould, N. P., Tung, R. D., Freidinger, R. M., Williams, P. D., and Veber, D. F., *J. Org. Chem.*, 57, 4394, 1992.
170. Carpino, L. A., Sadat-Aalee, D., Chao, H. G., and DeSelms, R., *J. Am. Chem. Soc.*, 112, 9651, 1990.
171. Bertho, J.-N., Loffet, A., Pinel, C., Reuther, F., and Sennyey, G., *Tetrahedron Lett.*, 32, 1303, 1991.
172. Carpino, L. A. and El-Faham, A., *J. Am. Chem. Soc.*, 117, 5401, 1995.
173. Benoiton, N. L., *Int. J. Pept. Protein Res.*, 44, 399, 1994.
174. Antonovics, I. and Young, G. T., *J. Chem. Soc. (C)*, 595, 1967.
175. Bodanszky, M. and Bodanszky, A., *J. Chem. Soc. Chem. Commun.*, 591, 1967.
176. Carpino, L. A., *J. Org. Chem.*, 53, 875, 1988.
177. Wenschuh, H., Beyermann, M., Haber, H., Seydel, J. K., Krause, E., Bienert, M., Carpino, L. A., El-Faham, A., and Albericio, F., *J. Am. Chem. Soc.*, 60, 405, 1995.

178. Pearson, A. J. and Lee, K., *J. Org. Chem.*, 60, 7153, 1995.
179. Boger, D. L., Borzilleri, R. M., and Nukui, S., *Bioorg. Med. Chem. Lett.*, 5, 3091, 1995.
180. Iselin, B., Feurer, M., and Schwyzer, R., *Helv. Chim. Acta*, 38, 1508, 1955.
181. Kovacs, J., Mayers, G. L., Johnson, R. H., and Ghatak, U. R., *J. Chem. Soc. Chem. Commun.*, 1066, 1968.
182. Mayers, G. L. and Kovacs, J., *J. Chem. Soc. Chem. Commun.*, 1145, 1970.
183. Kovacs, J., Mayers, G. L., Johnson, R. H., Cover, R. E., and Ghatak, U. R., *J. Org. Chem.*, 35, 1810, 1970.
184. Barber, M. and Jones, J. H., in *Peptides 1976. Proceedings of the 14th European Peptides Symposium*, Loffet, A., Ed., Editions de l'Université de Bruxelles, Bruxelles, 1976, 109.
185. Barber, M., Jones, J. H., and Witty, M. J., *J. Chem. Soc. Perkin Trans. 1*, 2425, 1979.
186. Kovacs, J., Holleran, E. M., and Hui, K. Y., *J. Org. Chem.*, 45, 1060, 1980.
187. Schnabel, E., *Hoppe-Seyler's Z. Physiol. Chem.*, 314, 114, 1959.
188. Bohak, Z. and Katchalski, E., *Biochemistry*, 2, 228, 1963.
189. Bergmann, M. and Zervas, L., *Biochem. Z.*, 280, 1928.
190. Goodman, M. and Levine, L., *J. Am. Chem. Soc.*, 86, 2918, 1964.
191. Crisma, M., Valle, G., Moretto, V., Formaggio, F., and Toniolo, C., *Pept. Res.*, 8, 187, 1995.
192. Grahl-Nielsen, O., *J. Chem. Soc. Chem. Commun.*, 1588, 1971.
193. Kemp, D. S. and Chien, S. W., *J. Am. Chem. Soc.*, 89, 2745, 1967.
194. Benoiton, N. L., *Biopolymers (Pept. Sci.)*, 40, 245, 1996.
195. Jones, J. H. and Witty, M. J., *J. Chem. Soc. Chem. Commun.*, 281, 1977.
196. Benoiton, N. L. and Chen, F. M. F., *Can. J. Chem.*, 59, 384, 1981.
197. O' Brien, J. L. and Niemann, C., *J. Am. Chem. Soc.*, 79, 80, 1957.
198. Mohr, E. and Geis, T., *Ber. Dtsch. Chem. Ges.*, 41, 798, 1908.
199. du Vigneaud, V. and Meyer, C. E., *J. Biol. Chem.*, 98, 295, 1932.
200. du Vigneaud, V. and Meyer, C. E., *J. Biol. Chem.*, 99, 143, 1933.
201. Carter, H. E. and Stevens, C. M., *J. Biol. Chem.*, 133, 117, 1940.
202. Bell, E. A., *J. Chem. Soc.*, 2423, 1958.
203. Cornforth, J. W. and Elliott, D. F., *Science*, 112, 534, 1950.
204. Goodman, M. and Stueben, K. C., *J. Am. Chem. Soc.*, 84, 1279, 1962.
205. Williams, M. W. and Young, G. T., *J. Chem. Soc.*, 3701, 1964.
206. McDermott, J. R. and Benoiton, N. L., *Can. J. Chem.*, 51, 2562, 1973.
207. Davies, J. S. and Mohammed, A. K., *J. Chem. Soc. Perkin Trans. 1*, 2982, 1981.
208. Weygand, F., Prox, A., and König, W., *Chem. Ber.*, 99, 1446, 1966.
209. Benoiton, N. L., Kuroda, K., and Chen, F. M. F., *Int. J. Pept. Protein Res.*, 13, 403, 1979.
210. Benoiton, N. L. and Kuroda, K., *Int. J. Pept. Protein Res.*, 17, 197, 1981.
211. Le Nguyen, D., Dormoy, J.-R., Castro, B., and Prevot, D., *Tetrahedron*, 37, 4229, 1981.
212. Manning, J. M. and Moore, S., *J. Biol. Chem.*, 243, 5591, 1968.
213. Weygand, F., König, W., Prox, A., and Burger, K., *Chem. Ber.*, 99, 1443, 1966.
214. Meienhofer, J., *Biopolymers*, 20, 1761, 1981.
215. Toniolo, C., Bonora, G. M., Mutter, M., and Pillai, V. N. R., *Makromol. Chem.*, 182, 1997, 1981.
216. Toniolo, C., Bonora, G. M., Mutter, M., and Pillai, V. N. R., *Makromol. Chem.*, 182, 2007, 1981.
217. Narita, M., Fukunaga, T., Wakabayashi, A., and Ishikawa, K., *Int. J. Pept. Protein Res.*, 23, 306, 1984.
218. Narita, M., Nagasawa, S., Chen, J.-Y., Sato, H., and Tanaka, Y., *Makromol. Chem.*, 185, 1069, 1984.
219. Haack, T. and Mutter, M., *Tetrahedron Lett.*, 33, 1589, 1992.
220. Quibell, M., Turnell, W. G., and Johnson, T., *J. Org. Chem.*, 59, 1745, 1994.
221. Benoiton, N. L., Kuroda, K., and Chen, F. M. F., *Int. J. Pept. Protein Res.*, 20, 81, 1982.
222. Jones, J. H., in *The Peptides. Analysis, Synthesis, Biology*, Vol. 1, *Major Methods of Peptide Bond Formation*, Gross, E. and Meienhofer, J., Eds., Academic Press, New York, 1979, 65.
223. Meienhofer, J., in *The Peptides. Analysis, Synthesis, Biology*, Vol. 1, *Major Methods of Peptide Bond Formation*, Gross, E. and Meienhofer, J., Eds., Academic Press, New York, 1979, 197.
224. Weygand, F., Prox, A., and König, W., *Chem. Ber.*, 99, 1451, 1966.
225. Carpino, L. A., El-Faham, A., and Albericio, F., *J. Org. Chem.*, 60, 3561, 1995.
226. Steinauer, R., Chen, F. M. F., and Benoiton, N. L., *J. Chromatogr.*, 325, 111, 1985.

227. Fischer, P. M., Solbakken, M., and Undheim, K., *Tetrahedron*, 50, 2277, 1994.
228. Fruton, J. S. and Bergmann, M., *J. Biol. Chem.*, 145, 253, 1942.
229. Künzi, H. and Studer, R. O., *Helv. Chim. Acta*, 58, 139, 1975.
230. Gunnarsson, K. and Ragnarsson, U., *Acta Chem. Scand.*, 44, 944, 1990.
231. Camble, R., Garner, R., and Young, G. T., *Nature (London)*, 217, 247, 1968.
232. Camble, R., Garner, R., and Young, G. T., *J. Chem. Soc. (C)*, 1911, 1969.
233. Garner, R. and Young, G. T., *J. Chem. Soc. (C)*, 50, 1971.
234. Fletcher, G. A., and Young, G. T., *J. Chem. Soc. Perkin Trans. 1*, 1867, 1972.
235. Macrae, R., and Young, G. T., *J. Chem. Soc. Perkin Trans. 1*, 1185, 1975.
236. Beyerman, H. C., de Leer, E. W. B., and Floor, J., *Recl. Trav. Chim. Pays-Bas*, 92, 481, 1973.
237. van Zon, A. and Beyerman, H. C., *Helv. Chim. Acta*, 56, 1729, 1973.
238. Beyerman, H. C., Rammeloo, T., Renirie, J. F. M., Syrier, J. L. M., and van Zon, A., *Recl. Trav. Chim. Pays-Bas*, 95, 143, 1976.
239. Izeboud, E. and Beyerman, H. C., *Recl. Trav. Chim. Pays-Bas*, 97, 1, 1978.
240. van Zon, A. and Beyerman, H. C., *Helv. Chim. Acta*, 59, 1112, 1976.
241. Kisfaludy, L., in *The Peptides. Analysis, Synthesis, Biology*, Vol. 2, *Special Methods in Peptide Synthesis,* Part A, Gross, E. and Meienhofer, J., Eds., Academic Press, New York, 1979, 417.
242. Beyermann, M., Bienert, M., Niedrich, H., Carpino, L. A., and Sadat-Aalee, D., *J. Org. Chem.*, 55, 721, 1990.
243. Carpino, L. A., Sadat-Aalee, D., and Beyermann, M., *J. Org. Chem.*, 55, 1673, 1990.
244. Mutter, M., Hagenmaier, H., and Bayer, E., *Angew. Chem. Int. Ed. Engl.*, 10, 811, 1971.
245. Bayer, E. and Mutter, M., *Nature (London)*, 237, 512, 1972.
246. Mutter, M. and Bayer, E., in *The Peptides. Analysis, Synthesis, Biology*, Vol. 2, *Special Methods in Peptide Synthesis,* Part A, Gross, E. and Meienhofer, J., Eds., Academic Press, New York, 1979, 285.
247. Mutter, M. and Bayer, E., *Angew. Chem.*, 86, 101, 1974.
248. Bonora, G. M., Palumbo, M., Toniolo, C., and Mutter, M., *Makromol. Chem.*, 180, 1293, 1979.
249. Toniolo, C., Bonora, G. M., and Mutter, M., *J. Am. Chem. Soc.*, 101, 450, 1979.
250. Bonora, G. M., Toniolo, C., Pillai, V. N. R., and Mutter, M., *Gazz. Chim. Ital.*, 110, 503, 1980.
251. Rahman, S. A. E., Anzinger, H., and Mutter, M., *Biopolymers*, 19, 173, 1980.
252. Pillai, V. N. R. and Mutter, M., *Acc. Chem. Res.*, 14, 122, 1981.
253. Toniolo, C., Bonora, G. M., Anzinger, H., and Mutter, M., *Macromolecules*, 147, 1983.
254. Gregory, H., Hardy, P. M., Jones, D. S., Kenner, G. W., and Sheppard, R. C., *Nature (London)*, 204, 931, 1964.
255. Anderson, J. C., Barton, M. A., Gregory, R. A., Hardy, P. M., Kenner, G. W., MacLeod, J. K., Preston, J., Sheppard, R. C., and Morley, J. S., *Nature (London)*, 204, 933, 1964.
256. Anderson, J. C., Kenner, G. W., MacLeod, J. K., and Sheppard, R. C., *Tetrahedron*, 22, Suppl. 8, 39, 1966.
257. Anderson, J. C., Barton, M. A., Hardy, P. M., Kenner, G. W., Preston, J., and Sheppard, R. C., *J. Chem. Soc., (C)*, 108, 1967.
258. Kimura, T., Morikawa, T., Takai, M., and Sakakibara, S., *J. Chem. Soc. Chem. Commun.*, 340, 1982.
259. Kimura, T., Takai, M., Yoshizawa, K., and Sakakibara, S., *Biochem. Biophys. Res. Commun.*, 114, 493, 1983.
260. Gutte, B. and Merrifield, R. B., *J. Am. Chem. Soc.*, 91, 501, 1969.
261. Gutte, B. and Merrifield, R. B., *J. Biol. Chem.*, 246, 1922, 1971.
262. Denkewalter, R. G., Veber, D. F., Holly, F. W., and Hirschmann, R., *J. Am. Chem. Soc.*, 91, 502, 1969.
263. Jenkins, S. R., Nutt, R. F., Dewey, R. S., Veber, D. F., Holly, F. W., Paleveda, W. J., Lanza, T., Strachan, R. G., Schoenwaldt, E. F., Barkemeyer, H., Dickinson, M. J., Sondey, J., Hirschmann, R., and Walton, E., *J. Am. Chem. Soc.*, 91, 505, 1969.
264. Veber, D. F., Varga, S. L., Milkowski, J. D., Joshua, H., Conn, J. B., Hirschmann, R., and Denkewalter, R. G., *J. Am. Chem. Soc.*, 91, 506, 1969.
265. Hirschmann, R., Nutt, R. F., Veber, D. F., Vitali, R. A., Varga, S. L., Jacob, T. A., Holly, F. W., and Denkewalter, R. G., *J. Am. Chem. Soc.*, 91, 507, 1969.

266. Fujii, N. and Yajima, H., *J. Chem. Soc. Perkin Trans. 1*, 789, 1981.
267. Fujii, N. and Yajima, H., *J. Chem. Soc. Perkin Trans. 1*, 797, 1981.
268. Fujii, N. and Yajima, H., *J. Chem. Soc. Perkin Trans. 1*, 804, 1981.
269. Fujii, N. and Yajima, H., *J. Chem. Soc. Perkin Trans. 1*, 811, 1981.
270. Fujii, N. and Yajima, H., *J. Chem. Soc. Perkin Trans. 1*, 819, 1981.
271. Fujii, N. and Yajima, H., *J. Chem. Soc. Perkin Trans. 1*, 831, 1981.
272. Yajima, H. and Fujii, N., *Biopolymers*, 20, 1859, 1981.
273. Yajima, H. and Fujii, N., *J. Am. Chem. Soc.*, 103, 5867, 1981.

Chapter 4

Convergent Approaches to the Synthesis of Large Peptides and Proteins

Despite certain notable successes, the chemical strategies discussed up to now are not really adequate for the routine synthesis of large peptides and proteins in a pure state. The development of methods that allow the preparation of such target molecules, however, remains one of the major goals of modern peptide chemistry. The synthesis of a protein by classical solution methods is a formidable undertaking, requiring considerable investment in time and effort and, although still carried out occasionally,[1-3] is the domain of specialized research groups. Linear SPPS on the other hand, despite its success in a wide variety of applications, presents certain limitations that make the assembly of large peptides particularly challenging.

Foremost among these is the problem of nonquantitative yields in amino acid-coupling steps and in N^α-deprotection reactions. This leads to the formation of deletion peptides and truncated sequences, all of which may be present in the crude mixture after cleavage from the solid support.[4] Since some, if not all, of the undesired peptides will very closely resemble the target molecule, its isolation in a pure state represents a difficult separation problem that can usually only be solved using high-resolution methods. For some large peptides even these may be inadequate and it may not be possible to obtain the desired material pure. Furthermore, peptides having difficult sequences[5-8] (see Section 2.4.2.2.2), in which low coupling yields are observed for a series of amino acids, often cannot be satisfactorily synthesized by standard SPPS and require either special protocols or a different type of synthetic strategy.

Such problems, and others, conspire against the preparation of large peptides and proteins by linear solid-phase methods even though some very impressive examples have been reported (see Section 2.6). Homogeneous large target molecules are probably best produced chemically by convergent strategies that rely on the union of peptide segments. These can be purified and characterized, providing a series of checks on the nature of the material being synthesized and facilitating the purification of the final product. Most modern convergent peptide synthesis strategies are based to some extent on the use of solid-phase methods because of the advantages in economics and rapidity that these present. The most promising are the subject of this chapter.

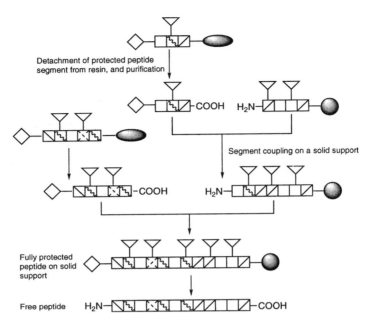

SCHEME 4.1 CSPPS. Amino acids are represented by squares, the protected α-amino groups by diamonds, and the protected side chain functional groups by triangles.

4.1 CONVERGENT SOLID-PHASE PEPTIDE SYNTHESIS

The problems caused by the poor solubility of the protected intermediates when peptides are synthesized in solution tend to become more acute as the synthesis progresses and the size of the peptides increases. One of the methods investigated in order to alleviate such problems was the coupling of the protected segments on solid supports. Several successful syntheses of larger peptides were carried out using this type of mixed approach in which peptide segments, synthesized in solution by conventional means, were coupled on a polymeric carrier.[9-12]

If the protected peptide segments themselves are also synthesized using solid-phase methods, then the advantages of the rapidity of SPPS and those of the purification and characterization of the intermediates of classical synthesis in solution are combined. This strategy has been termed *convergent solid-phase peptide synthesis* (CSPPS)[13-15] and is represented diagrammatically in Scheme 4.1.

In order to synthesize protected peptide segments using solid-phase methods some modification of the chemistry involved is required since standard SPPS usually provides the free peptide, as discussed in detail in Chapter 2. Once synthesized, each segment must then be purified to homogeneity prior to its coupling. When construction of the target peptide is complete, cleavage and purification procedures are carried out as in linear SPPS.

4.1.1 Solid-Phase Synthesis of Protected Peptide Segments

The first stage in a CSPPS strategy is the solid-phase synthesis of the various protected peptide segments corresponding to the amino acid sequence of the target molecule. Protected, as opposed to free, peptide segments are required because, generally speaking, in order to couple them in an unambiguous manner, they must be protected at all reactive functional groups except those required for amide bond formation. (See Section 4.3, however, for alternative approaches that allow unprotected peptide segments to be coupled.) The acidolytic conditions normally used to detach peptides from the solid support in standard linear SPPS are not compatible with the synthesis of protected peptide segments since the majority of side chain–protecting groups are not designed to withstand them.

Protected peptide segments are normally required with a carboxylic acid at the *C*-terminus, in order to allow coupling to another segment, and consequently the peptide–resin anchorage is usually an ester. This being so, the standard solid supports used for linear SPPS can be used to produce protected peptides if cleavage is brought about by saponification or transesterification. However, neither method is very satisfactory despite being quite well documented (see Section 4.1.1.2.3). A much more powerful alternative is to modify the peptide–resin anchorage so that the bond that links the peptide to the resin can be cleaved selectively. This requires a very fine adjustment of the chemistry involved, but this aspect of peptide research has been extensively investigated and many different types of anchorage have been described. Stability to all of the conditions necessary to effect synthesis of the peptide is a prerequisite, but at the same time it must be possible to detach the protected segment from the solid support in high yield, under mild conditions, and without racemization at the *C*-terminal amino acid.

A variety of different chemical processes can be used for the cleavage of protected peptide segments from solid supports, but there must be a high degree of compatibility between the protecting groups of the segment and the peptide–resin anchorage. If protected peptides are to be produced by Boc/Bzl synthesis, acidolysis cannot be used for cleavage from the resin. Photolysis or base-mediated cleavage are two options, with allyl transfer providing a third. For Fmoc/*t*Bu synthesis, cleavage usually cannot be performed under basic conditions; mild acidolysis is the method of choice here. The type of cleavage reaction used depends upon the peptide segments to be synthesized and the protection scheme used. The most important methods are discussed below.

4.1.1.1 Acidolytic Cleavage of Protected Peptides from the Solid Support

If acidolysis is to be used for cleavage, then, obviously, all of the protecting groups of the peptide must be stable to the conditions used. This means that strong acids such as liquid hydrogen fluoride are out of the question and that cleavage must be brought about using the mildest acidolysis possible. The development of CSPPS has stimulated the search for ever more acid-labile peptide–resin anchorages. Modification of the chemical structure of the solid support can increase its lability to acid, and a series of highly acid-labile resins and linkers is now available. These are only compatible with Fmoc/tBu SPPS since the use of acidic conditions is avoided throughout: highly acid-labile solid supports are not compatible with Boc/Bzl synthesis. The most commonly used handles (see Section 2.3) and resins that are cleaved under mild acidolytic conditions are discussed below.

4.1.1.1.1 Highly Acid-Labile Handles and Resins

Handles and resins that can be cleaved by extremely mild acidolysis currently appear to be the most-promising option for the solid-phase preparation of protected peptide segments. The usefulness of such solid supports is illustrated by the examples in Section 4.1.4, below. Most highly acid-labile solid supports are formed by introducing electron-donating substituents into the benzyl groups that serve as the peptide–resin anchor. The first handle of this type, 4-hydroxymethyl-3-methoxyphenoxyacetic acid (HMPA) **1**, was introduced by Sheppard and Williams.[16,17] A closely related, and more widely used, alternative is 4-(4-hydroxymethyl-3-methoxyphenoxy)butyric acid[18-20] (HMPB) **2**.

The difference between these handles resides in the presence of two extra CH_2 groups between the carboxyl group and the oxygen atom of the aryl-alkyl ether in **2**. These diminish the electron-withdrawing effect of the carboxyl group at the ether oxygen, allowing it to function as a more powerful electron donor, increasing the lability of the handle to acid. This results in even milder detachment conditions. Such handles, when attached to suitable resins provide highly acid-sensitive solid supports for peptide synthesis.[21] Fmoc/tBu-protected peptides can be cleaved, in the case of **2**, by repetitive treatments with 1% trifluoroacetic acid in dichloromethane for 6 to 8 min at room temperature. These conditions do not affect *tert*-butyl-based protecting groups.

Two highly acid-labile resins that have found widespread application are **3**, known as SASRIN (Super Acid Sensitive ResIN)[22,23] and the 2-chlorotrityl resin **4**. Cleavage of peptides from **3** can be brought about, in high yield, by treatment of the peptide–resin with 1% trifluoroacetic acid in dichloromethane.

The trityl-based resin **4** was developed specifically for use in CSPPS and is among the most acid-labile of all solid supports.[24-27] Cleavage can be accomplished in excellent yield by treatment with less than 1% trifluoroacetic acid or a mixture of acetic acid and trifluoroethanol in dichloromethane. An alternative cleavage reagent is hexafluoroisopropanol in dichloromethane,[28] which has the advantage of avoiding the contamination of the protected peptide segment with a carboxylic acid. Such contamination can lead to capping of the free N^{α}-amino group of protected peptide segments in coupling reactions.[29]

Several trityl-based handles, which can be attached to a range of functionalized solid supports, have also been described.[30-36] A representative example is **5**, which has been used for the synthesis of a 94-residue protein[37] (see Section 4.1.4.4). A number of other highly acid-labile handles, such as the hyper-acid-labile linker (HAL) **6**, are also available for the synthesis of protected peptide segments.[38,39]

Protected peptide segments are often required with Gly or Pro at the *C*-terminus in order to reduce the risk of epimerization on segment coupling (see Section 3.3.2.1 and Section 4.1.3.5 below). The presence of Pro in this position tends to favor the formation of diketopiperazines (see Section 2.4.2.1.1) on incorporation of the third amino acid of the peptide to be synthesized, with the concomitant drop in the level of resin functionalization that this implies. This unwanted cyclization can normally be controlled using special Fmoc deprotection protocols. The use of shorter exposure times to piperidine in dimethylformamide[40] or of tetrabutylammonium fluoride quenched with methanol[41] has been reported to alleviate the problem. The steric hindrance to cyclization normally ensures that diketopiperazine formation is minimal in trityl-based resins and handles, even when Pro is at the *C*-terminus.

4.1.1.1.2 The Wang Resin

The *p*-alkoxybenzyl alcohol resin **7** was designed and introduced into peptide synthesis by Wang[42] (see Section 2.3). It is mainly used for the synthesis of free peptides by the Fmoc/*t*Bu approach. Cleavage from the Wang resin is brought about by treatment with high concentrations of trifluoroacetic acid in the presence of scavengers.

This support is not useful for the preparation of Boc/Bzl- or Fmoc/tBu-protected peptide segments. In the former the acidolyses required to remove the N^α-protecting group would lead to cleavage of the peptide from the resin, and in the latter the side chain–protecting groups would not survive the cleavage conditions. However, **7** can be used for the synthesis of protected peptide segments if other protecting group combinations are used. One possibility is to use benzyl-based side chain protection in combination with the Fmoc N^α-protecting group. Cleavage with trifluoroacetic acid then gives Fmoc/Bzl-protected peptides.[13,43-46] Similar methods to those described above are sometimes required for avoiding diketopiperazine formation on incorporation of the third amino acid with this resin, especially when Pro is at the C-terminus.

4.1.1.2 Nucleophile- and Base-Mediated Cleavage of Protected Peptides from the Solid Support

In the same way that highly acid-labile resins and handles are only compatible with Fmoc/tBu synthesis, base-labile solid supports can normally only be used for the Boc/Bzl approach. The conditions necessary for Fmoc group removal would lead to the cleavage of a base-labile handle or resin.

4.1.1.2.1 *The Kaiser Oxime Resin*

The base-labile resin that has been most used in peptide synthesis is, without doubt, the *p*-nitrobenzophenone oxime resin **8** developed by Kaiser.[47-49] This resin has been used extensively for the synthesis of Boc/Bzl-protected peptide segments.

Various methods have been used to cleave protected peptides from the resin, including hydrazinolysis, ammonolysis, or aminolysis using a suitable amino acid ester.[50-54] Probably the most useful procedure, however, is transesterification of the peptide–resin with hydroxypiperidine. This initially forms the hydroxypiperidine ester of the protected peptide, which after treatment with zinc in acetic acid furnishes the corresponding free carboxylic acid. This method does not lead to epimerization at the C-terminal amino acid nor to loss of acid-sensitive protecting groups. Cleavage yields are usually high.

The main problem associated with the use of the Kaiser oxime resin is its lability to nucleophiles. Loss of peptide from the resin can be provoked by the free N^α-amino group of the growing peptide chain on neutralization with base, after acidolytic removal of the N^α Boc group. Chain elongation on this resin is usually carried out with *in situ* neutralization, in order to reduce such loss. This is particularly important in the coupling of the third amino acid, where the possibility of diketopiperazine formation must also be borne in mind.

4.1.1.2.2   Cleavage of the Peptide—Resin Anchorage by a β-Elimination Reaction

Base-mediated cleavage of peptides from solid supports incorporating the fluorene-based handles such as **9**, **10**, and **11** occurs by a β-elimination mechanism. Solid supports incorporating such handles are completely stable to acidic conditions. Prolonged exposure to diisopropylethylamine in dimethylformamide, however, leads to some cleavage of the peptide from the resin.

This is especially problematic[55] in **9**, but less so in **10** since the extra methylene group between the fluorene nucleus and the carbonyl group renders it somewhat more stable to base-catalyzed cleavage.[56] Premature cleavage of peptide from the resin could become troublesome when these supports are used to synthesize longer peptides. The problem is exacerbated if a separate neutralization step with diisopropylethylamine is carried out after removal of the N^α-Boc group. Resins incorporating these handles should be neutralized *in situ,* only after the addition of the next activated amino acid of the peptide sequence. Quantitative cleavage of peptides from both **9** and **10** can be brought about by treatment with 15% piperidine in dimethylformamide for 5 min.[57]

The handle **11** can be used to form solid supports that are more stable than either **9** or **10** with respect to premature cleavage during synthesis. This is probably because the link between the fluorene nucleus and the side chain is an electron-donating N-amide group. Resins incorporating **11** are much more stable to solutions of diisopropylethylamine in dimethylformamide, so that loss of peptide chains from the support during Boc/Bzl synthesis is avoided. However, lability to secondary amines such as piperidine is maintained, allowing protected segments to be released from the resin in high yield using 20% morpholine in dimethylformamide for 2 h.[58,59]

Several other handles and resins allow cleavage by base-catalyzed β-elimination reactions.[60-63] The 2-(2-nitrophenyl)ethyl handle **12** has been used for the synthesis of protected peptides and of nucleopeptides.[64-67] Cleavage can be brought about in high yield on treatment with 0.1 M 1,8-diazabicyclo[5.4.0]undec-7-ene in dioxane, or 20% piperidine in dimethylformamide, for 2 h at room temperature.

4.1.1.2.3 Miscellaneous Methods

Ammonolysis, hydrazinolysis and aminolysis. If the peptide is attached to the polymer by an ester linkage, then, in principle, ammonolysis[68-71] or hydrazinolysis[72-76] can be used to cleave protected peptide segments from a range of solid supports. This gives rise either to the protected peptide C-terminal amides or hydrazides. These latter compounds may be converted into the azides for subsequent coupling reactions.

Transesterification and saponification. Protected peptides having ester linkages to the solid support may also, in principle, be detached by transesterification.[77-81] This produces protected peptide esters that can be transformed into the azides, via the corresponding hydrazides, for subsequent coupling to other segments. Alternatively, the peptide esters can be saponified giving the free carboxylic acids, although epimerization and other side reactions may intervene. Reaction rates for the transesterification reaction depend upon the nature of the C-terminal amino acid. Lower rates are observed for Pro and for other sterically hindered amino acids such as Val or Ile. The cleavage of protected peptides from solid supports may also be achieved directly by saponification.[82-85] The cleavage yields can be very high, but again care must be taken to avoid epimerization and other side reactions.

None of the nucleophilic cleavage techniques discussed in this last section are compatible with protected peptides containing Asp or Glu residues since the protecting groups of these are not stable to the nucleophiles involved. The methods, therefore, are of limited utility.

4.1.1.3 Photolytic Cleavage of Protected Peptides from the Solid Support

Photolysis[86] of the peptide–resin bond is a mild, noninvasive technique that is, in principle, compatible with both the Boc/Bzl and Fmoc/tBu approaches making it a very attractive cleavage technique for use in CSPPS. Although a range of possibilities exists for introducing photolability into a solid support, the most widely used in peptide synthesis are nitrobenzyl- or phenacyl-anchoring linkages.

4.1.1.3.1 Nitrobenzyl Resins

Several nitrobenzyl-based solid supports have been employed for the synthesis of protected peptide segments.[87-93] The most useful are those resins incorporating the 4-bromomethyl-3-nitrobenzoic acid handle **13**. A typical example is 3-nitro-4-bromomethylbenzhydrylamido-polystyrene **14** (known as Nbb–resin), which is formed by attaching **13** to a benzhydrylamine resin. This support is fully compatible with the Boc/Bzl SPPS and partially compatible with the Fmoc/tBu approach. The peptide-handle bond is not completely stable to treatment with piperidine, so that the synthesis of long peptides by Fmoc/tBu synthesis is not advisable.[94]

In order to use **14**, the first amino acid must be incorporated as its cesium carboxylate,[95] displacing the benzyl bromide by nucleophilic substitution. This can be done either by esterifying the amino acid onto the handle in solution to give a preformed handle[96] or by attaching it to the handle previously incorporated onto the resin (see Section 2.3). The coupling of the third amino acid of the sequence (regardless of whether Boc/Bzl or Fmoc/tBu approaches are used) must be carried out under conditions that minimize the production of diketopiperazines.[97] Formation of these cyclic dipeptides is more of a problem in Fmoc/tBu synthesis and can only be avoided on **14** by coupling the second and third amino acids as a protected dipeptide.[40] In the Boc/Bzl approach, the problem can be minimized by using specially designed coupling protocols for the third amino acid, involving *in situ* neutralization.[98,99] The coupling of all other amino acids is carried out using standard protocols.

Photolysis of a suspension of the peptide–resin in a mixture of dichloromethane and trifluoroethanol by irradiation at 360 nm then detaches the peptide from the solid support in moderate-to-good yields. All commonly used protecting groups are stable to the cleavage conditions.

4.1.1.3.2 *Phenacyl Resins*

An alternative to nitrobenzyl-based resins is provided by resins such as **15** or handles such as **16** or **17**. These allow the formation of a photolabile α-methylphenacyl ester[100] anchoring linkage, on treatment with the cesium carboxylate of the first amino acid of the peptide to be synthesized. Phenacyl-based resins[101-105] are, however, only compatible with the Boc/Bzl approach since the peptide–resin anchor is not stable to the basic conditions used in Fmoc/tBu synthesis.

Phenacyl-based solid supports are subject to the same side reactions that occur when the phenacyl ester is used as a *C*-terminus-protecting group in synthesis in solution (see Section 3.2.3.1.2). Racemization of the first amino acid can occur, especially in the case of Pro. Incorporation of the second amino acid can be hampered by cyclization of the free amino group of the first onto the carbonyl group of the resin, forming Schiff bases. Diketopiperazine formation may also compete with the coupling of the third amino acid, although this can normally be overcome using similar methods to those used for nitrobenzyl resins. Occasionally, in difficult cases, these side reactions can only be avoided by incorporating previously prepared di- or even tripeptides onto the resin.[102]

Photolysis is carried out as for nitrobenzyl resins and cleavage yields of protected peptides are similar for both types of support.

4.1.1.4 Cleavage of Protected Peptides from Allyl-Functionalized Resins

Allyl-based solid supports are potentially very useful because the mild cleavage conditions are, in principle, compatible with both Boc/Bzl and Fmoc/tBu peptide synthesis. Furthermore, when used with the latter approach, they provide an orthogonal[96,106,107] scheme for the preparation of protected peptide segments. Useful allyl-functionalized polymers are provided by the incorporation of the handles **18**, **19**, or **20** into suitable resins.[108-110]

The general requirements for the cleavage of peptides are a palladium catalyst and a several-fold excess, relative to the degree of allyl substitution of the resin, of a suitable nucleophile or "allyl acceptor." Two general procedures have been developed, the first of which makes use of tetrakis(triphenylphosphine) palladium, $(Ph_3P)_4Pd$, as catalyst and 1-hydroxybenzotriazole, morpholine, or N,N-dimethyl-barbituric acid as nucleophile.[111,112] For Fmoc-protected peptides the nucleophile used as an allyl acceptor must be carefully chosen so as not to provoke the removal of this group; N-methylaniline can give good results.[113,114] An alternative hydrostannolytic cleavage procedure,[115,116] in which the peptide–resin is treated with palladium dichloride, $PdCl_2$, and tributyltin hydride in the presence of a proton donor, can also be used for cleavage. This method is particularly useful in the case of Fmoc-containing peptides since the conditions involved do not provoke its loss. For both cleavage procedures the conditions are mild and yields are usually high, although careful control of the reaction conditions is sometimes required.

4.1.2 Purification of Protected Peptide Segments

The purification of the protected peptide intermediates is an important aspect of the CSPPS strategy since it ensures that they are homogeneous molecular species, free from single-residue deletion peptides and other impurities. However, a prerequisite for any chromatographic purification is adequate solubility of the material to be purified, in a solvent compatible with the purification procedure. Protected peptides exhibit unpredictable, but generally poor, solubility in most of the commonly used organic solvents, which makes them difficult to purify and causes some of the most serious problems in CSPPS.

4.1.2.1 Enhancement of the Solubility of Protected Peptide Segments

The low solubility of protected peptide segments has prompted the investigation of ways in which it can be enhanced in order to facilitate purification and segment-coupling reactions. Poor solubility is thought to be due, in the main, to intermolecular association by hydrogen bonding, leading to the formation of β-sheet-like secondary structures,[117] in much the same way that such structures are thought to be the cause of difficult sequences in linear SPPS.[7] Disruption of the hydrogen bonding that stabilizes these secondary structures is necessary in order to solubilize protected peptides. Similar approaches to those that are useful for the synthesis of difficult sequences (see Section 2.4.2.2.2) can also be useful for solubility enhancement. On the one hand, improved solubility can be achieved when certain solvents or solvent mixtures are used. Alternatively, the structure of the protected peptide segments themselves may be modified in such a way as to reduce the formation of β-sheet-like structures, rendering them more soluble.

4.1.2.1.1 Structural Modification

Model studies on peptides containing Pro **21** demonstrated that the presence of tertiary amide bonds at central positions in the segment led to better solubility.[118] For target peptides lacking a regular distribution of Pro, pseudoprolines **22**, may be incorporated into the segments by the formation of oxazolidines or thiazolidines from Ser, Thr, or Cys.[119-122] Regeneration of the original residue is easy in the case of Ser- and Thr-derived oxazolidines, although it can be more difficult Cys-derived thiazolidines. Another alternative is to introduce tertiary amides by the protection of peptide bonds.[123-130] Hmb amide bond protection **23** has been used to great effect in CSPPS. It confers much-improved solubility upon protected peptide segments, significantly facilitating their purification, as has been demonstrated by the synthesis of the Tau protein of Alzheimer's disease[37,131-133] (see Section 4.1.4.4).

This type of structural modification, however, is not always a practical proposition since it tends to depend, to some extent at least, on the presence of specific residues in the peptide to be synthesized. As a consequence, efforts have been devoted to developing more generally applicable strategies based upon the protection of the C- or N-terminals or the side chains with "solubilizing" groups. The first attempts in this direction were those of Young and Macrae[134] who reported on the use of picolyl esters as *C*-terminus protection. Another example is the "liquid-phase" peptide synthesis of Bayer and Mutter[135] where the solubility of a peptide in many organic solvents can be increased dramatically by attaching its *C*-terminus to a polyethylene glycol polymer (see Section 3.5.3).

The Sulfmoc group **24**, described by Merrifield and Bach,[136] was the first of a series of solubilizing N^α-protecting groups that have been evaluated for use in the purification of peptides. Other removable chromatographic probes, based on derivatives of the Fmoc group, have been reported,[137-139] as have several non-Fmoc-based groups.[140-142]

Various possibilities are available for the enhancement of solubility by modification of side chain protection. The presence of oxidized or unoxidized[143] Met in a protected peptide segment, for example, can have important consequences for its purification since Met(O) can increase the overall polarity of the molecule appreciably.[29,144] This extra polarity can be beneficial in the purification of Met-containing segments. Picolyl groups may be used as side chain protection instead of the more common benzyl groups. This gives rise to protected peptides **25** that are, in general, more polar, have lower high-performance liquid chromatography (HPLC) retention times, and are more soluble in acetic acid.[145] Such peptides may be amenable to purification by cation exchange chromatography, as may those containing Lys residues, which can be reversibly protected.[146,147]

Another possibility is to work with semiglobal protection, leaving certain amino acid residues without protection at their side chains. This often leads to improved solubility in water–acetonitrile or water–methanol mixtures that can facilitate purification by standard reversed-phase chromatographic techniques. A drawback, however, is that side reactions are more prone to occur especially during the long coupling times that may be necessary for large peptide segments.

4.1.2.1.2 Use of Special Solvents or Additives to Enhance Solubility

The structural modification of peptides or the use of special protection schemes in order to enhance solubility are promising approaches and are topics of considerable current interest. However, the use of special solvent systems and/or additives has the advantage that it can allow the initially synthesized, unmodified protected peptide segment to be dissolved. Protected peptide segments are usually only poorly soluble in aqueous media and in most of the commonly used organic solvents, but many are at least moderately soluble in dipolar aprotic solvents, such as dimethylformamide, dimethylsulfoxide, or *N*-methylpyrrolidinone. In addition to these, attention has also focused on the use of trifluoroethanol and hexafluoroisopropanol[148] and of mixed solvent systems such as hexamethylphosphorotriamide–dimethylsulfoxide, hexafluoroisopropanol–ethanol–dichloromethane, and hexafluoroisopropanol–dimethylformamide.[149-153] The solubility of peptides in nonpolar solvents such as tetrahydrofuran is improved by the addition of inorganic salts.[154] This has been used in the purification of protected peptide segments[155] and is an interesting and potentially useful new development in this area of research.

4.1.2.2 Purification Methods

Even in favorable cases, the purification of a synthetic peptide can take considerable amounts of time and effort. The purification of protected peptide intermediates, complicated as it often is by problems of poor solubility, is often laborious. Assuming that the protected peptide can be dissolved in some suitable solvent, purification can be carried out using one or more of the chromatographic techniques that are applicable to peptide molecules. Frequently, however, quite drastic modifications of the normal operating conditions are necessary if a protected peptide is to be purified to homogeneity.[29,144,156] Two techniques, in particular, deserve special mention because of the frequency with which they are applied in the purification of both free and protected peptide molecules.

4.1.2.2.1 Gel Filtration

Gel filtration[157,158] separates molecules on the basis of molecular size — larger molecules elute more rapidly than smaller ones, which are retarded by the stationary phase to a greater extent.[159-161] The technique is used routinely in the purification of peptides, but its resolving power is somewhat limited. Although gel filtration efficiently removes low-molecular-weight impurities from a crude mixture of peptides after a synthesis, it is normally not sufficient on its own for separating the target peptide from closely related peptidic impurities. For the purification of free peptides gel filtration is usually carried out in aqueous solvent systems. The purification of protected peptides, however, often requires the use of dipolar aprotic solvents such as dimethylformamide, dimethylsulfoxide, or N-methylpyrrolidinone.

4.1.2.2.2 Reversed-Phase High-Performance Liquid Chromatography

The importance of HPLC to the peptides field cannot be overemphasized and the development and refinement of this technique has exerted a profound influence upon peptide analysis, characterization, and synthesis.[162] The major difference between HPLC and other chromatographic systems is that it is carried out with equipment designed to produce and withstand very high pressures. This allows smaller particle sizes to be used in the columns, leading to an increase in resolution. The most common variant used today, reversed-phase HPLC, is so called because the elution conditions are essentially the reverse of those used in normal-phase chromatography.

Retention of the sample by the column occurs through hydrophobic interaction with the column support. Elution is carried out in such a way as to decrease the ionic nature, or to increase the hydrophobicity, of the eluant so that it competes for the hydrophobic groups on the column.

Column supports in reversed-phase HPLC are normally alkane chains bound to a silica matrix. The most common lengths are chains of 4, 8, or 18 carbon atoms (referred to as C-4, C-8, and C-18 columns, respectively), although other functionalized silica supports can also be used, such as those containing nitrile or phenyl groups. Which column to use is, of course, determined by the type of molecule to be analyzed. The beaded particles making up the support have an average size in any one column, referred to as the mesh size. Modern HPLC columns are usually available with mesh sizes of between 3 and 10 μm. Smaller particle sizes normally give better resolution but increase the operating pressure. In addition, the particles are not solid but contain pores, which has the effect of increasing their surface area and, consequently, their interaction with the peptide. For a molecule to be able to interact effectively with the column support, it must be able to penetrate the pores of the packing material. Pore sizes of between 80 and 100 μm might be the optimum for smaller peptides, but for larger ones a pore size of 300 μm is more effective.

A two-eluant system, which is often 0.1% trifluoroacetic acid in water (eluant A) and 0.1% trifluoroacetic acid in acetonitrile (eluant B), is used. Normally gradient elutions are carried out, so that elution is commenced with a low percentage of B in A (typically 0 to 10%), which is then taken over a suitable analysis time (20 to 30 min usually) to higher proportions of eluant B (normally 60 to 100%). Of course, any number of variations on these conditions can be used in practice, depending upon the nature of the peptide to be analyzed or purified. Detection is usually carried out by measuring the ultraviolet absorbance of the effluent.

The power of HPLC is such that it is often sufficient alone for the purification of many peptides and it has, to some extent, superseded all other purification methods in the peptides field. Preparative HPLC is broadly similar to the analytical technique and, as long as the necessary equipment is available, gram quantities of peptides can, in favorable cases, be purified in a single run. Many column sizes are now available commercially as are pumping systems which can provide flow rates of many hundreds of milliliters per minute. For typical laboratory-scale applications, purifications are most commonly carried out at the hundreds of milligrams scale. The only drawback to preparative HPLC purifications of peptides is that it may be necessary to use quite large volumes of solvents in a purification.

The analysis of protected peptides by reversed-phase HPLC often requires that they be injected into the system as solutions in a dipolar aprotic solvent, such as dimethylformamide. Perfectly acceptable analytical chromatograms are usually obtained when this is done, with the loading solvent appearing as a broad peak at the beginning of the chromatogram. For the purification of protected peptide segments at the preparative level the addition of another solvent, such as isopropanol or dimethylformamide, to the eluants may be necessary to ensure adequate solubility.[29,144,156,163,164]

Other purification techniques that can be useful for the purification of protected peptide segments, depending upon the case, are normal-phase column chromatography on silica gel and medium-pressure liquid chromatography (MPLC), both in normal- and reversed-phase modes.[14,15]

4.1.2.3 Determination of Covalent Structure

The characterization of a peptide after purification is necessary to establish whether or not the desired peptide, and not some structural modification of it, has in fact been isolated. Characterization of peptide molecules is not always straightforward owing to the particular type of structural complexity that these molecules present. Amino acid analysis is used very extensively for the analysis of both free and protected peptides. In the latter case, however, it gives no information on whether or not protecting groups are still present. More rigorous characterizations are provided by mass spectrometry, especially in the fast atom bombardment (FAB) mode[165-168] and by NMR spectroscopy. In this latter technique, the interpretation of the NMR spectra of large protected peptides can be complicated. One-dimensional spectra are not normally sufficient on their own, and more sophisticated two- and even three-dimensional experiments are usually required for rigorous assignments.[169] Extensive use should be made both of mass spectrometric and NMR techniques for a comprehensive characterization of protected peptide segments.

4.1.3 Solid-Phase Coupling of Protected Peptide Segments

Although the coupling of a protected peptide segment on a solid support is analogous to that of a single amino acid, it is more demanding for several reasons. The risks of epimerization at the *C*-terminus must always be borne in mind, and consequently protected peptides are often chosen so as to have either Gly or Pro at this position (see Section 3.3.2.2.1). In linear SPPS, excesses of urethane-protected amino acids can be used in order to drive the coupling reaction to completion. In CSPPS, on the other hand, the use of excesses of a protected peptide segment whose synthesis may have required considerable investment in time and effort is something that must be considered carefully. Finally, there is the question of solubility. It is often difficult to achieve acceptably high concentrations of peptide segment in the coupling medium. This tends to lower the yield of the coupling reaction making it necessary to repeat it until coupling is complete, with the consumption of segment that this implies.

These difficulties notwithstanding, protected peptide segments can be coupled on solid supports with good results, and in order to maximize the efficiency of the process several factors must be taken into account.

4.1.3.1 The Solid Support

The nature of the solid support can have a significant effect on the outcome of segment-coupling reactions. Not surprisingly, the solid supports that give the best results are those that also perform best in standard SPPS. Studies have shown that polystyrene and polyacrylamide resins give higher segment coupling yields and shorter reaction times than is the case when other types of support, such as controlled pore glass, are used.[170] Polyethylene glycol-grafted polystyrene also gives similarly good results in peptide segment couplings.[93]

In addition to the resin type, the degree of functionalization is also important. The optimum level for segment coupling is lower than that normally used in linear SPPS, with values in the range 0.04 to 0.2 meq g^{-1} giving the best results.

4.1.3.2 Synthesis Strategy

In addition to normal chain elongation in the C to N direction, solid-phase segment coupling has also been carried out by N to C elongation.[171-175] For stepwise SPPS N to C chain elongation suffers from the disadvantage that the risks of epimerization of the C-terminus of the resin-bound peptide in the amino acid–coupling steps are much greater (see Section 3.3.2.1.2). However, in CSPPS, although activation of the resin-bound carboxyl groups may be correspondingly more difficult, epimerization of the activated C-terminal amino acid is no more of a risk than it is in standard C to N elongation. One of the possible advantages of N to C chain elongation in segment coupling is that, since the activated C-terminus of the peptide remains attached to the resin, it might be possible to recycle the excess of protected peptide in solution. Nonetheless, N to C chain elongation has not established itself and is rarely used. Almost all SPPS is carried out in the C to N direction.

When designing a convergent solid-phase synthesis strategy for a particular peptide, one of the factors that must be considered is the size of the segments to be coupled. If the criterion of choosing only those protected peptide segments having Pro or Gly at the C-terminus is strictly adhered to, then, inevitably, some segments will be rather long. The danger of working with long protected peptide segments is that they may well be too insoluble to be purified or to provide sufficiently concentrated solutions for the coupling reaction.[176] A compromise must be struck between working with a longer, poorly soluble segment having Gly or Pro at the C-terminus, on the one hand, and using a shorter, more soluble segment that, since it lacks C-terminal Pro or Gly, will be much more prone to epimerize on coupling, on the other. A related consideration is whether peptide segments should be chosen to have Pro at the C-terminus or in the middle. As has been discussed (see Section 4.1.2.1.1 above), Pro in the middle of a protected peptide segment can confer better solubility upon it, whereas at the C-terminus it is resistant to epimerization. One possibility is to choose segments having Pro at the C-terminus and to introduce tertiary amide bonds into the peptide using Hmb peptide bond protection (see Section 4.1.4.4).

4.1.3.3 Incorporation of the First Segment

The first segment of the target molecule can, of course, be synthesized directly on the resin by linear SPPS. Alternatively, a previously synthesized, purified, and characterized peptide segment can be attached to the resin. The advantage of stepwise synthesis is that it is rapid, but a drawback is that the peptide cannot be purified. If the segment is quite long, then the possibility of having deletion or truncated peptides in the C-terminal region is increased. If the segment is relatively short, however, modern SPPS usually allows it to be synthesized in an essentially pure state. The attachment of previously synthesized protected peptide segments to a solid support, on the other hand, means that the C-terminal segment can be rigorously purified and characterized.

Protected peptide segments can usually be attached to solid supports by amide bond formation in good yield, but their incorporation by esterification is somewhat more difficult. This means that when peptide C-terminal acids are required, the first segment might best be incorporated by stepwise synthesis on the resin, rather than by attempted esterification of a protected segment. An alternative here is to synthesize the first segment with a handle already incorporated at the C-terminus. This then allows the segment–handle to be purified and attached to the solid support by amide bond formation. An example of this method is to be found in Section 4.1.4.4.

Irrespective of how the first segment is incorporated, it is often done in such a way as to reduce the substitution level of the resin.[9,54] For stepwise synthesis the first amino acid can be loaded using less than the amount required to react with all of the active sites on the resin. The resin is then acetylated to block the remaining amino groups, and synthesis is continued under normal conditions. If a protected peptide is to be loaded, then the substitution level of the resin can be reduced when an internal reference amino acid is incorporated.

4.1.3.4 Coupling Methods

Even if a reasonable excess of the segment to be coupled can be used and, further, even if solubility is not a problem, a segment-coupling reaction may require many hours or even several days to go to completion. It is important, therefore, to treat each one on a case-by-case basis. Careful attention should be paid to the different variables, such as reagents, reaction time, concentration of the soluble components, temperature, and efficiency of the agitation of the resin, if good results are to be obtained.

Coupling methods for protected peptide segments have evolved in an analogous manner to those used for the coupling of single amino acids. Early procedures included the azide and oxidation–reduction protocols. These were superseded by the use of carbodiimides, usually in the presence of additives such as HOSu or HOBt. Contemporary methods for effecting the solid-phase coupling of peptide segments are based on the use of phosphonium or uronium salts in the presence of HOBt. New reagents based upon 1-hydroxy-7-azabenzotriazole[177,178] such as N-[(dimethylamino)-1H-1,2,3-triazolo[4,5b]pyridin-1-ylmethylene]-N-methylmethanaminium hexafluorophosphate N-oxide (HATU) show much promise[179] and have been used successfully in solid-phase segment couplings (see Section 4.1.4.5).

4.1.3.5 Side Reactions

Perhaps the most important side reaction is epimerization at the activated C-terminal amino acid of the segment (see Section 3.3.2.1.2). When this is Gly, epimerization is not possible, and when it is Pro, epimerization is thought to be minimal under normal coupling conditions. For all other amino acids, however, epimerization may occur to a greater or lesser extent.[180] Peptide segments with *C*-terminal amino acids that are neither Gly nor Pro must, therefore, be coupled in such a way that epimerization is minimized. Furthermore, analytical methods that are sufficiently sensitive to detect the presence of diastereomeric peptides in quite small quantities are necessary, in order to be able to quantify the amount of epimerization that has occurred in a given segment coupling.[181-183]

Other side reactions may have their origins in the instability of side chain-protecting groups during coupling reactions that may take considerable lengths of time to go to completion. One example is the formation of pGlu when Glu is the *N*-terminal amino component. The stability of side chain-protecting groups to the coupling conditions should be carefully checked.

4.1.3.6   Monitoring of the Coupling Reaction

Monitoring of solid-phase segment-coupling reactions is not necessarily straightforward (see Section 2.4.3). The qualitative ninhydrin test is useful for determining whether or not unreacted amino groups remain on the resin. As the length of the peptide chain increases, however, the test becomes less sensitive. Amino acid analysis can also be used for determining the extent of segment-coupling reactions. But again, as the peptide attached to the solid support becomes longer, the information that it can provide becomes more limited. It can be difficult to judge the extent of incorporation of a new segment if it contains residues that are already present, in the peptide. If several of these residues are already present, the problem is correspondingly more difficult. When neither of these techniques gives clear results, solid-phase Edman sequencing can be a useful and accurate alternative for determining the yields of segment couplings.[93,184,185] Since, during the course of a synthesis on a solid support, the weight of the peptide–resin increases, this can, in principle, be used to monitor segment couplings. The method has the advantage of simplicity, and instances of its use are documented.[186,187] However, it is often not a reliable or accurate guide to coupling yields. It should be used only as supporting evidence, when the coupling yield is determined using other techniques. The most straightforward method of monitoring a segment-coupling reaction is, of course, to remove an aliquot of the resin and to cleave the peptide from it. The product can then be analyzed by one or more physical techniques such as HPLC, capillary electrophoresis, mass spectrometry, or NMR.

4.1.4   Examples of Convergent Solid-Phase Peptide Synthesis

4.1.4.1   Rat Atrial Natriuretic Factor

The 26-residue peptide corresponding to the 8-33 sequence of rat atrial natriuretic factor was synthesized[188] by a convergent solid-phase strategy on a polystyrene resin cross-linked with 1% *p*-divinylbenzene. The protected peptide segments required

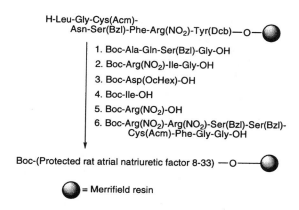

H-Leu-Gly-Cys(Acm)-
    Asn-Ser(Bzl)-Phe-Arg(NO$_2$)-Tyr(Dcb)—O—⬤

1. Boc-Ala-Gln-Ser(Bzl)-Gly-OH
2. Boc-Arg(NO$_2$)-Ile-Gly-OH
3. Boc-Asp(OcHex)-OH
4. Boc-Ile-OH
5. Boc-Arg(NO$_2$)-OH
6. Boc-Arg(NO$_2$)-Arg(NO$_2$)-Ser(Bzl)-Ser(Bzl)-
    Cys(Acm)-Phe-Gly-Gly-OH

Boc-(Protected rat atrial natriuretic factor 8-33) —O——⬤

⬤ = Merrifield resin

SCHEME 4.2   Convergent solid-phase synthesis of rat atrial natriuretic factor.

for the synthesis, Boc-Ala-Gln-Ser(Bzl)-Gly-OH, Boc-Arg(NO$_2$)-Ile-Gly-OH, and Boc-Arg(NO$_2$)-Arg(NO$_2$)-Ser(Bzl)-Ser(Bzl)-Cys(Acm)-Phe-Gly-Gly-OH, were synthesized on the same type of resin and were detached by transesterification, using methanolic solutions of triethylamine,[77,78] affording the methyl esters. These were purified by crystallization before saponification to give the corresponding peptide carboxylic acids. In addition to the protected peptide segments, single amino acids were also coupled to the peptide–resin in this synthesis, which is outlined in Scheme 4.2.

The *C*-terminal octapeptide was incorporated onto the resin by stepwise synthesis. Segment coupling reactions were carried out using an excess of protected peptide, DCC, and HOBt in dimethylformamide. Coupling times were between 18 and 24 h. Yields were judged to be very high by HPLC analysis of peptide cleaved from the resin at several points during the synthesis. The coupling of the final octapeptide segment was more problematical and required repetitive couplings and greater excesses of segment in order to achieve an acceptable yield. The peptide–resin was transformed, by subsequent synthetic operations, into the desired atrial natriuretic factor peptide, which exhibited full biological activity. It is worth emphasizing the scale at which the synthesis was carried out: more than 10 g of the desired final peptide was obtained, illustrating that CSPPS can be useful for the provision of multigram quantities of complex peptides.

4.1.4.2   Prothymosin α

Prothymosin α, a protein consisting of 109 residues, is the largest peptide that has been synthesized to date using a CSPPS strategy.[189] Its synthesis demonstrates the potential of CSPPS for the provision of large peptide and proteins. The segments spanning the 1-75 sequence of the protein were synthesized on trityl resin **4** and, after cleavage, were purified by chromatography on silica gel, eluting with chloroform–methanol mixtures. The *C*-terminal segment, consisting of 34 amino acid residues and corresponding to the 76-109 sequence of the protein, was synthesized by stepwise Fmoc/*t*Bu SPPS. The initial functionalization level of resin **4** was

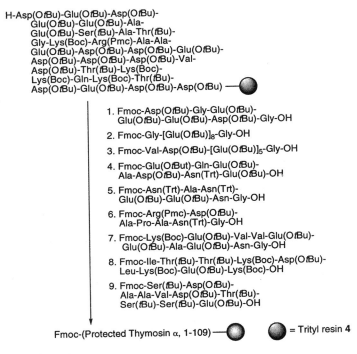

H-Asp(OtBu)-Glu(OtBu)-Asp(OtBu)-
Glu(OtBu)-Glu(OtBu)-Ala-
Glu(OtBu)-Ser(tBu)-Ala-Thr(tBu)-
Gly-Lys(Boc)-Arg(Pmc)-Ala-Ala-
Glu(OtBu)-Asp(OtBu)-Asp(OtBu)-Glu(OtBu)-
Asp(OtBu)-Asp(OtBu)-Asp(OtBu)-Val-
Asp(OtBu)-Thr(tBu)-Lys(Boc)-
Lys(Boc)-Gln-Lys(Boc)-Thr(tBu)-
Asp(OtBu)-Glu(OtBu)-Asp(OtBu)-Asp(OtBu) ⬤

1. Fmoc-Asp(OtBu)-Gly-Glu(OtBu)-
   Glu(OtBu)-Glu(OtBu)-Asp(OtBu)-Gly-OH

2. Fmoc-Gly-[Glu(OtBu)]₈-Gly-OH

3. Fmoc-Val-Asp(OtBu)-[Glu(OtBu)]₅-Gly-OH

4. Fmoc-Glu(OtBut)-Gln-Glu(OtBu)-
   Ala-Asp(OtBu)-Asn(Trt)-Glu(OtBu)-OH

5. Fmoc-Asn(Trt)-Ala-Asn(Trt)-
   Glu(OtBu)-Glu(OtBu)-Asn-Gly-OH

6. Fmoc-Arg(Pmc)-Asp(OtBu)-
   Ala-Pro-Ala-Asn(Trt)-Gly-OH

7. Fmoc-Lys(Boc)-Glu(OtBu)-Val-Val-Glu(OtBu)-
   Glu(OtBu)-Ala-Glu(OtBu)-Asn-Gly-OH

8. Fmoc-Ile-Thr(tBu)-Thr(tBu)-Lys(Boc)-Asp(OtBu)-
   Leu-Lys(Boc)-Glu(OtBu)-Lys(Boc)-OH

9. Fmoc-Ser(tBu)-Asp(OtBu)-
   Ala-Ala-Val-Asp(OtBu)-Thr(tBu)-
   Ser(tBu)-Ser(tBu)-Glu(OtBu)-OH

Fmoc-(Protected Thymosin α, 1-109) ⬤   ⬤ = Trityl resin **4**

SCHEME 4.3   Convergent solid-phase synthesis of prothymosin-α.

between 0.4 and 0.6 meq/g. The protected peptide segments were then coupled to the 76-109 peptide–resin, as shown in Scheme 4.3.

A fivefold excess of each peptide segment–HOBt–DCC (1:1.5:1) in dimethyl sulfoxide was used and reactions were complete after 6 to 18 h. The segment corresponding to the 1-10 sequence of the protein could only be coupled in 55% yield. All other segment couplings, however, proceeded in very high yields as judged by HPLC analysis of material obtained by cleavage of aliquots of peptide–resin taken throughout the synthesis. Prothymosin α was obtained in 11% overall yield, and the synthetic material had identical biological activity to the natural protein.

4.1.4.3 β-Amyloid Protein

The very sparingly soluble β-amyloid protein, found in amyloid plaques in sufferers of Alzheimer's disease, has a very strong tendency to aggregate both in solution and on solid supports. This makes it very difficult to synthesize by standard linear SPPS, although pure protein has been obtained by modified methods that use Hmb amide bond protection at selected residues[132,190] (see Section 2.6.4). An earlier synthesis of this protein in a pure state was, however, carried out using a convergent solid-phase strategy.[185,191,192]

The requisite protected peptide segments were all prepared on the oxime resin **8** and were detached by transesterification with 1-hydroxypiperidine, followed by

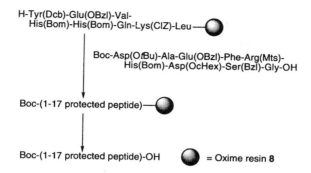

SCHEME 4.4   Convergent solid-phase synthesis of the 1-17 segment of β-amyloid protein.

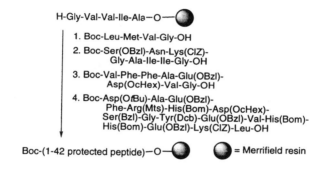

SCHEME 4.5   Convergent solid-phase synthesis of β-amyloid protein.

treatment with zinc in acetic acid. The segments were purified by reversed-phase HPLC. The 1-17 sequence of the protein was itself synthesized by segment coupling as shown in Scheme 4.4.

The *C*-terminal peptide corresponding to the 10-17 sequence was built up on the oxime resin in a stepwise manner. The 1-9 segment was then coupled to it using BOP reagent in dimethylformamide at 0 to 4°C for 14 h giving the 1-17 sequence in a yield of 59%. This was then detached from the resin and used in the synthesis of the β-amyloid protein, outlined in Scheme 4.5.

The synthesis was carried out on a standard Merrifield resin on which the *C*-terminal pentapeptide was incorporated by stepwise synthesis. Each peptide segment was dissolved in dimethylformamide and coupled using BOP reagent and HOBt. The coupling yields for the segments were 95%, 70 to 90%, and 80 to 90%, respectively. The coupling of the 1-17 segment was, however, somewhat more difficult. After deprotection of the Boc group and neutralization of the peptide–resin, the 1-17 segment was coupled in 50 to 85% yield by four successive room-temperature coupling reactions, each using a progressively smaller amount of segment (a total of 2.3 equivalents).

Model studies indicated that the amount of epimerization in this last coupling step was about 8%. After treatment with hydrogen fluoride and purification, synthetic β-amyloid protein was obtained in an overall yield of 42%. This yield is lower than that (approximately 66%) that would be obtained in an efficient linear synthesis, assuming that a yield of 99.9% were possible in each coupling step, but since the material is devoid of single amino acid deletion impurities, the final purification is facilitated.

4.1.4.4   The 3-Repeat Region of Human Tau-2

The Tau protein is also associated with Alzheimer's disease although its precise role in the onset and development of the condition is unknown. The 3-repeat region of this protein, from Asp[158] to Leu[251], a 94-residue peptide, has been synthesized by CSPPS, using the versatile Hmb amide bond-protecting group (see Sections 2.4.2.2 and 2.6.4). In the same way that the protection of selected amide bonds can dramatically reduce the level of on-resin aggregation, protected peptide segments with Hmb-derived protection at strategic points are also much more soluble than those lacking it. This increase in solubility has two important consequences. First of all, the purification of the protected peptide segments is facilitated since conventional chromatographic methods can be used. Second, but equally importantly, the increased solubility allows more concentrated solutions to be achieved in segment-coupling reactions. This promotes more efficient couplings, with shorter reaction times and higher yields.

The problem of the incorporation of the first segment of the protein onto the solid support in a pure state was solved in a very elegant manner as outlined in Scheme 4.6. Fmoc-β-Ala was esterified onto the pentafluorophenol active ester of handle **6**, and this preformed handle was attached to a polyamide resin giving the solid support **26**. After removal of the Fmoc group, the HMPAA handle,[193] incorporating Fmoc-protected leucine was attached giving the double-handle-containing solid support **27**. The first segment was then elaborated on this support, incorporating Hmb amide bond protection at Lys[245]. When chain elongation was complete, the peptide–resin was acetylated giving **28**. Treatment with 0.75% trifluoroacetic acid in dichloromethane selectively cleaved the trityl-based handle giving the protected peptide HMPAA ester **29**. Good solubility of this species was ensured by its Hmb amide bond protection. After purification by chromatography on silica gel, it was attached to a polyamide resin giving **30**, upon which the rest of the synthesis was carried out. The substitution level of the peptide resin at this point was 0.048 mmol peptide per gram of peptide–resin.

All of the other protected peptide segments required for the synthesis were also prepared on trityl-based solid supports of the type **26** (except that, instead of β-alanine, it incorporated the *C*-terminal amino acid of the segment in question). In each case, after chain elongation was complete the peptide–resins were acetylated, so that backbone amide protection was in the form of AcHmb groups. Cleavage of the protected peptides occurred in high yield on treatment of the peptide–resins with 0.75% trifluoroacetic acid in dichloromethane.

Removal of the Fmoc group of peptide–resin **30** with piperidine also brought about deacetylation of the resin-bound backbone-protecting groups giving peptide resins of the type **31**. If Fmoc removal is done in such a way as to leave the AcHmb

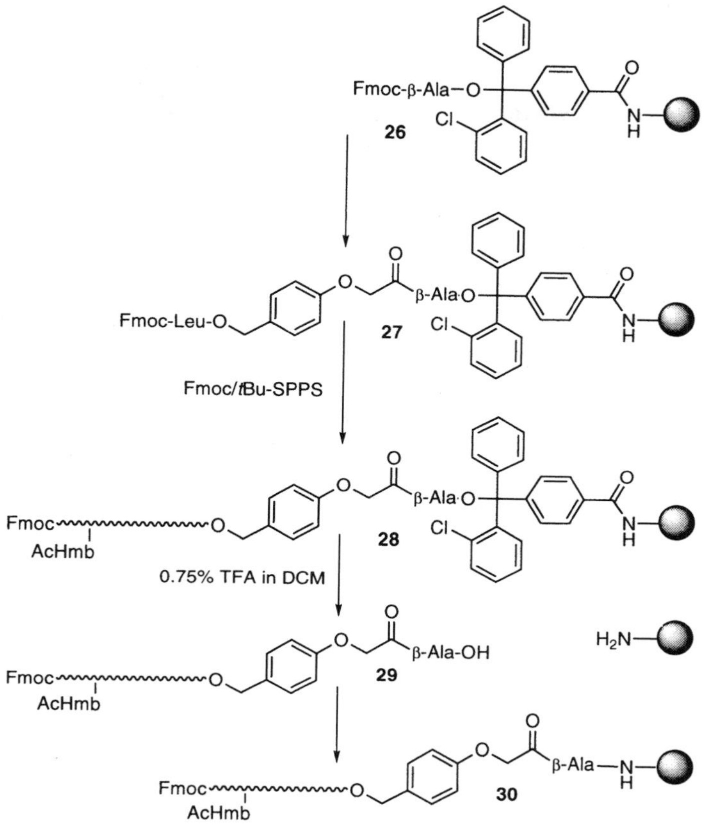

SCHEME 4.6 Attachment of the first protected peptide to the resins in the synthesis of the 3-repeat region of human Tau-2.

groups unaffected, *O*- to *N*-acyl migration can occur on segment coupling, effectively capping the resin-bound N^α-amino group and causing low yields. Segment couplings were carried out with two equivalents of peptide segment using BOP reagent in a minimum volume of dimethylformamide and were usually complete within 6 h. Even quite large (up to 21 residues) protected peptide segments could be coupled in high yield. After each coupling step the resin was acetylated, capping any residual resin-bound amino groups and reacetylating the resin-bound Hmb groups. Once chain assembly was complete, treatment of the peptide–resin **32** with piperidine bought about removal of the N^α-Fmoc group and deacetylation of all backbone protection. Acidolytic cleavage with trifluoroacetic acid containing scavengers then gave the crude protein fragment **33** that was sufficiently soluble to allow its purification by reversed-phase HPLC in water and acetonitrile mixture. The synthesis is outlined in Scheme 4.7.

The use of backbone-amide bond protection in CSPPS provides a possible solution to the problem of poor segment solubility, allowing the facile purification

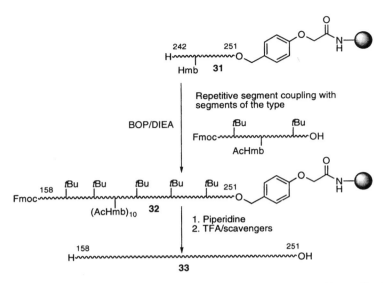

SCHEME 4.7 Convergent solid-phase synthesis of the 3-repeat region of human Tau-2.

of protected peptide segments and improving the yields in coupling reactions involving them. These factors clearly outweigh the slight disadvantage of the modified coupling protocols required for the incorporation of Hmb-protected amino acids into peptides.

4.1.4.5 The *N*-Terminal Repeat Region of γ-Zein

Finally, an example from the authors' own laboratory[179] that illustrates the particular suitability of CSPPS for the preparation of peptides with repetitive sequences. The *N*-terminal region of the maize γ-zein protein consists of eight conserved repeats of the sequence Val-His-Leu-Pro-Pro-Pro. If the *N*-terminal region of this protein is synthesized by CSPPS, then only the one protected peptide segment, corresponding to the monomer, need be prepared. Sequential couplings of the monomer then give the desired protein fragment. Its synthesis is outlined in Scheme 4.8.

The peptide was elaborated on a 4-methylbenzhydrylamine resin incorporating the highly acid-labile HMPB handle **2**. Three residues of Phe were incorporated between the solid support and the handle, as an internal standard. The first Pro residue was anchored by esterification using DIPCDI in the presence of DMAP, in a yield greater than 98%. The second two Pro residues were incorporated as the Fmoc-Pro-Pro-OH dipeptide since sequential incorporation led to unacceptably high amounts of diketopiperazine formation[194-197] (see Section 2.4.2.1.1). The remaining amino acids were added using standard coupling protocols. Treatment of peptide–resin **34** with 1% trifluoroacetic acid in dichloromethane led to cleavage of peptide **35** in 91% yield. This peptide was purified by reversed-phase MPLC using water and acetonitrile eluants containing 0.1% pyridine in order to avoid deprotection of the His(Trt) group. Sequential couplings of this protected segment then allowed

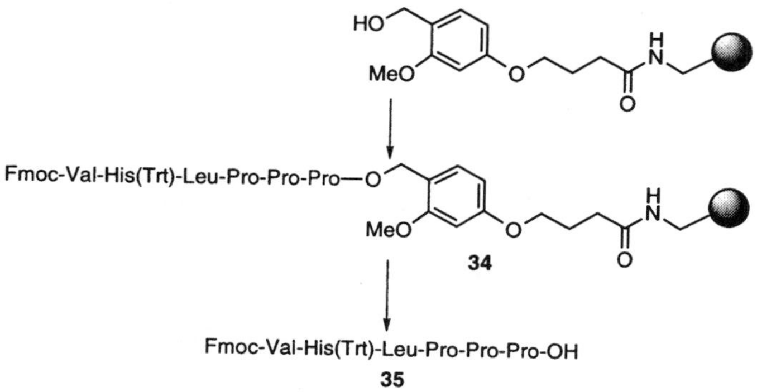

Fmoc-Val-His(Trt)-Leu-Pro-Pro-Pro-O
34

Fmoc-Val-His(Trt)-Leu-Pro-Pro-Pro-OH
35

SCHEME 4.8   Solid-phase synthesis of the Fmoc-Val-His(Trt)-Leu-Pro-Pro-Pro-OH monomer.

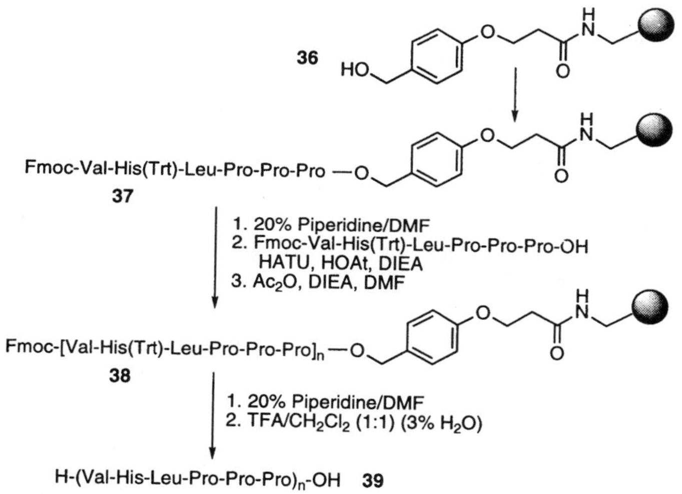

36

Fmoc-Val-His(Trt)-Leu-Pro-Pro-Pro —O
37

1. 20% Piperidine/DMF
2. Fmoc-Val-His(Trt)-Leu-Pro-Pro-Pro-OH
   HATU, HOAt, DIEA
3. Ac$_2$O, DIEA, DMF

Fmoc-[Val-His(Trt)-Leu-Pro-Pro-Pro]$_n$ —O
38

1. 20% Piperidine/DMF
2. TFA/CH$_2$Cl$_2$ (1:1) (3% H$_2$O)

H-(Val-His-Leu-Pro-Pro-Pro)$_n$-OH   **39**

SCHEME 4.9   Convergent solid-phase synthesis of the *N*-terminal repeat region of γ-zein.

the repetitive sequence of the *N*-terminal region of maize γ-zein to be synthesized, as shown in Scheme 4.9.

The Val-His(Trt)-Leu-Pro-Pro-Pro sequence was first synthesized on an aminomethyl resin incorporating the 3-(4-hydroxymethyl)phenoxypropionic acid handle[21] and three Phe residues as internal standard, **36**. Peptide–resin **37** was synthesized using the same methods as those for the protected segment **35**. Removal of the *N*-terminus Fmoc group from **37**, followed by HATU-mediated segment coupling and a capping step, then gave the dimer **38** (n = 2). Higher oligomers were synthesized by repetition of these steps until the desired octamer had been produced. Removal of the Fmoc group and cleavage with trifluoroacetic acid in dichloromethane (1:1) containing 3% water then furnished the desired repetitive protein fragment, **39**.

4.2 COUPLING OF PROTECTED PEPTIDE SEGMENTS IN SOLUTION

Although many protected peptide segments can be coupled successfully on solid supports, problems are sometimes encountered in achieving acceptably high yields, even if excesses of segment are used to try to drive the reaction to completion. This has led some workers to advocate an approach in which protected peptide segments are synthesized using solid-phase methods but are coupled in solution.[19,198] However, experience gained in classical peptide synthesis demonstrates that the coupling of protected peptide segments in solution is itself not always straightforward, in particular because of problems of solubility and of the low molar concentrations of the species involved. Modern approaches have attempted to overcome these problems as much as possible.

4.2.1 Lipophilic Segment-Coupling Strategy

The protecting groups used in Fmoc/tBu synthesis tend to give rise to rather lipophilic protected segments that are relatively soluble in organic solvents, such as tetrahydrofuran, chloroform, or ethyl acetate. This observation has been the basis for an approach to complex peptide synthesis developed and advocated by the Ciba-Geigy group.[19,20,199] The lipophilicity of the peptides can be increased by using maximal protection schemes so that His, Asn, and Gln are always protected with the trityl group, Arg is always protected with the Pmc group, and Trp with the Boc group. Cys can either be protected with the trityl group or with the acetamidomethyl group, if an acid-stable alternative is required. *tert*-Butyl protection is used for all other amino acid side chains. The greater solubility of such protected peptides means that they can be purified more easily and higher concentrations of segments can be achieved in coupling reactions. The strategy is outlined below in Scheme 4.10.

The segments are prepared by Fmoc/tBu solid-phase synthesis on a support incorporating the highly acid-labile HMPB handle **2** and are chosen so as to have either Gly or Pro at the *C*-terminus, where possible. Cleavage is brought about by treatment of the peptide–resin with 1% trifluoroacetic acid in dichloromethane, conditions that leave all side chain–protecting groups unaffected. The segment that is to serve as the amino component **40** is protected at its *C*-terminus as the *tert*-butyl ester **41**. The *N*-terminus of the last segment to be coupled **42** is protected with the Boc group so that, when the target peptide **43** has been constructed, acidolysis with trifluoroacetic acid in the final deprotection step gives the free peptide **44**; the Fmoc group would be stable to this treatment. The segments are purified by either normal-phase or reversed-phase chromatography, and segment coupling is normally done in *N*-methylpyrrolidinone, using phosphonium or uronium reagents as coupling reagents. After segment coupling, all protecting groups are removed with trifluoroacetic acid providing the crude peptide, which is then purified. The lipophilic segment coupling strategy has been successfully applied to the synthesis of several medium-sized peptides, including calcitonin and human neuropeptide Y.

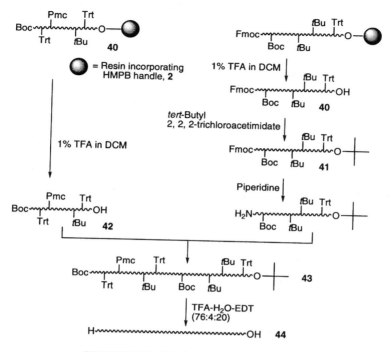

SCHEME 4.10   The lipophilic coupling strategy.

4.3   COUPLING OF MINIMALLY PROTECTED PEPTIDES IN AQUEOUS SOLUTION

The problems caused by the poor solubility of protected peptide segments, inherent both in the CSPPS strategy and in classical synthesis in solution, stimulated the investigation of other convergent strategies based upon the coupling of unprotected or minimally protected peptide segments. These are much more soluble in aqueous solvent systems, allowing them to be purified to homogeneity using the standard chromatographic techniques used for the purification of peptides. The coupling of minimally protected peptide segments is a delicate and demanding undertaking since the multitude of free functional groups can lead to all kinds of side reactions competing with the desired amide bond formation. In order to couple two unprotected peptides chemically, it must be possible to activate the C-terminal carboxylic acid of one of the peptides without affecting any of the other carboxyl groups present in either peptide. Additionally, all other amino groups except the one involved in coupling must be either blocked or deactivated in some way, otherwise they will take part in acylation reactions, leading to the formation of branched peptides. However, methods are available for accomplishing this and the approach is one of the most promising general strategies for the synthesis of proteins currently available.

4.3.1   Activation of C-Terminal Thiocarboxyl or Thioester Groups

The origins of the coupling of minimally protected segments as an approach to large peptide synthesis can be traced back to the use of semiglobal protecting schemes to enhance solubility in classical peptide synthesis in solution (see Section 3.2.2). However, the methods currently receiving the most attention were pioneered by Blake, Li, and Yamashiro[200-207] in the early 1980s. They are based on the coupling of minimally protected peptide segments by specific activation of a C-terminal thiocarboxyl group.[200,201]

The basic strategy adopted by Blake, Li, and Yamashiro was to divide the target molecule into two (sometimes more) large segments, one of which has a thiocarboxyl group at the C-terminus. Both of these segments can be synthesized using Boc/Bzl SPPS; thioester peptide–resin linkages are stable to the conditions used. They are then detached from the resin with strong acid so that most side chain–protecting groups are also removed, ensuring that even quite large segments are soluble in aqueous solvent systems. Specific activation of the thiocarboxyl group using silver salts or by acyl disulfide formation then allows the two segments to be joined by an amide bond in a chemoselective manner, giving the desired target molecule.

Thioglycine is most commonly chosen as the C-terminus since racemization on segment coupling is not possible and yields tend to be higher. However, other thiocarboxyl amino acids can be used if desired. Boc-thiocarboxylglycine, easily prepared from Boc-Gly-OH, is first made to react with handle **45** in solution, forming the preformed handle **46**. This is then attached to a suitable resin giving **47**, as illustrated in Scheme 4.11; similar procedures apply in the cases of other thiocarboxyl amino acids.

Standard Boc/Bzl synthesis is then carried out on **47** and the peptide detached from the resin by treatment with liquid hydrogen fluoride, removing most protecting groups. That of the N-terminus, however, may be chosen so as to withstand the cleavage conditions. N^α-protecting groups that are not removed by strong-acid cleavage

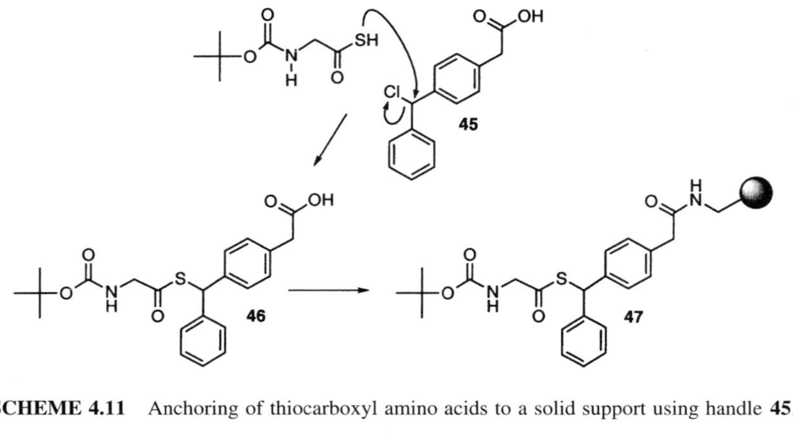

SCHEME 4.11   Anchoring of thiocarboxyl amino acids to a solid support using handle **45**.

under normal conditions include the acetyl (Ac) **48**, Fmoc **49**, trifluoroacetyl (Tfa) **50**, 2-(methylsulfonyl)ethoxycarbonyl[208] (Msc) **51**, Troc **52**, and isonicotinoyloxycarbonyl[209] (iNoc) **53** groups.

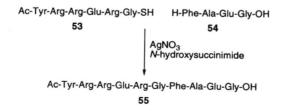

Structures are drawn to include the nitrogen atom of the amino acid.

The Fmoc **49**, Tfa **50**, and Msc **51** groups are all base labile while the Troc **52** and iNoc **53** groups can both be removed by treatment with zinc dust in acetic acid. The N-acetyl **48** group usually cannot be removed selectively from peptides since the conditions required are too vigorous and lead to the rupture of amide bonds. It is used when the amino group in question is to remain permanently blocked, as is the case in several natural peptides and proteins.

The second segment to be coupled is prepared by more-or-less standard Boc/Bzl SPPS and is detached from the resin by hydrogen fluoride treatment. After purification of the segments in aqueous medium, a process facilitated by the absence of side chain protection, the thiocarboxyl-terminus peptide is activated with silver ion and coupled with the free N^α-amino group of the second minimally protected peptide. Such silver ion activation is specific for the thiocarboxyl group and allows peptides having free carboxyl groups in their side chains to be coupled.

The potential of the method was initially demonstrated in model studies.[200] Peptide segment **54** was synthesized on solid support **47** and peptide **55** on a standard Merrifield resin. Both were cleaved with liquid hydrogen fluoride. After purification, the peptides were coupled in aqueous solution, using silver nitrate to activate the thiocarboxyl C-terminus of **54**. Model peptide **55** was produced in 40% yield, as shown in Scheme 4.12.

Ac-Tyr-Arg-Arg-Glu-Arg-Gly-SH   H-Phe-Ala-Glu-Gly-OH
53   **54**

AgNO₃
N-hydroxysuccinimide

Ac-Tyr-Arg-Arg-Glu-Arg-Gly-Phe-Ala-Glu-Gly-OH
55

SCHEME 4.12   Synthesis of a model peptide by silver ion activation of a peptide C-terminal thiocarboxylic acid.

Model peptide **55**, however, contains no Lys so that amide bond formation between **53** and **54** is not complicated by competition with another nucleophilic amine: the side chains of the Arg residues, or those of Asn, Gln, His, or Trp were they to be present, are not effective nucleophiles under normal conditions. Lysine on the other hand has a highly nucleophilic amine side chain that can undergo amide bond formation with the activated thiocarboxyl *C*-terminus leading to chain branching. If the strategy is to be applicable in more complex syntheses, a method for blocking the side chain amino groups of Lys is required. The option adopted by Blake, Li, and Yamashiro was to treat Lys-containing peptides, produced after cleavage from the solid support, with citraconic anhydride **56**. This gives rise to peptides protected as in **57**. The citraconyl group can be removed, when necessary, by mild acidic hydrolysis in 25% aqueous acetic acid.[210]

The *N*-terminus of the thiocarboxyl peptide must always be blocked during segment coupling. However it not always necessary to differentiate between *N*-terminus protection and that of any Lys residues it may contain. This being so, all its amino groups can be protected with citraconyl anhydride. The *N*-terminus of the peptide that is to act as amino component, on the other hand, requires that Lys side chain protection and that of the N^α-amino group be distinguished chemically. This is because it must be possible to liberate the *N*-terminus selectively, for coupling. Differential protection is normally achieved by synthesizing the segment with one of the protecting groups **49** to **53** (but not **48**) at the *N*-terminus. Protection of the internal Lys residues with the citraconyl group then allows the N^α-amino group to be liberated selectively.

Such an approach to the management of Lys was first tried in the synthesis of [Gly¹⁷]-β-endorphin, a peptide hormone analogue consisting of 31 amino acid residues.[201] The 18-31 segment, which is to act as amino component in the segment-coupling reaction, was prepared by Boc/Bzl synthesis on a Merrifield resin. The last amino acid was incorporated as the hydrogen fluoride-stable Fmoc derivative. After cleavage from the solid support, peptide **54** was treated with citraconic anhydride to protect the Lys N^ε amino groups. The Fmoc group was then removed with piperidine, providing the amino component peptide segment **55**, ready for coupling, as shown in Scheme 4.13.

The 1-17 segment, having a thiocarboxyl *C*-terminus, was synthesized on support **47**. After cleavage from the resin and purification, peptide **56** was treated with citraconic anhydride to protect both the amino groups of the Lys residue at position 9 and the *N*-terminus, giving **57**, as shown in Scheme 4.14.

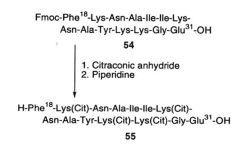

SCHEME 4.13 Synthesis of the 18-31 segment of [Gly^{17}]-β-endorphin.

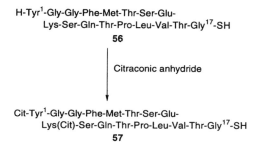

SCHEME 4.14 Synthesis of the 1-17 segment of [Gly^{17}]-β-endorphin.

The two segments **55** and **57** were coupled by activation of the thiocarboxyl group of **57** with silver nitrate, in the presence of *N*-hydroxysuccinimide to give β-endorphin in 40% yield, after purification. With this synthesis the fundamental Blake–Li–Yamashiro strategy was established and was successfully applied, with certain refinements (see Section 4.3.1.1), in the preparation of several complex large peptides and proteins. It has been very influential and several more-recent approaches to large peptide synthesis are based on the coupling or chemoselective "ligation" (see Section 4.4) of unprotected or minimally protected peptide segments having thiocarboxyl or thioester groups.

A conceptually similar, more recent variation[211-217] makes use of the specific activation of *C*-terminus thioesters of minimally protected peptide segments for differentiating the carboxyl group of the *C*-terminus from those of the side chains. The approach is outlined in Scheme 4.15.

The *S*-alkyl thioester of glycine **59** is synthesized by reaction of the thioacid **58** with Boc-Gly-OH and is then loaded onto a *p*-methylbenzhydrylamine resin using conditions similar to those used for loading a normal amino acid. Boc/Bzl peptide synthesis is then carried out on **60** until the desired sequence has been assembled. The last amino acid is incorporated with *N*$^{\alpha}$-Troc **52** or -*i*Noc **53** protection, giving the peptide–resin **61**. After hydrogen fluoride-mediated cleavage the peptide *S*-alkylthioester **62** is treated with *tert*-butyl succinimidyl carbonate (BocOSu) to protect the side chains of any Lys residues it might contain, giving minimally protected peptide **63**. This can then be selectively activated at the *C*-terminus using

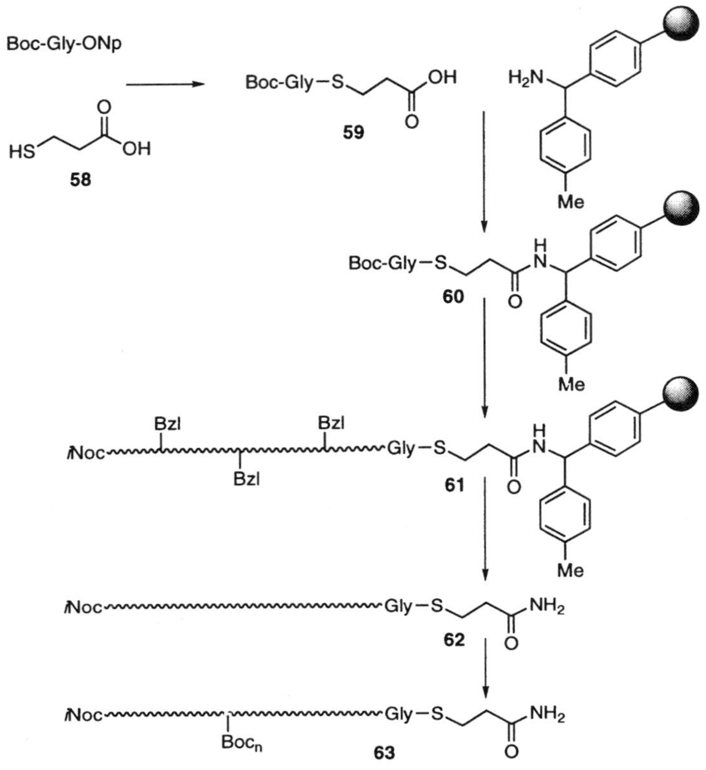

SCHEME 4.15 Synthesis of partially protected peptide *C*-terminal thioesters.

silver salts and converted to the corresponding *p*-nitrophenyl active ester for coupling with the N^α-amino group of another minimally protected peptide segment.

4.3.1.1 Examples of Synthesis Using *C*-Terminus Thiocarboxyl or Thioester Activation

4.3.1.1.1 Bovine [Cys(Cam)¹⁴, ¹⁷]-Apocytochrome C

This 104-amino acid protein was chosen by Blake[204] both because it was a challenging synthetic target and because the natural protein was commercially available, providing a convenient standard against which the synthetic material could be judged. The native protein, cytochrome c, has a heme group bound to the two Cys residues at positions 14 and 17. In order to simplify matters somewhat, this was removed by treatment with silver nitrate and the Cys residues protected as Cys(Cam) derivatives, giving the apocytochrome target molecule shown in Scheme 4.16.

The target protein was divided into the three segments spanning the regions 1-23, 24-60, and 61-104. The corresponding peptides, Ac-[Cys(Cam)¹⁴,¹⁷, GlyS²³]-apoCyt c-(1-23) **64**, Tfa-[GlyS⁶⁰]-apoCyt c-(24-60) **65**, and Tfa-apoCyt c-(61-104) **66** were all were synthesized on solid supports incorporating the handle **45**. Peptides

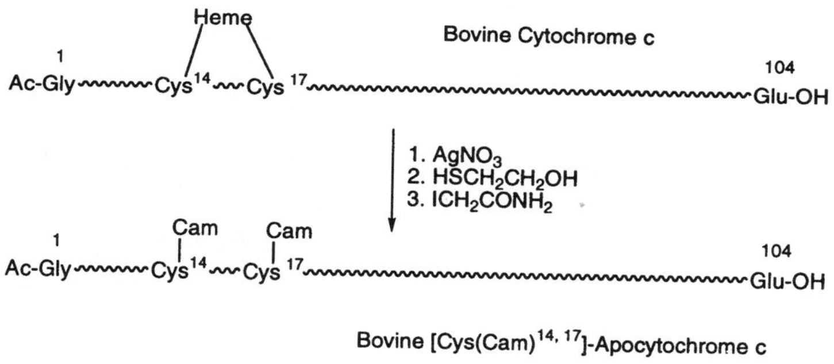

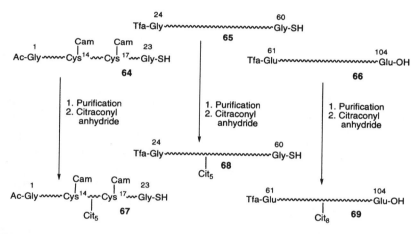

SCHEME 4.16 The preparation of bovine [Cys(Cam)[14, 17]]-apocytochrome c from natural bovine cytochrome c.

SCHEME 4.17 The preparation of the partially protected peptide segment required for synthesis of bovine [Cys(Cam)[14,17]]-apocytochrome c.

64 and **65** had thioglycine at the *C*-terminus, and in peptides **65** and **66** the last amino acid was incorporated as its N^α-trifluoroacetyl derivative so that it would remain blocked after hydrogen fluoride–mediated cleavage from the solid support (see Scheme 4.17).

After purification by reversed-phase chromatography, each of the segments was reacted with citraconyl anhydride to block the side chain amino groups of the Lys residues giving the minimally protected peptides **67**, **68**, and **69**. The peptide corresponding to the *C*-terminal region **69** was then treated with a 10% aqueous hydrazine solution to remove its N^α-protecting group, and coupling between **68** and **70** was brought about, in an estimated yield of between 10 and 20%, by treatment of an equimolar solution of both peptides in aqueous dimethylformamide with silver nitrate and *N*-hydroxysuccinimide. After isolation of **71**, its N^α-trifluoroacetyl group was removed by treatment with 10% hydrazine and coupling between **67** and **72** was carried out, using a threefold excess of the latter peptide segment, again with silver nitrate and *N*-hydroxysuccinimide, as shown in Scheme 4.18.

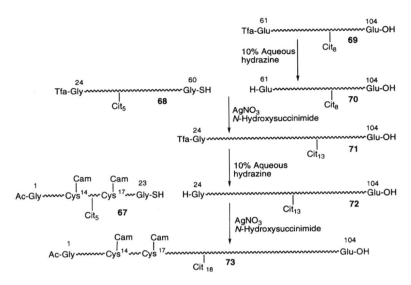

SCHEME 4.18   The synthesis of bovine [Cys(Cam)^{14,17}]-apocytochrome c.

After chromatography to remove uncoupled segment **67**, the crude protected peptide **73** was treated with 33% aqueous acetic acid to remove the citraconyl groups and the resulting material was then subjected to further, extensive chromatography. The synthetic target protein was isolated in 0.6% overall yield and was identical to a sample of the same molecule derived from native bovine cytochrome c.

4.3.1.1.2 α-*Inhibin-92*

Blake, Li, and Yamashiro carried out two separate syntheses of this 92-amino acid protein, using a similar strategy to that described above for bovine apocytochrome c. In the first of these,[205] the three protected peptides Cit-[Lys(Cit)^{14,25,28}, GlyS^{34}]-α-IB-92-(1-34) **79**, Tfa-[Lys(Cit)^{43,51,53,64}GlyS^{65}]-α-IB-92-(35-65) **76**, and Msc-[Lys(Cit)^{67,86}]-α-IB-92-(66-92) **74** were prepared on solid supports incorporating handle **45** and their Lys side chains were protected as the citraconyl derivatives as described above. The α-inhibin-92 sequence was then constructed as outlined in Scheme 4.19.

The base-labile Msc group **51** was removed from peptide **74** by treatment with 0.1 M aqueous sodium hydroxide. The resulting peptide **75** was coupled to peptide **76** (R = Tfa) in 37% yield, using silver ion activation of the thiocarboxyl group in the presence of *N*-hydroxysuccinimide. After purification, the *N*-terminal Tfa group of peptide **77** (R = Tfa) was removed by treatment with 10% aqueous hydrazine and peptide **78** was then coupled to peptide **79** in 22% yield, again using silver ion activation in the presence of *N*-hydroxysuccinimide. Treatment of peptide **80** with aqueous acetic acid to remove the citraconyl groups followed by purification then gave material identical with the natural protein.

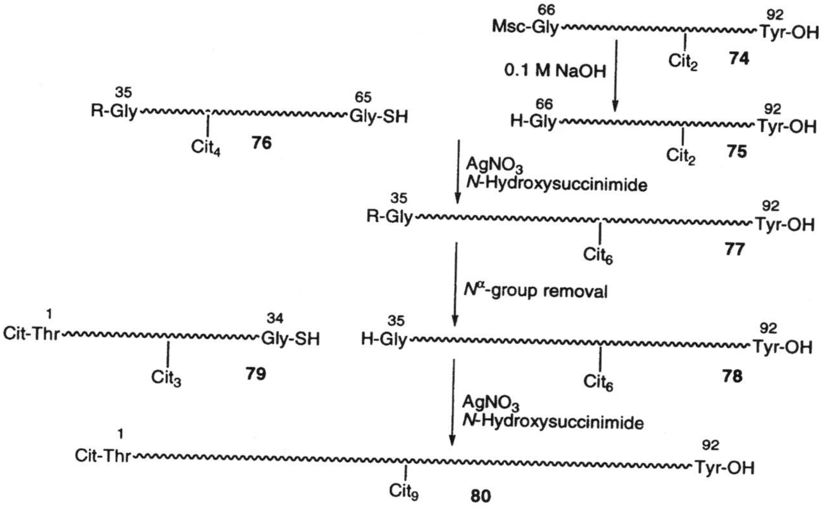

SCHEME 4.19   The synthesis of α-inhibin-92.

In the second synthesis of this molecule[206] certain refinements were introduced. First of all, the handle **81**, which is easier to synthesize than **45**, was used for the preparation of the peptide segments, **74**, **76** (R = Msc), and **79**.

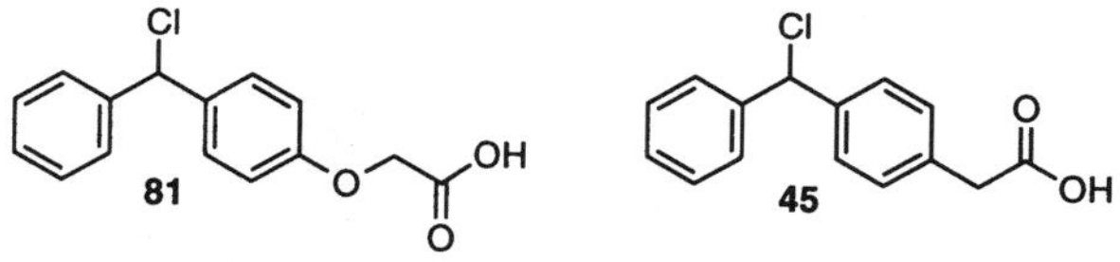

As an alternative to the silver ion-catalyzed coupling reaction of thiocarboxyl-terminus peptides, acyldisulfides were used as activated derivatives. Model experiments showed that the thiocarboxyl group of Boc-thioalanine reacts with diaryl disulfides to give unsymmetrical acyl disulfides in dimethylformamide or symmetrical diacyl disulfides in aqueous dimethylformamide as shown in Scheme 4.20.

Both disulfides underwent efficient amide bond formation with the N^α-amino group of resin-bound leucine. Exploratory experiments indicated that the most useful procedure for segment coupling was use of 2,2'-dipyridyldisulfide in dimethylsulfoxide, in the presence of hydroxysuccinimide. Comparative studies indicated that the coupling of **74**, **76** (R = Msc), and **79** proceeded in higher overall yield when acyl disulfide, rather than silver ion, activation of the thiocarboxyl group was applied. The overall yield, based on starting resin, for the production of α-inhibin-92 using the three-segment strategy with acyldisulfide activation was 8%. A two-segment strategy, involving the synthesis and coupling of segments **77** and **79** only, gave an

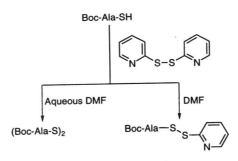

SCHEME 4.20 Diaryl disulfide activation of amino acid thioesters.

improved overall yield of 11%. These are both superior to the 4.5% yield produced in a linear synthesis of the same molecule.

4.3.1.1.3 *DNA-Binding Protein of* **Bacillus stearothermophilus**

The synthesis of the HU-type DNA-binding protein (HBs), consisting of 90 amino acids was carried out using the variation of the Blake–Li–Yamashiro strategy developed by Hojo and Aimoto.[213] The three minimally protected peptides Boc-[Lys(Boc)[3]]-HBs-(1-15) **90**, *i*Noc-[Lys(Boc)[18,19,23,28]]-HBs-(16-39) **87**, and *i*Noc-[Lys(Boc)[41,59]]-HBs-(40-60) **84** were synthesized on the solid support **60**, as outlined in Scheme 4.15. The segment *i*Noc-[Lys(Boc)[75,80,83,86,90]]-HBs-(61-90) was synthesized on a phenylacetamidomethyl (Pam) resin, detached from the support, and treated with BocOSu in order to protect the Lys side chains, giving **82**. These segments were coupled together as outlined in Scheme 4.21.

The *i*Noc N^α-protecting group was removed from **82** using zinc dust in acetic acid. Coupling of **84** and **83** was brought about by treatment of a solution containing one equivalent of each, with silver nitrate in the presence of *N*-hydroxysuccinimide. After isolation and purification of **85**, the *i*Noc N^α-protecting group was removed by treatment with zinc dust in acetic acid. Coupling of **86** with **87** was carried out by treating **86** with a twofold excess of **87** and silver nitrate in the presence of *N*-hydroxysuccinimide. After isolation and purification, the N^α-protecting group of **88** was removed. Coupling of **89** to **90** was again achieved by treating **89** with a twofold excess of **90** and *N*-hydroxysuccinimide in the presence of silver ion. The desired, partially protected sequence **91** was then purified and treated with trifluoroacetic acid to give the desired protein.

Segment-coupling times were between 1 and 2 days in each case, longer than those needed for the coupling of thiocarboxyl terminal peptides, because of the lower reactivity of *C*-terminus thioesters. The overall yield for this synthesis was 8.2%, based on segment **82**. Comparative studies showed that the best overall yield (16%) was obtained by using a similar strategy to that outlined in Scheme 4.21, but without isolation of the various intermediates after segment coupling. Linear SPPS of the protein gave a lower overall yield and less-pure crude product.

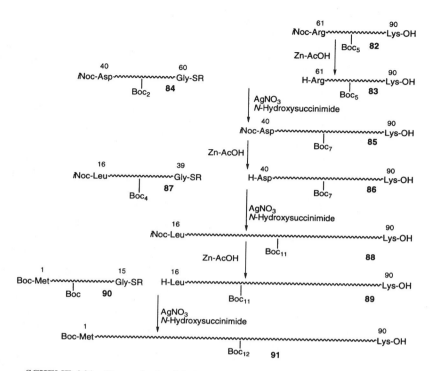

SCHEME 4.21 The synthesis of the DNA-binding protein of *B. stearothermophilus.*

4.4 CHEMICAL LIGATION OF PEPTIDE SEGMENTS

While the chemical synthesis of "native" proteins (those having an all peptide-bond backbone) is still some way from being routine, the successes that have been achieved have encouraged a more in-depth study of the molecular basis of protein structure and function. In addition to refining and extending methods for the synthesis of the naturally occurring molecules, chemists today are also concerned with probing the chemical behavior of proteins by the synthesis of analogues. These may be ones in which given amino acid residues are replaced by other proteinogenic residues, or they may be ones into which some artificial element has been introduced. The study of analogues can provide valuable insights into the structural or mechanistic behavior of natural proteins in biological systems.

A very influential approach to the synthesis of nonnative analogues of proteins has been to join peptide segments together by bonds that are not amides. This is often called *chemical ligation*[218] although, somewhat confusingly, *ligation* or *chemical ligation* is also used to describe the union of free peptides by amide bond formation, especially when this is done enzymatically. In this book we will take *chemical ligation* to mean the covalent linking together of peptide segments by any bond that is not an amide. The chemical ligation of peptide segments can be done in such a way as to form stable nonnatural structures. In these, the peptide backbone, while still consisting predominantly of amide bonds, incorporates some other covalent linkage at intervals, coinciding with the ligation sites. Such structures are said to be *backbone-engineered* and can, depending upon the case, exhibit the full biological activity of the natural protein.

4.4.1 Chemical Ligation to Give Backbone-Engineered Protein Analogues

The controlled formation of an amide bond between two peptide segments normally requires the use of protecting groups, to a greater or lesser extent. This is because the *C*-terminal carboxyl group of one of the peptides must be differentiated from all other carboxyl groups present in either of the peptides. Furthermore, all amino groups other than the one involved in amide bond formation must be rendered nonreactive in some way (see Section 4.3). If, however, the bond used to join two peptides is not an amide, then the need for protecting groups is not so pressing. If each of the segments has a unique and complementary reactive functional group, they can, in principle, be left completely unprotected. Chemical ligation can then be brought about chemoselectively by a specific chemical reaction between them, although very strict selectivity requirements are placed upon such a reaction.

Methods for linking two different proteins, or proteins to peptides or other types of molecules, by the formation of non-amide covalent bonds have been known for some time.[219-221] Even though the chemistry used often gives rise to heterogeneous products, the preparation of such protein conjugates is useful in immunological or other biochemical studies. These methods have been influential in the development of chemical ligation strategies, although a higher degree of refinement is necessary if homogeneous products of defined structure are to be produced. The most important chemical ligation strategies are considered below.

4.4.1.1 Chemical Ligation by Thioester or Thioether Formation

The first method for the chemical ligation of large unprotected peptides was developed by Kent[222,223] and owes something to the Blake–Li–Yamashiro approach to the coupling of minimally protected peptide segments in aqueous solution (see Section 4.3). Kent's method is, essentially, thioester formation between the two peptides, one of which, **92**, has a thiocarboxyl group at its *C*-terminus (often chosen to be thiocarboxyl glycine) and the other, **93**, a bromoacetyl group at its *N*-terminus.[224,225] Nucleophilic attack of the thiocarboxyl group at the carbon atom bearing the bromine atom then leads to thioester formation between the segments as in **94**. The overall effect at the ligation site is to replace a glycine residue by a 2-thioacetic acid residue. This means that the –NH– group of the amide bond in question has been replaced by a sulfur atom, as shown in Scheme 4.22.

The reaction is conducted at acid pH so that all free amino groups are rendered nonnucleophilic by protonation. The thiocarboxyl and bromoacetyl groups are

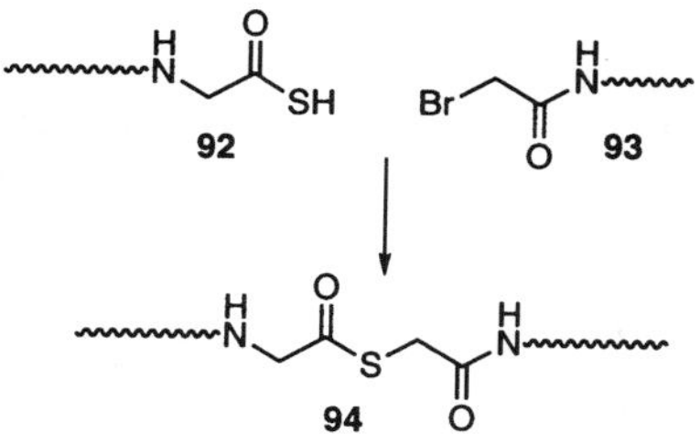

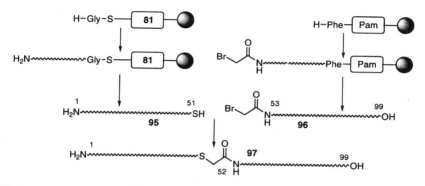

SCHEME 4.22 Chemical ligation by thioester formation.

SCHEME 4.23 Synthesis of a backbone-engineered HIV-1 protease analogue by thioester chemical ligation.

therefore exclusively mutually reactive under the reaction conditions, which makes the ligation reaction efficient, in spite of requiring bond formation between two large peptides. This chemical ligation was first applied in the synthesis of an HIV-1 protease analogue, as outlined in Scheme 4.23.

The ligation site chosen was between the Gly[51] and Gly[52] residues of the monomer of the [Aba[67,95,167,195]] HIV-1 protease analogue, previously synthesized by Kent,[226-228] (see Section 2.6.2), for a number of reasons. First, since glycine is achiral, there is no possibility of loss of chiral integrity at the ligation site. Second, the two segments to be ligated would both be more or less equal in length, at around 50 amino acids each. Third, this region of the HIV-1 protease was known to be sensitive to changes in the amino acid sequence so that it provided an opportunity to assess the effect of backbone engineering.

The 1-50 sequence of the molecule, having thiocarboxyl glycine at the C-terminus was synthesized on a solid support incorporating handle **81**, whereas the N-bromoacetylated (53-99) section was synthesized on a standard Pam resin. Residue 52 (2-thioacetic acid) is created *in situ* by the chemical ligation reaction itself. Both the required peptides **95** and **96** were produced on cleavage from their solid supports by treatment with hydrogen fluoride. This also removed all protecting

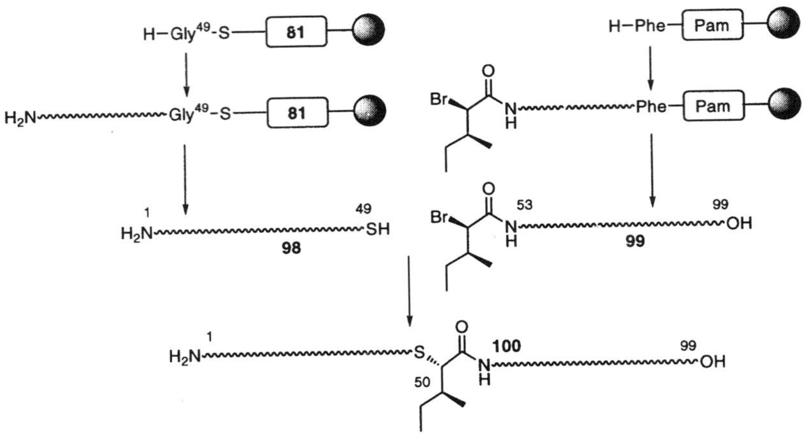

SCHEME 4.24 Synthesis of a second backbone-engineered HIV-1 protease analogue by thioester chemical ligation.

groups, ensuring excellent solubility in aqueous medium. After purification, chemical ligation by S_N2 substitution of the thiocarboxyl group of **95** at the bromoacetyl unit of **96** was brought about at pH 4.3, and was essentially complete in 3 h. The HIV-1 protease analogue **97** showed enzymatic activity comparable with that of the natural protein, implying that the NH group substituted by the sulfur atom at the ligation site was not functionally important. The enantiomer of **97** was also prepared, using an identical procedure but with D-amino acids.[229]

Similar chemistry has also been carried out on other Gly-Gly-containing proteins.[230] The method was extended, however, in the synthesis of another backbone-engineered HIV-1 protease analogue[231] in which the amide bond between the Gly[49] and Ile[50] residues was replaced by a thioester. The synthesis of this analogue, outlined in Scheme 4.24, demonstrates that the site for chemical ligation is not restricted solely to glycine but can, in principle, be applied at other points in the amino acid sequence. The limiting factor is the availability of the bromocarboxylic acid in question.

The 1-49 C-terminus thiocarboxyl peptide **98** was synthesized using handle **81**. The 51-99 segment **99**, however, must have (2R,3S)-2-bromo-3-methylvaleric acid at its N-terminus in order to give rise to the thioester analogue of the Gly[49]-Ile[50] amide bond on chemical ligation. This bromoacid is not available commercially and was prepared from D-allo-Ile (2R,3S Ile), a non-DNA-encoded isomer of Ile. The configuration at C-2 of residue 50 is changed to the (S) form on chemical ligation of **98** and **99** since the S_N2 substitution reaction takes place with inversion of configuration. The overall result in the HIV-1 protease analogue **100** is that DNA-encoded (2S,3S)-Ile is replaced with (2S,3S)-2-thio-3-methylvaleric acid at residue 50. Put more simply, the –NH– group of the Gly-Ile amide bond is replaced by a sulfur atom.

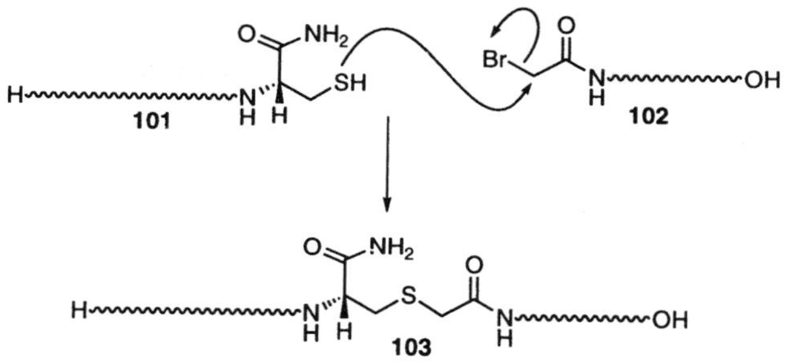

SCHEME 4.25 Chemical ligation by thioether formation.

These thioester protein analogues are stable in the pH range 3 to 6, but they do, however, undergo base-catalyzed hydrolysis above pH 7. An alternative use of sulfur nucleophiles in chemical ligation allows the formation of peptide and protein analogues that are stable at basic pH. This approach, also developed by Kent,[223] is based on thioether formation between the free thiol group of cysteine on one peptide and the bromoacetyl group of the second peptide to be ligated. Such thioether formation permits the preparation of a different class of backbone-engineered protein analogue in which an amide bond has now been replaced by a thioether. The general reaction is illustrated in Scheme 4.25 for a peptide **101**, having a C-terminal Cys amide. Chemical ligation with an N-terminal bromoacetyl peptide **102** then gives a backbone-engineered analogue **103**.

Care must be taken to avoid the formation of excessive amounts of the homodimer of the peptide having the cysteine residue, by disulfide bridge formation (see Chapter 5). Judicious choice of the conditions of the chemical ligation reaction can normally prevent this becoming severe. Unlike the thioester chemical ligation strategy that is constrained to have the thiocarboxyl group at the C-terminus of one of the peptides, chemical ligation by thioether formation can be brought about with a cysteine residue at any position in a peptide. This lends it an extra degree of flexibility for the design of nonnatural protein analogues.

Thioether chemical ligation was first applied to the synthesis of another HIV-1 protease analogue,[223] using the general strategy outlined in Scheme 4.25. However, thioether chemical ligation has also been used to produce different types of backbone-engineered proteins. The synthesis of a structural model for the cytoplasmic domain of a receptor for the integrins, cell-surface proteins thought to be involved in signal transduction,[232] provides an example. The model required was of the type **104**, in which two peptide chains, each having a high tendency to form helical structures are joined in a head-to-head fashion. Although such structures are not accessible by standard linear SPPS nor by biotechnological techniques, they can be prepared by the chemical ligation of two peptide chains at their N-termini. One possibility for doing this is thioether formation, as in **105**, by reaction between the thiol group of an N-terminal cysteine residue on one peptide and the N-terminal bromoacetyl group on the other.

The synthesis of model protein **105** is outlined in Scheme 4.26. Both the
N-terminal cysteine-containing peptide (1-50) **106** and the *N*-bromoacetylated (1-76)
peptide **107** were synthesized on a Pam resin and cleaved with hydrogen fluoride.
Chemical ligation of these requires careful control of the pH of the medium. It must
be high enough to produce sufficient quantities of the thiolate anion, while at the
same time being low enough to render any amino groups nonnucleophilic by protonation. Chemical ligation at pH 7.0 in an aqueous solvent system led to appreciable
amounts of the homodimer of peptide **106**, as a consequence of disulfide bridge
formation. This could be almost completely eliminated using 95% dimethylformamide containing aqueous 5% phosphate buffer as reaction medium, and by reducing
the concentration of the reactants. Chemical ligation was complete in under 3 h,
giving the desired protein model **105**, essentially without the competitive formation
of homodimer.

Kent has also demonstrated that chemical ligation by thioester ester formation
is compatible with the presence of free sulfhydryl groups in the peptide segments
to be ligated.[233] This is an important development and significantly extends the

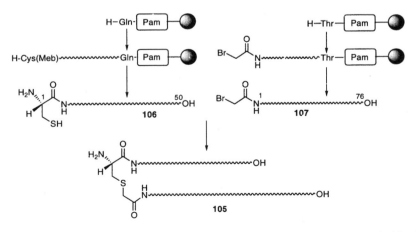

SCHEME 4.26 The synthesis of backbone-engineered protein model **105** by thioester chemical ligation.

usefulness of the methods. It is possible because of the difference in pK_a values between the thiocarboxylic acid (≈ 3) and the cysteine thiol (≈ 9), so that the thiocarboxylate anion can be generated selectively. Careful control of the pH of the ligation medium allows an exquisite differentiation of the two sulfur nucleophiles: at pH 3.2, thioester formation can be brought about smoothly without interference from free sulfhydryl groups. Disulfide bridge or thioether formation[234] can then be carried out on the ligated molecules by increasing the pH, taking care not to cause conditions that would lead to hydrolysis of the thioester.

Since their introduction, thioester and thioether chemical ligation strategies have also been adopted by other workers to prepare artificial four-helix bundle proteins[235] and other nonnatural structures.[236] Their usefulness has been demonstrated by the work outlined above, and they promise to be a very useful tool for the design and construction of artificial proteins to probe biochemical processes.

4.4.1.2 Chemical Ligation by Hydrazone or Oxime Formation

Other chemical ligation strategies for the construction of backbone-engineered protein analogues have been developed by Offord and Rose. The first of these relies upon the formation of hydrazones[237] between unprotected peptide segments.[238-241] The method requires a peptide C-terminal hydrazide, on the one hand, and an aldehyde or ketone, on the other. These are two complementary functional groups not normally found in proteins. The amino group of the peptide hydrazide (pK_a between 3 and 4) is appreciably less basic than the α-amino groups (pK_a values between 7 and 9). As a consequence, it is the only functional group present under the acid conditions of the chemical ligation reaction that can form a stable product with the aldehyde of the second peptide. Chemical ligation occurs rapidly at pH 4.6, and, although the hydrazone formed on ligation is somewhat sensitive to hydrolysis, a more stable hydrazide linkage between the peptide segments can be formed on reduction with sodium cyanoborohydride. The chemistry of this ligation strategy is outlined in Scheme 4.27.

The peptides to be ligated can be generated by chemical synthesis, by enzymatic methods,[242-244] or by a combination of the two. Peptide C-terminal hydrazides **106** can be produced enzymatically, using proteases to catalyze the formation of an amide bond between the peptide C-terminal carboxylic acid and the hydrazine derivative.[245-247] The introduction of an aldehyde function at the N-terminus of N-terminal serine- or threonine-containing peptides **107** can be done by mild periodate oxidation.[248] This gives peptides with an aldehyde at the N-terminus, as in **108**. Peptides **107** and **108** can be smoothly joined by chemical ligation at pH 4.6 giving backbone-engineered peptide hydrazone **109**. If desired, reduction with sodium cyanoborohydride then gives the more stable peptide hydrazide **110**. The application of the strategy to the synthesis of a 174-residue analogue of the granulocyte/colony-stimulating factor (G/CSF) protein[240,241] is outlined in Scheme 4.28.

The N-terminal (1-62) and C-terminal (76-174) peptide segments were produced by enzymatic digestion, with the protease *Achromobacter lyticus*, of two recombinant analogues of G/CSF, produced in *Escherichia coli*.[240] The chemically synthesized

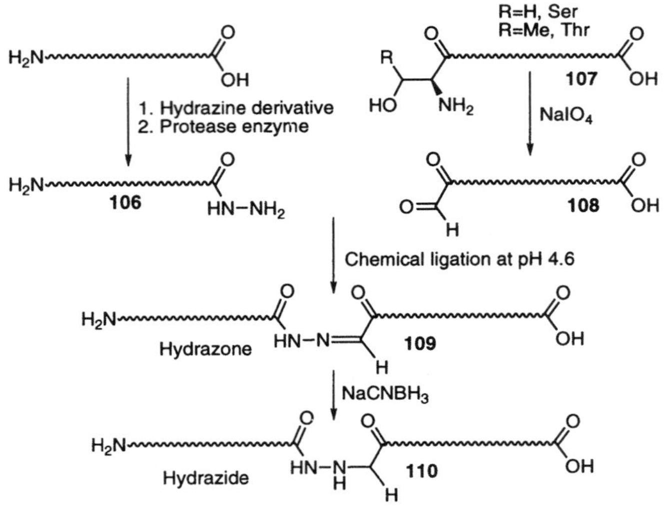

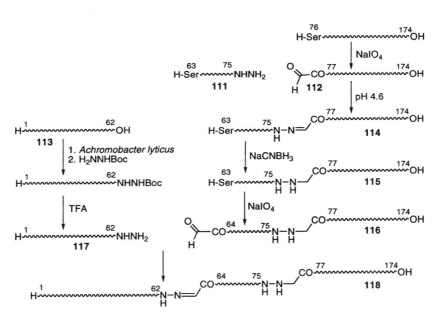

SCHEME 4.27   Chemical ligation by hydrazone formation.

SCHEME 4.28   Synthesis of a backbone-engineered analogue of the G/CSF protein, by hydrazone chemical ligation.

(63-75) segment was made by Fmoc/*t*Bu SPPS and cleaved from the resin by hydrazinolysis (see Section 4.1.1.2.3) giving the hydrazide **111**. Construction of the desired protein analogue began with sodium periodate-mediated oxidation of the

(76-174) peptide, which gave the *N*-terminal aldehyde **112**. Chemical ligation of **112** with the chemically synthesized hydrazide **111** was brought about in high yield at pH 5.3 using two equivalents of **111** relative to **112**. Reduction of the hydrazone **114** with sodium borohydride gave the peptide hydrazide **115**, which was oxidized to the *N*-terminal aldehyde **116**. The (1-62) peptide hydrazide **117** was produced by reverse proteolysis on treatment of the free (1-62) peptide **113** with Boc-hydrazide, in the presence of the enzyme *A. lyticus*, followed by treatment with trifluoroacetic acid. Condensation of **116** and **117** then gave the desired backbone-engineered protein analogue (1-62)-NHN=CHCO-(64-75)-NHNH-CH$_2$CO-(77-174) G/CSF, **118**.

The same study also showed how the three peptide segments could be ligated in reverse order, starting from the *N*-terminal segment and working toward the *C*-terminal one, although this did require the use of the base-labile Msc-protecting group **51** for the peptide *C*-terminal hydrazide **111**. Several different analogues of the same protein were constructed by the incorporation of systematically varied chemically synthesized (64-75) sections of the molecule. Additionally, it was possible to couple the (1-62) peptide **113** to the (63-75) peptide hydrazide **111** enzymatically by peptide bond formation. Chemical ligation of the (77-174) peptide *N*-terminal aldehyde by hydrazone formation then gave another analogue that had only one hydrazone linkage in its backbone. Several of the artificial proteins produced showed significant biological activity, in some cases comparable with that of the natural molecule.

Another ligation strategy, developed by the same group,[249-251] involves the use of peptide *O*-alkylated hydroxylamine **119** as the amino component. These can be chemically ligated to peptide aldehydes, derived from the oxidation of serine or threonine, giving peptide oximes **120**, as illustrated in Scheme 4.29.

The method was initially used to synthesize branched artificial proteins[252] (see Section 4.4.3.2). Very similar chemistry has been used for the preparation of backbone-engineered cyclic peptides, which incorporate an oxime bond between the *N*- and *C*-terminal amino acids of the linear precursor.[253]

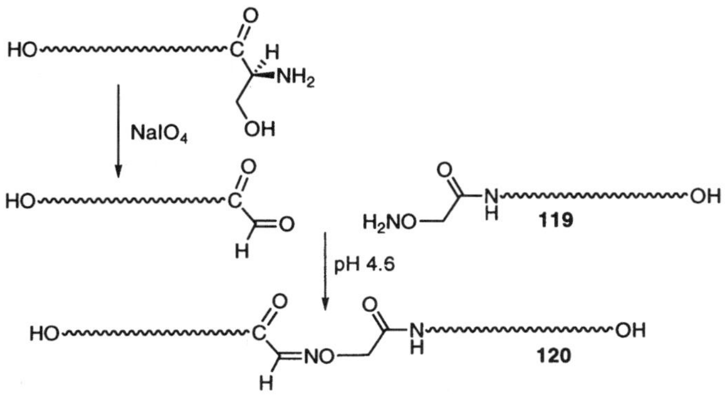

SCHEME 4.29 Chemical ligation by oxime formation.

4.4.1.3 Use of More than One Type of Chemical Ligation for Protein Analogue Synthesis

The synthesis of even more sophisticated artificial proteins is possible if two mutually compatible ligation chemistries are used together. This allows the condensation of several unprotected peptide segments in a specific manner. Kent[254] has applied both the thioester and the oxime chemical ligation chemistries to the synthesis of another protein analogue, that of the cMyc–Max transcription factor-related protein. This artificial molecule consists of 172 amino acid residues and is a heterodimer of two peptide chains of equal length, linked at their *C*-termini by an oxime bond. Each of the chains was synthesized from its two subunits by thioester chemical ligation.

The synthesis first required the construction of two different peptides, each with a functional group for subsequent chemical ligation at both the *N*- and *C*-termini. The syntheses of these peptides is shown in Scheme 4.30.

Attachment of Boc-Lys(Fmoc)-OH to a *p*-methylbenzhydrylamine resin provided the starting point for the synthesis, **121**. The desired sequences were built up and the Fmoc group removed, giving the two peptide–resins **122** and **123**. The first of these was converted to the *O*-alkylated hydroxylamine derivative **124**, while peptide–resin **123** was transformed into the *C*-terminus ketone-containing peptide **125**. After Boc group removal, a bromoacetyl group was introduced at the *N*-terminus of each. Hydrogen fluoride-mediated cleavage from the solid supports then gave the two peptides **126** and **127**.

The thiocarboxyl peptides **128** and **129** were synthesized on solid supports incorporating the handle **81**. With all four peptide chains in hand, construction of the protein analogue was accomplished as shown in Scheme 4.31. Chemical ligation between the 32-residue thiocarboxyl leucine *C*-terminus peptides **128** and **129** and each of the 53-residue bromoacetylated peptides **126** and **127**, respectively, by thioester formation occurred in less than 1 h at pH 4.7. The 86-residue products **130** and **131** were produced in good yield. Chemical ligation by oxime formation between **130** and **131** was also brought about at pH 4.7 in an aqueous solvent system, although the ligation reaction was slower and did not go to completion. The desired target molecule, the artificial cMyc–Max protein **132** was isolated and purified. It had biological activity similar to natural proteins of this type.

These chemical ligation strategies open up a whole new range of possibilities for the design and synthesis of backbone-engineered, artificial protein analogues having unusual architectures. Such structures cannot be prepared by conventional peptide chemistry or by biotechnological techniques. Current work is now directed toward the synthesis of tailor-made proteins, with specific properties.

4.4.2 Segment Coupling by Prior Chemical Ligation

Although the synthesis of nonnative proteins is currently a very active area of research, the refinement and development of methods for the synthesis of natural proteins, or of nonnaturally occurring proteins with "native" backbones, is still highly

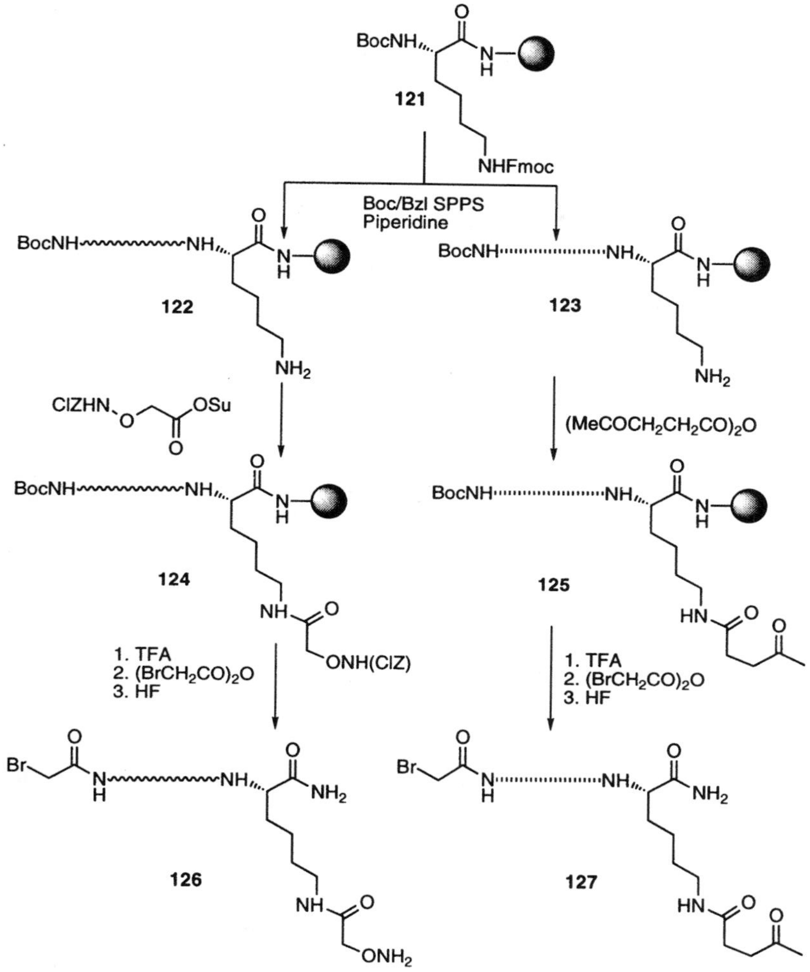

SCHEME 4.30 Synthesis of partially protected peptide segments required for the preparation of the cMyc–Max transcription factor-related protein.

desirable. The chemical ligation strategies discussed above have demonstrated that large unprotected peptides can be joined chemoselectively, in high yield giving chemically well-defined molecules. This has prompted the investigation of new methods for the formation of amide bonds between such peptides, so that native structures can be produced. Contemporary work has focused on the use of a prior chemical ligation step to bring the peptides together, before amide bond formation takes place. The original concept was outlined several decades ago, but it has only been fairly recently that it has been applied in complex peptide synthesis.

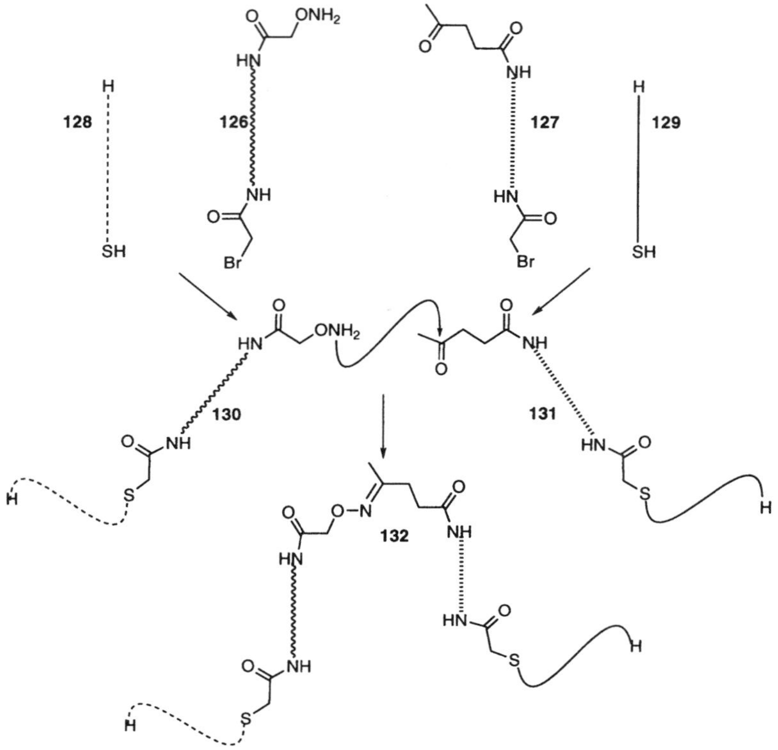

SCHEME 4.31 Synthesis of the cMyc–Max transcription factor-related protein, using two different, complementary chemical ligation chemistries.

4.4.2.1 Template-Assisted Coupling

It was recognized in the 1950s that amide bond formation between two large peptide segments would be more efficient if some way could be devised to bring, and hold, the N- and C-termini close enough together so that reaction could take place. This would allow amide bond formation between the segments to occur without formation of the highly reactive synthetic intermediates traditionally associated with chemical peptide synthesis. The use of excesses of one of the segments, to drive the coupling reaction to completion, would also be avoided. Pioneering studies into putting the idea into practice were carried out independently by Brenner[255-260] and Wieland.[261] Both elected to convert the bimolecular coupling reaction into an intramolecular one by bringing the amino and carboxyl components together on a "template" and allowing an intramolecular acyl transfer reaction to take place. The procedure is illustrated in Scheme 4.32 for the intramolecular coupling of phenylalanine with glycine on a salicylamide template, according to Brenner's scheme.[262]

    Z-Phe-OH was esterified onto the free hydroxyl group of salicoylglycine methyl ester **133** and, after hydrogenolytic removal of the N^α-protecting group, base-catalyzed intramolecular rearrangement[262] gave N-salicoylphenylalanyl-glycine methyl

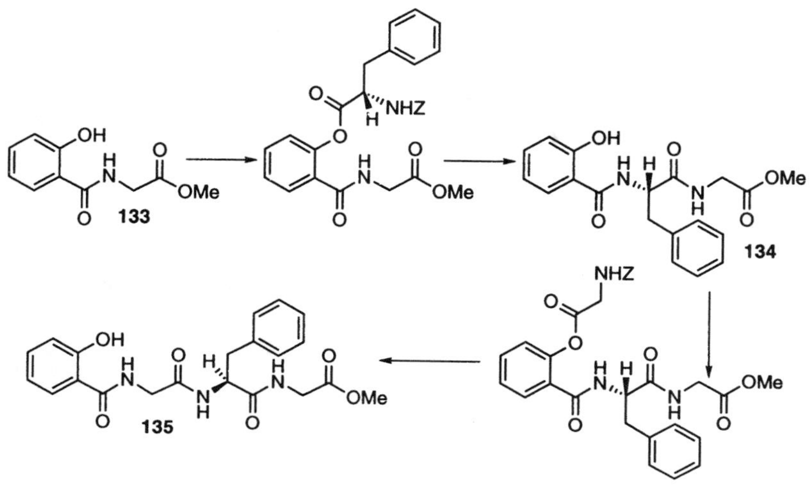

SCHEME 4.32 Brenner's "low-powered" approach to peptide synthesis.

ester **134**. The next amino acid, in this case, Z-Gly-OH, was then esterified onto the free phenolic hydroxy group of the template. Base-catalyzed intramolecular rearrangement then gave *N*-salicoylglycylphenylalanyl-glycine methyl ester **135**. Peptide synthesis can, in principle, be continued by the addition of other amino acids. However, although several model peptides were made using such methods, the conversion of *template-assisted* amide bond formation into a practical general method for peptide synthesis has proved to be very challenging. At present, it has not been achieved in an entirely satisfactory manner despite the application of various different approaches.[263-265] Among the problems that must be overcome if it is to be useful for the coupling of large peptide segments is that of bringing the components together on the template. This could conceivably be just as difficult to achieve as the coupling of the two segments under more conventional conditions.

The most-promising current variation of the strategy is that proposed by Kemp.[266] Known as *thiol capture ligation,* it revolves around amide bond formation between two segments, brought together on the 4-hydroxy-6-mercaptodibenzofuran template **136**, by chemical ligation through disulfide formation.[267,268] The first peptide to be coupled is synthesized on the template **136**, as shown in Scheme 4.33.

Template **136** is first attached, by disulfide bridge formation, to a solid support **137** incorporating the Cys derivative Z-Cys(Scm)-OH (see Section 5.1). Peptide synthesis is then carried out on **138**, using amino acids having Bpoc *N*α protection and *tert*-butyl side chain protection, giving peptide–resin **139**. This is then subjected to acidolysis with trifluoroacetic acid to remove the majority of protecting groups. Some, such as Cys, may be maintained protected at this point. Cleavage of the peptide-template ester **140** from the resin is brought about by treatment with triethyl phosphine.

The second peptide segment must have an *N*-terminal Cys residue in order to permit its chemical ligation by disulfide formation to the peptide–template **140**. It may be synthesized on the Wang resin **7** using either standard Fmoc/*t*Bu synthesis

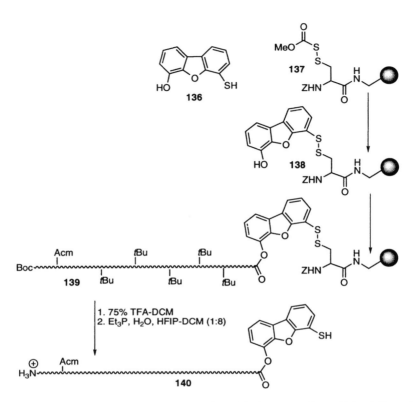

SCHEME 4.33 Synthesis of a partially protected peptide 4-mercapto-6-(peptidyloxy)dibenzofuran ester.

or amino acids having Bpoc N^α protection and *tert*-butyl-based side chain protection. The *N*-terminal Cys is incorporated as the Boc-Cys(Scm)-OH derivative (see Section 5.1), giving peptide–resin **141**. After trifluoroacetic acid-mediated cleavage from the solid support and purification, peptide **142** is then ligated to the 4-mercapto-6-(peptidyloxy)dibenzofuran **140** by formation of a disulfide bridge, giving **143**. This is shown in Scheme 4.34.

With the two peptides brought together on the template, as in **143**, base-catalyzed intramolecular acyl transfer can now take place, forming an amide bond between them, leading to the formation of the desired target sequence, **144**. Any remaining protecting groups, including the template itself, are then removed before the target molecule **145** is purified. The intramolecular acyl transfer is shown in Scheme 4.35.

Thiol capture ligation is a very elegant approach to complex peptide synthesis. It has been proved in model systems, including those with unprotected His, Lys, and Arg residues,[269,270] and in the construction of a complex 39-residue peptide.[271] Among its advantages over conventional segment-coupling methods are that, first, the solubility of the intermediates in aqueous solvent systems is improved owing to the lack of side chain protection. Second, the intramolecular acyl transfer reaction is fast, usually being complete within a few hours at most and sometimes within

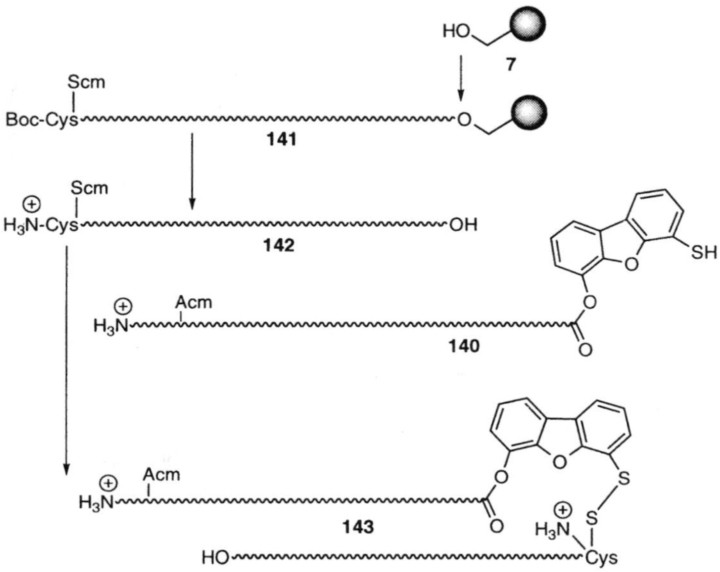

SCHEME 4.34 Chemical ligation of partially protected peptide 4-mercapto-6-(peptidyloxy)dibenzo-furan ester **140** and *N*-terminal cysteine-containing peptide **142** by disulfide formation.

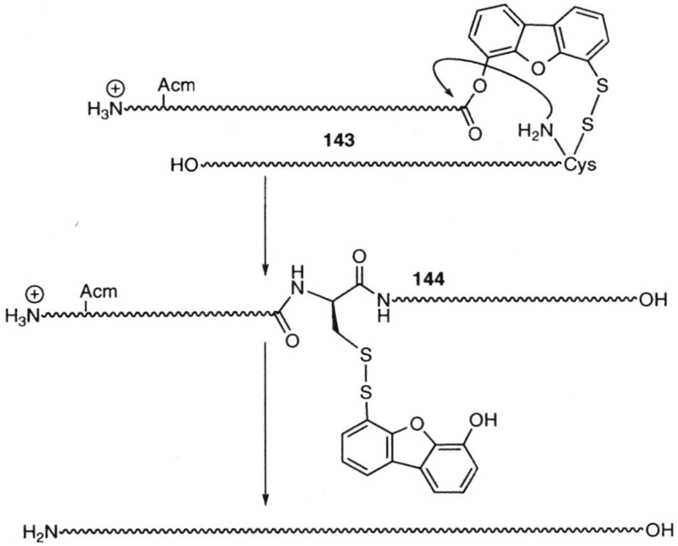

SCHEME 4.35 Template-assisted intramolecular acyl transfer to give an amide bond between the two peptide segments.

minutes. Third, the use of excesses of one of the segments is avoided. However, it does not constitute a general approach to peptide synthesis since it is only applicable in cases where there is a convenient distribution of Cys residues. Also, the chemistry

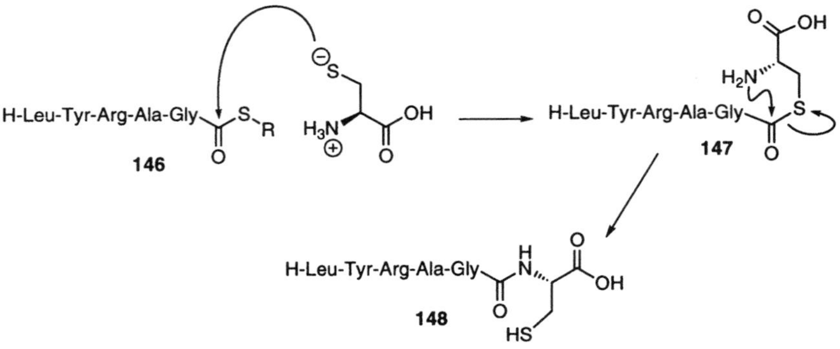

SCHEME 4.36 Native chemical ligation.

involved is rather intricate and careful selection of protecting groups is required. As it has been relatively little tried, it remains to be seen whether or not such template-assisted synthesis establishes itself as a general method for coupling peptide segments.

4.4.2.2 Native Chemical Ligation

Kemp's thiol capture ligation demonstrates that amide bond formation, even between large peptide segments with only moderate activation, can be rapid when the reacting amino and carboxyl termini are held in close proximity. An alternative approach, which does not rely upon the use of a template, has been devised by Kent.[272] It involves the chemical ligation of two unprotected segments by thioester formation, prior to the creation of an amide bond between them. For this to be possible, one of the peptides must have cysteine as its N-terminus and the other a thioester C-terminus. In model experiments, using H-Cys-OH and a pentapeptide thiobenzyl ester **146** (R = Bzl), initial chemical ligation gives cysteine thioester **147**. Amide bond formation to give hexapeptide **148** then takes place. The mechanism of such coupling is thought to be that shown in Scheme 4.36.

The initially formed chemical ligation product **147** spontaneously rearranges, giving an amide bond at the ligation site, hence the name "native chemical ligation." The rapid rearrangement of **147** is thought to be because of the favorable geometric arrangement of the α-amino group with respect to the thioester.

Such amide bond formation can also be applied to the coupling of larger peptides. The N-terminal segment is synthesized on a solid support incorporating handle **81** and has a thioglycine residue at its C-terminus. After cleavage, the thiocarboxyl peptide is made to react with an alkyl halide to give the required thioester. Model studies demonstrated that the use of better thioester leaving groups led to faster ligation reactions. The second segment, which must have Cys at its N-terminus, can be synthesized by standard solid-phase procedures. Once purified, the segments are ready for chemical ligation. Amide bond formation occurs spontaneously as in the model studies outlined above. Protection of functional groups other than the C-terminus thioester is unnecessary, and the method tolerates the presence of unprotected Lys or Cys residues in either of the segments to be coupled.

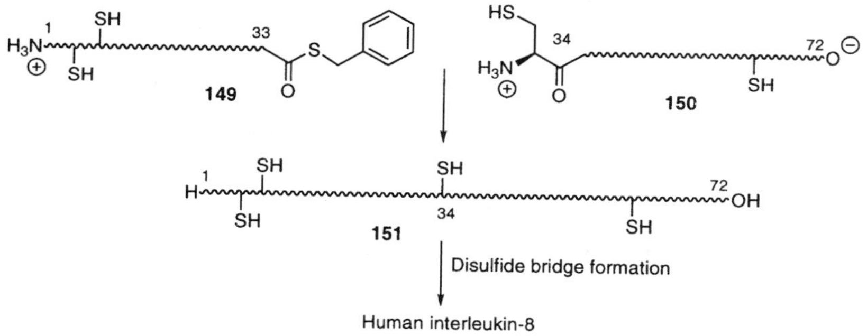

SCHEME 4.37 Synthesis of human interleukin-8 by native chemical ligation.

Native chemical ligation has been used to synthesize interleukin-8, a 72-amino acid protein.[272] The synthesis is outlined in Scheme 4.37.

The *N*-terminal peptide segment, which has two internal unprotected Cys residues in addition to unprotected Lys residues, was synthesized using handle **81**. After cleavage from the resin, treatment with benzyl bromide furnished thioester **149**. The second segment **150** was synthesized by standard SPPS and both peptides were purified by reversed-phase chromatography. Native chemical ligation of **149** and **150** was brought about at pH 7.6 in phosphate buffer in the presence of 6 M guanidine hydrochloride, affording the desired sequence **151**. Oxidation followed by purification then gave material identical with natural interleukin-8.

Tam[273-275] has also investigated native chemical ligation and has successfully applied it to the synthesis of complex peptides. Alternative possibilities for forming the initial chemical ligation product of the peptide segments are provided by nucleophilic attack of the thiocarboxyl group of peptide **152** on either the bromomethyl group of peptides, such as **153**, or the Npys group of peptides, such as **154**. Bromoalanine is easily derived from serine and can be incorporated at the *N*-terminus of peptides by standard methods. A more in-depth discussion of the Npys group is presented in Section 5.3.3.3. These alternative chemical ligations are outlined in Scheme 4.38.

In the case of bromoalanine peptide **153**, chemical ligation with **152** gives **155**, which undergoes rearrangement, forming the native ligation product **158**. A similar process occurs with peptide **154**, except that the disulfide **156** is formed on chemical ligation. This rearranges in a manner similar to **155**, giving the native chemical ligation product **157**, which must be reduced in order to provide the native-cysteine-containing peptide **158**.

4.4.2.3 Domain Ligation

A related method for coupling unprotected peptide segments, developed by Tam[276-280] relies on initial chemical ligation by thiazolidine or oxazolidine formation. The derivatives formed then undergo *O*- to *N*-acyl transfer giving an *N*-acyl thiazolidine or oxazolidine at the ligation site, but with the segments now linked by an amide

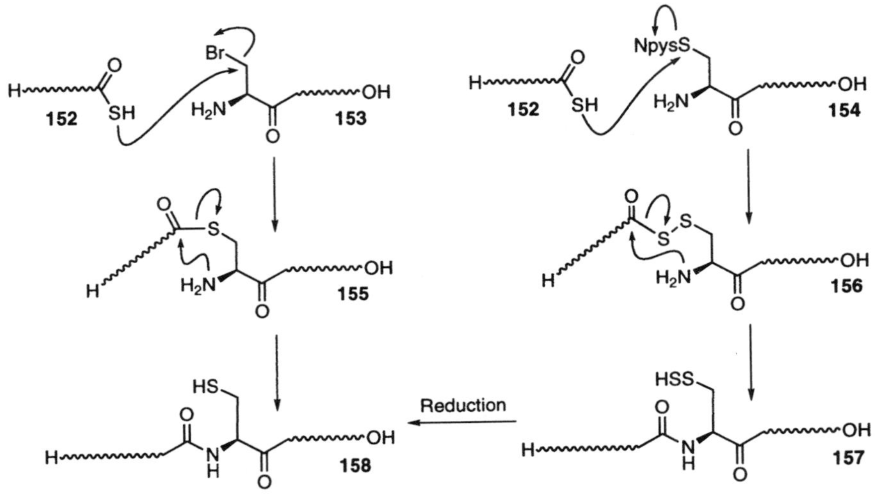

SCHEME 4.38 Alternative approaches to native ligation.

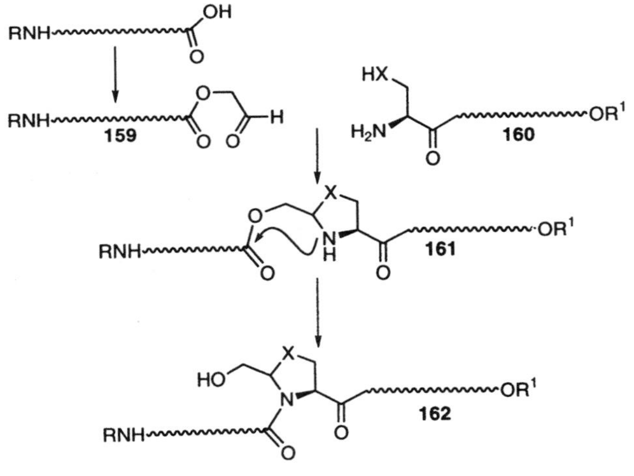

SCHEME 4.39 Domain chemical ligation.

bond. The *N*-acyl thiazolidine or oxazolidine can, in principle, be transformed back into the original Cys or Ser or Thr residue. The procedure is outlined in Scheme 4.39.

The segment that is acting as the carboxyl component must first be converted into a glycoaldehyde ester **159**. The peptide **160** that is to act as amino component must have Cys, Ser, or Thr as its *N*-terminus amino acid, since only these can form thiazolidine or oxazolidine intermediates. However, reaction rates are faster and yields better for *N*-terminal Cys peptides. Initial chemical ligation between the thiol and the aldehyde gives an addition product that then condenses with the α-amino group to form the cyclic thiazolidine ester **161** as a mixture of diastereomers.

Intermediates of the type **161** can normally be isolated and characterized. To avoid saponification of **161** or reaction of the aldehyde with reactive side chains that may be present in the segments, the chemical ligations are best conducted in aqueous acid. At pH 4 to 5 reaction is very rapid; at pH 2 it is somewhat slower.

Rearrangement of **161**, by *O*- to *N*-acyl migration, to form an amide bond occurs rapidly under basic conditions. However, ester hydrolysis can compete effectively with rearrangement so that the reaction is best conducted at around pH 7, or sometimes even lower. The weak basicity of the thiazolidine nitrogen atom means that *O*- to *N*-acyl migration still occurs at acidic pH, albeit slowly. In common with the esters **161**, the amides **162** also exist as mixtures of diastereomers that can usually be resolved by HPLC, if necessary. In peptides such as **162**, where X = S, although transformation into the original cysteine peptide is, in principle, possible, in practice it is difficult without bringing about the rupture of peptide bonds. With oxazolidines (**162**, X = O), the procedure would be much easier but the ligation and rearrangement yields for these are substantially lower than for the corresponding thiazolidines. Domain ligation is, however, still a very useful method since the thiazolidine can be considered to be a modified proline residue, with a thioether as the isoelectronic replacement for the methylene at position 4 and a hydroxymethyl substituent at position 5. Such modification probably does not alter the backbone conformation of proline-containing peptides, so that it provides an easy method for introducing this type of structural modification into peptides. More importantly, however, it allows the efficient coupling of large unprotected peptide segments to give nonnatural protein molecules. The domain ligation strategy has been applied to the synthesis of HIV-1 protease analogues.[281] The preparation of one of these is outlined in Scheme 4.40.

The key step in the synthesis of the peptide segment **166**, that of introducing the masked aldehyde function, was achieved by applying chemistry similar to that developed by Hojo and Aimoto[213] (see Section 4.3.1.1.3 above). Boc-Asn-S(CH$_2$)$_2$COOH was coupled to a Pam resin incorporating Gly giving **163**, and the peptide chain was elongated on this solid support. When complete, cleavage gave peptide **164**. Silver ion activation of the thioester and condensation with the masked aldehyde **165** then gave peptide **166**. Treatment of **166** with trifluoroacetic acid and condensation with the *N*-terminal Cys peptide **167**, synthesized on a Pam resin by conventional Boc/Bzl SPPS gave ligation product **168**. Rearrangement of this at pH 5 then gave the desired HIV-1 protease analogue **169**, in which the Pro^{39} residue has been replaced by a hydroxymethylthiaproline residue at the same position. The residue at position 39 is generated *in situ* by the chemical ligation reaction itself. Experimentally, the procedure is simple — acetal deprotection, ligation, and rearrangement all take place in the same reaction vessel and require only changes in pH to make them proceed in high yield.

An analogue having Ala instead of Leu at the 38 position was also synthesized using a very similar procedure. A modification of the method, however, allowed the backbone-engineered protein **172** to be prepared. This was formed by ligation of the peptide aldehyde equivalent **170** with *N*-terminal cysteine peptide **171**, as shown in Scheme 4.41.

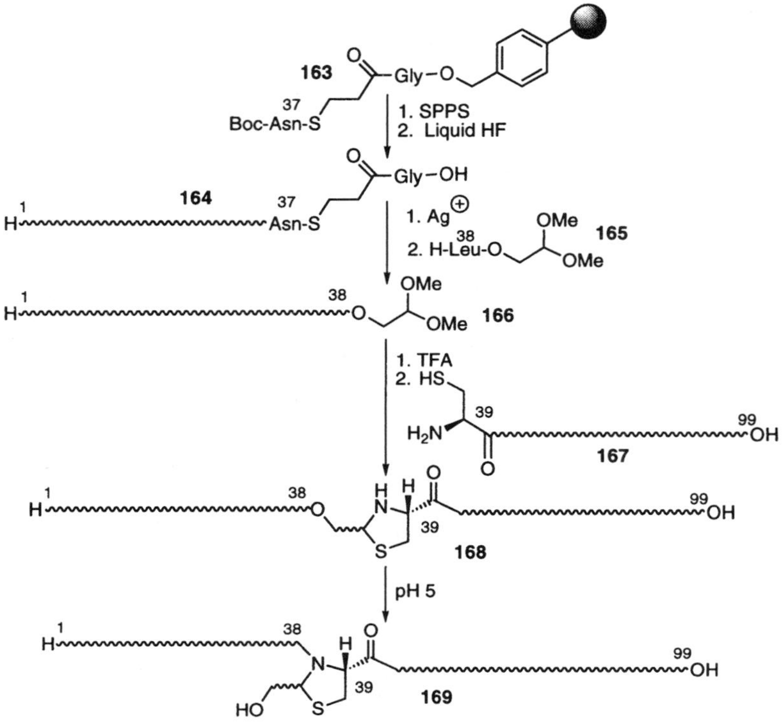

SCHEME 4.40 Synthesis of an HIV-1 protease analogue by domain ligation.

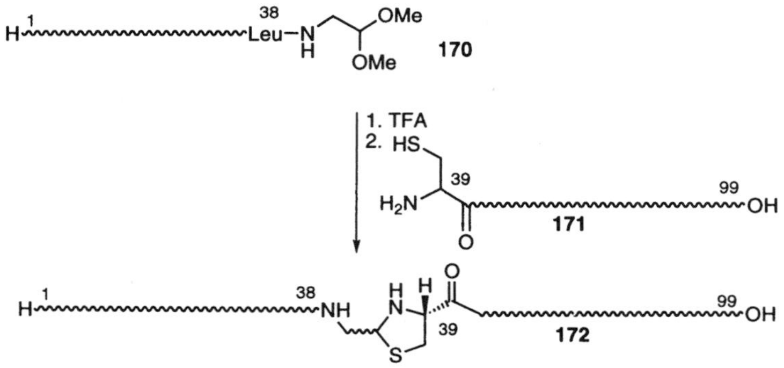

SCHEME 4.41 Synthesis of a backbone-engineered HIV-1 protease analogue by a variation of the domain ligation strategy.

In this case the chemical ligation product **172** has an amide rather than an ester bond between the peptide chain and the thiazolidine ring and so cannot rearrange to give peptides of the type **162**. This method provides another level of flexibility for the creation of other nonnatural protein analogues by chemically ligating unprotected peptides.

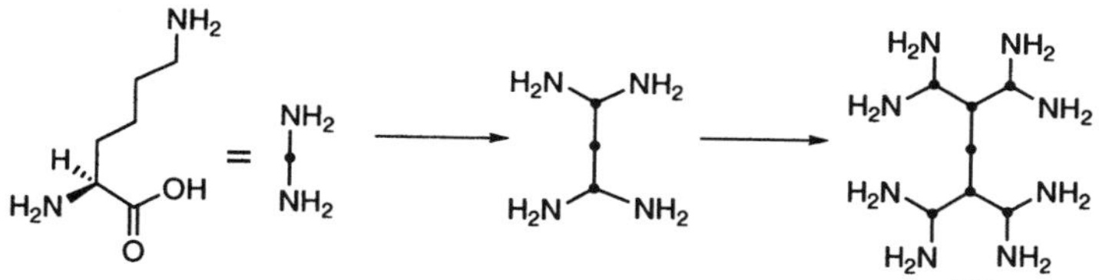

SCHEME 4.42 The lysine core.

4.4.3 Chemical Ligation for the Synthesis of Branched Artificial Proteins

Although branched proteins are unknown in nature, artificial branched peptides have been used both for immunological studies and for the construction of topologically novel protein-like structures. Branched peptides can be synthesized by conventional SPPS as well as by the coupling of protected peptide segments. However, the chemical ligation of unprotected peptides provides another alternative, which often has the advantage of improved solubility of the synthetic intermediates.

4.4.3.1 Multiple Antigenic Peptides

The use of peptide molecules in immunological studies often requires their attachment (or "conjugation") to a known protein or a synthetic polymeric carrier. Although this can give good results from an immunological point of view, the techniques normally used for conjugation give rise to heterogeneous materials whose compositions are ambiguous. In order to improve on this and to produce chemically well-defined molecules for immunological studies, Tam[282-286] introduced the multiple antigenic peptide (MAP) concept. This consists of attaching the peptides to be studied to an amino acid "core" **173**, built up of seven lysine residues, as shown in Scheme 4.42.

Synthesis of lysine core **173** can be accomplished by standard SPPS. Boc-Lys(Boc)-OH is anchored to a Pam resin, and treatment with trifluoroacetic acid solution liberates both amino groups, to which Boc-Lys(Boc)-OH is again coupled. Deprotection, coupling, and deprotection cycles are then carried out sequentially, giving **173**. The peptides whose immunological properties are to be investigated are synthesized on each of the eight branches of this core. When completed, hydrogen fluoride–mediated acidolysis detaches the branched structure (a peptide dendrimer) from the resin, removing all protecting groups in the process. After purification the MAP **174** can be injected into the animal in question and the immunological response determined. The synthetic procedure is outlined in Scheme 4.43.

If the final lysines of the core are incorporated, not as Boc-Lys(Boc)-OH, but rather as Fmoc-Lys(Boc)-OH, then two different peptides can be synthesized on the same core. The first peptide is synthesized after trifluoroacetic acid–mediated removal of the lysine Boc groups. Once completed, the second peptide can then be synthesized on the same peptide core, after piperidine-mediated removal of the lysine Fmoc group.[283]

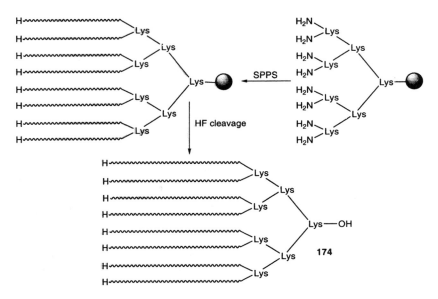

SCHEME 4.43 The synthesis of an MAP by SPPS.

A drawback of the MAPs approach, is that SPPS of the desired peptides on the core is not always straightforward. The simultaneous synthesis of eight closely spaced peptide chains can sometimes lead to poor resin swelling and to aggregation of the peptide chains on the solid support. This can result in low coupling yields and the production of deletion peptides. Such problems can be overcome to a large extent by chemical ligation of the components — any of the methods described above could, in principle, be used for the preparation of MAPs. Several approaches have, in fact, been reported, including disulfide formation[287] and thioether and oxime formation.[277,278,280,288] The application of a variation of the domain chemical ligation strategy (see Section 4.4.2.3) to the preparation of MAPs is outlined in Scheme 4.44.

The lysine scaffolding with eight aldehyde groups **175** was prepared from core **173**, first by coupling serine residues to it and then by subjecting the product to sodium periodate oxidation.[248] The peptide to be ligated, **176**, which must have

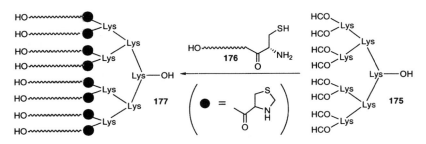

SCHEME 4.44 The synthesis of an MAP by chemical ligation.

N-terminal cysteine, was prepared by SPPS. Chemical ligation of **175** to **176** in aqueous solution gave the desired peptide dendrimer **177** in good yield, after 18 h. This chemical ligation has allowed the formation of MAPs having molecular weights in excess of 24,000; these molecules are consequently among the largest artificial proteins ever made. Despite their size, reasonable solubility in aqueous solvent systems is maintained, as a result of the lack of peptide side chain protection.

4.4.3.2 Template-Assembled Synthetic Proteins

The template-assembled synthetic protein (TASP) molecules designed by Mutter[289–291] represent another important class of synthetic branched peptides. The underlying objective of TASP is to gain a better understanding of the relationship between protein structure and function. The question of how a linear sequence of amino acids folds into the three-dimensional arrangement represented by the native protein molecule is one of the most intensely investigated in contemporary science. Despite the enormous amount of work that has been done, the protein-folding problem has not been solved, although a reasonably clear picture of the factors involved has emerged. The three-dimensional arrangements found in naturally occurring proteins can be mimicked using branched, rather than linear peptides. TASP molecules consist of peptide sequences having a high propensity to adopt secondary structures such as α-helices or β-sheets, linked to a peptide template; under the right conditions they can adopt protein-like tertiary structures. The study of such molecules can help gain an understanding of how natural proteins function. A TASP molecule that has the overall topology of a protein four-helix bundle is represented diagrammatically as **178**.

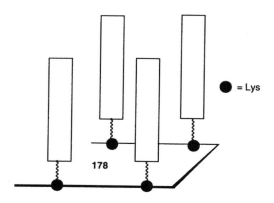

= Lys

A TASP molecule mimicking a protein four-helix bundle.

TASP molecules can be synthesized by standard SPPS. One possibility is to incorporate lysine residues into the template so that their side chains can serve as branch points. The peptides that are to adopt the secondary structure under

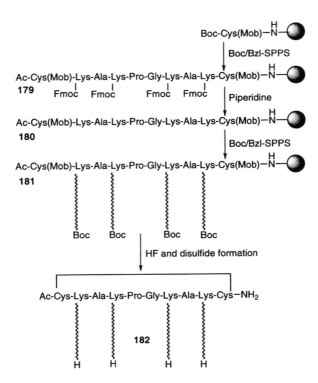

SCHEME 4.45 The solid-phase synthesis of a TASP molecule.

consideration are then synthesized on the lysine side chains. This strategy requires that the side chain protection of lysine be orthogonal to the chemistry used for preparation of the template. The solid-phase synthesis of a TASP molecule, using this approach is outlined in Scheme 4.45.

The template, in this case, is a cyclic disulfide peptide having the sequence Ac-Cys-Lys-Ala-Lys-Pro-Gly-Lys-Ala-Lys-Cys-NH$_2$. Lysine is incorporated with Fmoc side chain protection, which is stable to the acidolysis required for the elaboration of the linear precursor of the template, **179**. Treatment with piperidine at the appropriate moment removes the Fmoc groups, giving **180**, upon which the branches are now synthesized. When chain assembly is complete, treatment of **181** with hydrogen fluoride leads to cleavage from the solid support. Intramolecular disulfide bridge formation then gives TASP molecule **182**. In this TASP all four branches are identical, but a whole range of different structures can be synthesized, if so desired. The templates can be linear or cyclic, of varying sizes, and with varying numbers of branch point residues. The side chains of the lysine residues can be differentially protected so that different peptides can be attached to the same template. In addition, other amino acid residues, such as glutamic acid, can be used as branch points. In this case the direction of synthesis of the branched peptide is changed from $C{\rightarrow}N$

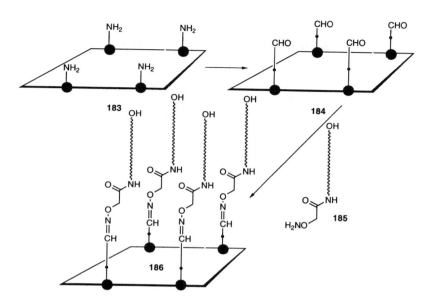

SCHEME 4.46 The synthesis of a TASP molecule by chemical ligation.

to $N{\rightarrow}C$. There are very many possibilities available for constructing arrays of peptides with predetermined three-dimensional arrays, and the TASP concept promises to have a profound effect on the *de novo* design of proteins.

However, the branched TASP molecules are not always easy to synthesize by linear SPPS, and similar difficulties to those encountered in the case of MAPs led to the investigation of convergent strategies for their preparation. This has been done by coupling protected peptide segments to the template in solution,[292-294] but chemical ligation of unprotected peptides is more promising, offering as it does the advantage of improved solubility in aqueous solvent systems. The preparation of TASP molecules using thioether,[295] thioester,[296] and oxime[297,298] formation has been reported. The chemical ligation of unprotected peptides to the template by oxime formation is outlined in Scheme 4.46.

The lysine-containing template **183** is converted to the tetra-aldehyde derivative **184** by coupling serine, followed by sodium periodate-mediated oxidation. *N*-terminus oxime peptide **185** can then be ligated in aqueous solution in good yield, giving the TASP molecule **186**.

Oxime chemical ligation has also been used by Rose[252] to prepare another type of branched artificial protein, as outlined in Scheme 4.47.

The polyaldehyde peptide template **188** was produced by sodium periodate oxidation of the corresponding serine-containing peptide **187**. Reaction with an excess of the *N*-terminal oxime peptide **189** in aqueous solution at pH 4.6 then gave branched artificial proteins of the general structure **190** in over 90% yield.

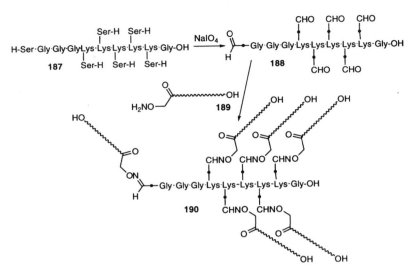

SCHEME 4.47 The synthesis of branched artificial proteins by chemical ligation.

REFERENCES

1. Sakakibara, S., *Biopolymers (Pept. Sci.)*, 37, 17, 1995.
2. Inui, T., Kubo, S., Bódi, J., Kimura, T., and Sakakibara, S., in *Peptides 1994. Proceedings of the 23rd European Peptides Symposium*, Maia, H. L. S., Ed., ESCOM, Leiden, 1995, 36.
3. Inui, T., Bódi, J., Kubo, S., Nishio, H., Kimura, T., Kojima, S., Maruta, H., Muramatsu, T., and Sakakibara, S., *J. Pept. Sci.*, 2, 28, 1996.
4. Bayer, E., Eckstein, H., Hägele, K., König, W. A., Brüning, W., Hagenmaier, H., and Parr, W., *J. Am. Chem. Soc.*, 92, 1735, 1970.
5. Kent, S. B. H., in *Peptides. Structure and Function. Proceedings of the 9th American Peptides Symposium*, Deber, C. M., Hruby, V. J., and Kopple, K. D., Eds., Pierce Chemical Company, Rockford, IL, 1985, 407.
6. Kent, S. B. H., *Annu. Rev. Biochem.*, 57, 957, 1988.
7. Milton, R. C. d. L., Milton, S. C. F., and Adams, P. A., *J. Am. Chem. Soc.*, 112, 6039, 1990.
8. Bedford, J., Hyde, C., Johnson, T., Jun, W., Owen, D., Quibell, M., and Sheppard, R. C., *Int. J. Pept. Protein Res.*, 40, 300, 1992.
9. Yajima, H. and Kiso, Y., *Chem. Pharm. Bull.*, 22, 1087, 1974.
10. Yajima, H., Kiso, Y., Okada, Y., and Watanabe, H., *J. Chem. Soc. Chem. Commun.*, 106, 1974.
11. Protein Synthesis Group, *Sci. Sin. Ser. B (Engl. Ed.)*, 18, 745, 1975.
12. Wong, C. H., Chen, S. T., Ho, C. L., and Wang, K. T., *Biochim. Biophys. Acta*, 536, 376, 1978.
13. Pedroso, E., Grandas, A., Saralegui, M. A., Giralt, E., Granier, C., and van Rietschoten, J., *Tetrahedron*, 38, 1183, 1982.
14. Lloyd-Williams, P., Albericio, F., and Giralt, E., *Tetrahedron*, 49, 11065, 1993.
15. Benz, H., *Synthesis*, 337, 1994.
16. Sheppard, R. C. and Williams, B. J., *J. Chem. Soc. Chem. Commun.*, 587, 1982.
17. Sheppard, R. C. and Williams, B. J., *Int. J. Pept. Protein Res.*, 20, 451, 1982.
18. Flörsheimer, A. and Riniker, B., in *Peptides 1990. Proceedings of the 21st European Peptides Symposium*, Giralt, E. and Andreu, D., Eds., ESCOM, Leiden, 1991, 131.
19. Riniker, B., Flörsheimer, A., Fretz, H., Sieber, P., and Kamber, B., *Tetrahedron*, 49, 9307, 1993.
20. Riniker, B., Flörsheimer, A., Fretz, H., and Kamber, B., in *Peptides 1992. Proceedings of the 22nd European Peptides Symposium*, Schneider, C.H. and Eberle, A.N., Eds., ESCOM, Leiden, 1993, 34.

21. Ibericio, F. and Barany, G., *Int. J. Pept. Protein Res.*, 26, 92, 1985.
22. Mergler, M., Nyfeler, R., Tanner, R., Gosteli, J., and Grogg, P., *Tetrahedron Lett.*, 29, 4005, 1988.
23. Mergler, M., Nyfeler, R., Tanner, R., Gosteli, J., and Grogg, P., *Tetrahedron Lett.*, 29, 4009, 1988.
24. Barlos, K., Gatos, D., Kapolos, S., Papaphotiu, G., Schäfer, W., and Wenqing, Y., *Tetrahedron Lett.*, 30, 3947, 1989.
25. Barlos, K., Gatos, D., Kallitsis, J., Papaphotiu, G., Sotiriu, P., Wenqing, Y., and Schäfer, W., *Tetrahedron Lett.*, 30, 3943, 1989.
26. Barlos, K., Chatzi, O., Gatos, D., and Stavropoulos, G., *Int. J. Pept. Protein Res.*, 37, 513, 1991.
27. Barlos, K., Gatos, D., Papaphotiou, G., and Schäfer, W., *Liebigs Ann. Chem.*, 215, 1993.
28. Bollhagen, R., Schmiedberger, M., Barlos, K., and Grell, E., *J. Chem. Soc. Chem. Commun.*, 2559, 1994.
29. Gairí, M., Lloyd-Williams, P., Albericio, F., and Giralt, E., *Int. J. Pept. Protein Res.*, 46, 119, 1995.
30. van Vliet, A., Smulders, R. H. P. H., Rietman, B. H., and Tesser, G. I., in *Innovation and Perspectives in Solid Phase Synthesis. Peptides, Polypeptides and Oligonucleotides*, Epton, R., Ed., Intercept Ltd, Andover, 1992, 475.
31. van Vliet, A., Smulders, R. H. P. H., Rietman, B. H., Eggen, I. F., van der Werben, G., and Tesser, G. I., in *Peptides 1992. Proceedings of the 22nd European Peptides Symposium*, Schneider, C. H. and Eberle, A. N., Eds., ESCOM, Leiden, 1993, 279.
32. Zhang, L., Rapp, W., Goldhammer, C., and Bayer, E., in *Innovation and Perspectives in Solid Phase Synthesis. Peptides, Proteins and Nucleic Acids. Biological and Biomedical Applications*, Epton, R., Ed., Mayflower Worldwide Ltd, Birmingham, 1994, 717.
33. White, P., in *Innovation and Perspectives in Solid Phase Synthesis. Peptides, Proteins and Nucleic Acids. Biological and Biomedical Applications*, Epton, R., Ed., Mayflower Worldwide Ltd, Birmingham, 1994, 701.
34. Rietman, B. H., Smulders, R. H. P. H., Eggen, I. F., van Vliet, A., van de Werben, G., and Tesser, G. I., *Int. J. Pept. Protein Res.*, 44, 199, 1994.
35. Zikos, C. C. and Ferderigos, N. G., *Tetrahedron Lett.*, 35, 1767, 1994.
36. van Vliet, A., Rijkers, D. T. S., and Tesser, G. I., in *Peptides 1994. Proceedings of the 23rd European Peptides Symposium*, Maia, H. L. S., Ed., ESCOM, Leiden, 1995, 267.
37. Quibell, M., Packman, L. C., and Johnson, T., *J. Am. Chem. Soc.*, 117, 11656, 1995.
38. Rink, H., *Tetrahedron Lett.*, 28, 3787, 1987.
39. Albericio, F. and Barany, G., *Tetrahedron Lett.*, 32, 1015, 1991.
40. Pedroso, E., Grandas, A., de las Heras, X., Eritja, R., and Giralt, E., *Tetrahedron Lett.*, 27, 743, 1986.
41. Ueki, M. and Amemiya, M., *Tetrahedron Lett.*, 28, 6617, 1987.
42. Wang, S. S., *J. Am. Chem. Soc.*, 95, 1328, 1973.
43. Wang, S. S. and Kulesha, I. D., *J. Org. Chem.*, 40, 1227, 1975.
44. Chang, C.-D., Felix, A. M., Jimenez, M. H., and Meienhofer, J., *Int. J. Pept. Protein Res.*, 15, 485, 1980.
45. Grandas, A., Pedroso, E., Giralt, E., Granier, C., and van Rietschoten, J., *Tetrahedron*, 42, 6703, 1986.
46. Sabatier, J.-M., Tessier-Rochat, M., Granier, C., van Rietschoten, J., Pedroso, E., Grandas, A., Albericio, F., and Giralt, E., *Tetrahedron*, 43, 5973, 1987.
47. DeGrado, W. F. and Kaiser, E. T., *J. Org. Chem.*, 47, 3258, 1982.
48. Kaiser, E. T., *Acc. Chem. Res.*, 22, 47, 1989.
49. Kaiser, E. T., Mihara, H., Laforet, G. A., Kelly, J. W., Walters, L., Findeis, M. A., and Sasaki, T., *Science*, 243, 187, 1989.
50. Scarr, R. B. and Findeis, M. A., *Pept. Res.*, 3, 238, 1990.
51. Nakagawa, S. H. and Kaiser, E. T., *J. Org. Chem.*, 48, 678, 1983.
52. Nakagawa, S., Lau, H. S. H., Kézdy, F. J., and Kaiser, E. T., *J. Am. Chem. Soc.*, 107, 7087, 1985.
53. Sasaki, T. and Kaiser, E. T., *J. Am. Chem. Soc.*, 111, 380, 1989.
54. Sasaki, T., Findeis, M. A., and Kaiser, E. T., *J. Org. Chem.*, 56, 3159, 1991.
55. Mutter, M. and Bellof, D., *Helv. Chim. Acta*, 67, 2009, 1984.
56. Liu, Y.-Z., Ding, S.-H., Chu, J.-Y., and Felix, A. M., *Int. J. Pept. Protein Res.*, 35, 95, 1990.
57. Blaha, I., Nemec, J., Tozser, J., and Oroszlan, S., *Int. J. Pept. Protein Res.*, 38, 453, 1991.

58. Rabanal, F., Giralt, E., and Albericio, F., *Tetrahedron Lett.*, 33, 1775, 1992.
59. Rabanal, F., Giralt, E., and Albericio, F., *Tetrahedron*, 51, 1449, 1995.
60. Tesser, G. I., Buis, J. T. W. A. R. M., Wolters, E. T. M., and Bothé-Helmes, E. G. A. M., *Tetrahedron*, 32, 1069, 1976.
61. Buis, J. T. W. A. R. M., Tesser, G. T., and Nivard, R. J. F., *Tetrahedron*, 32, 2321, 1976.
62. Schwyzer, R., Felder, E., and Failli, P., *Helv. Chim. Acta*, 67, 1316, 1984.
63. Katti, S. B., Misra, P. K., Haq, W., and Mathur, K. B., *J. Chem. Soc. Chem. Commun.*, 843, 1992.
64. Robles, J., Pedroso, E., and Grandas, A., *J. Org. Chem.*, 59, 2482, 1994.
65. Robles, J., Pedroso, E., and Grandas, A., *Nucleic Acids Res.*, 23, 4151, 1995.
66. Eritja, R., Robles, J., Fernadez-Forner, D., Albericio, F., Giralt, E., and Pedroso, E., *Tetrahedron Lett.*, 32, 1511, 1991.
67. Albericio, F., Giralt, E., and Eritja, R., *Tetrahedron Lett.*, 32, 1515, 1991.
68. Bodanszky, M. and Sheehan, J. T., *Chem. Ind.*, 1423, 1964.
69. Wright, D. E., Agarwal, N. S., and Hruby, V. J., *Int. J. Pept. Protein Res.*, 15, 271, 1980.
70. Niu, C. I., Yang, S. Z., Ma, A. Q., Chen, Y. N., Jiang, Y. Q., and Huang, W. T., *Biopolymers*, 20, 1833, 1981.
71. Blake, J. and Li, C. H., *Int. J. Pept. Protein Res.*, 3, 185, 1971.
72. Kessler, W. and Iselin, B., *Helv. Chim. Acta*, 49, 1330, 1966.
73. Ohno, M. and Anfinsen, C. B., *J. Am. Chem. Soc.*, 89, 5994, 1967.
74. Visser, S., Roeloffs, J., Kerling, K. E. T., and Havinga, E., *Recl. Trav. Chim. Pays-Bas*, 87, 559, 1968.
75. Murakami, Y., Nakano, A., Matsumoto, K., and Iwamoto, K., *Bull. Chem. Soc. Jpn.*, 51, 2690, 1978.
76. Kaufmann, K. D., Kunzek, H., Dölling, R., Halatsch, W.-R., Nieke, E. M., Rose, K. B., Bauschke, S., and Schönherr, C., *Z. Naturforsch.*, 20, 99, 1980.
77. Bodanszky, M. and Sheehan, J. T., *Chem. Ind.*, 1597, 1966.
78. Beyerman, H. C., Hindicks, H., and de Leer, E. W. B., *J. Chem. Soc. Chem. Commun.*, 1668, 1968.
79. Beyerman, H. C., Kranenburg, P., and Syrier, J. L. M., *Recl. Trav. Chim. Pays-Bas*, 90, 791, 1971.
80. Penke, B. and Birr, C., *Liebigs Ann. Chem.*, 1999, 1974.
81. Reddy, G. L., Bikshapathy, E., and Nagaraj, R., *Tetrahedron Lett.*, 26, 4257, 1985.
82. Marshall, G. R. and Merrifield, R. B., *Biochemistry*, 4, 2394, 1965.
83. Giralt, E., Celma, C., Ludevid, M. D., and Pedroso, E., *Int. J. Pept. Protein Res.*, 29, 647, 1987.
84. Celma, C., Albericio, F., Pedroso, E., and Giralt, E., *Pept. Res.*, 5, 62, 1992.
85. Anwer, M. K. and Spatola, A. F., *Tetrahedron Lett.*, 33, 3121, 1992.
86. Pillai, V. N. R., *Synthesis*, 1, 1980.
87. Rich, D. H. and Gurwara, S. K., *J. Chem. Soc. Chem. Commun.*, 610, 1973.
88. Rich, D. H. and Gurwara, S. K., *J. Am. Chem. Soc.*, 97, 1575, 1975.
89. Giralt, E., Albericio, F., Andreu, D., Eritja, R., Martin, P., and Pedroso, E., *An. Quim.*, 77C, 120, 1981.
90. Bayer, E., Dengler, M., and Hemmasi, B., *Int. J. Pept. Protein Res.*, 25, 178, 1985.
91. Ajayaghosh, A. and Pillai, V. N. R., *J. Org. Chem.*, 52, 5714, 1987.
92. Ajayaghosh, A. and Pillai, V. N. R., *Tetrahedron*, 44, 6661, 1988.
93. Kneib-Cordonier, N., Albericio, F., and Barany, G., *Int. J. Pept. Protein Res.*, 35, 527, 1990.
94. Albericio, F., Nicolás, E., Josa, J., Grandas, A., Pedroso, E., Giralt, E., Granier, C., and van Rietschoten, J., *Tetrahedron*, 43, 5961, 1987.
95. Gisin, B. F., *Helv. Chim. Acta*, 56, 1476, 1973.
96. Barany, G. and Albericio, F., *J. Am. Chem. Soc.*, 107, 4936, 1985.
97. Giralt, E., Eritja, R., and Pedroso, E., *Tetrahedron Lett.*, 22, 3779, 1981.
98. Suzuki, K., Nitta, K., and Endo, N., *Chem. Pharm. Bull.*, 23, 222, 1975.
99. Gairí, M., Lloyd-Williams, P., Albericio, F., and Giralt, E., *Tetrahedron Lett.*, 31, 7363, 1990.
100. Sheehan, J. C. and Umezawa, K., *J. Org. Chem.*, 38, 3771, 1973.
101. Wang, S. S., *J. Org. Chem.*, 41, 3258, 1976.
102. Tjoeng, F. S., Tam, J. P., and Merrifield, R. B., *Int. J. Pept. Protein Res.*, 14, 262, 1979.
103. Tjoeng, F. S. and Heavner, G. A., *J. Org. Chem.*, 48, 355, 1983.
104. Bellof, D. and Mutter, M., *Chimia*, 39, 317, 1985.

105. Abraham, N. A., Fazal, G., Ferland, J. M., Rakhit, S., and Gauthier, J., *Tetrahedron Lett.*, 32, 577, 1991.
106. Barany, G. and Merrifield, R. B., *J. Am. Chem. Soc.*, 99, 7363, 1977.
107. Albericio, F. and Barany, G., *Int. J. Pept. Protein Res.*, 30, 177, 1987.
108. Kunz, H. and Dombo, B., *Angew. Chem. Int. Ed. Engl.*, 27, 711, 1988.
109. Blankemeyer-Menge, B., and Frank, R., *Tetrahedron Lett.*, 29, 5871, 1988.
110. Guibé, F., Dangles, O., Balavoine, G., and Loffet, A., *Tetrahedron Lett.*, 30, 2641, 1989.
111. Kunz, H., in *Innovation and Perspectives in Solid Phase Synthesis and Related Technologies. Peptides, Polypeptides and Oligonucleotides. Macro-Organic Reagents and Catalysts*, Epton, R., Ed., SPCC (U.K.) Ltd, Birmingham, 1990, 371.
112. Lloyd-Williams, P., Jou, G., Albericio, F., and Giralt, E., *Tetrahedron Lett.*, 32, 4207, 1991.
113. Kates, S. A., Daniels, S. B., and Albericio, F., *Anal. Biochem.*, 212, 303, 1993.
114. Lloyd-Williams, P., Merzouk, A., Guibé, F., Albericio, F., and Giralt, E., *Tetrahedron Lett.*, 35, 4437, 1994.
115. Dangles, O., Guibé, F., Balavoine, G., Lavielle, S., and Marquet, A., *J. Org. Chem.*, 52, 4984, 1987.
116. Loffet, A., Galeotti, N., Jouin, P., Castro, B., Guibé, F., Dangles, O., and Balavoine, G., in *Peptides. Chemistry, Structure and Biology. Proceedings of the 11th American Peptides Symposium*, Rivier, J. E. and Marshall, G. R., Eds., ESCOM, Leiden, 1990, 1015.
117. Toniolo, C., Bonora, G. M., Heimer, E. P., and Felix, A. M., *Int. J. Pept. Protein Res.*, 30, 232, 1987.
118. Toniolo, C., Bonora, G. M., Mutter, M., and Pillai, V. N. R., *Makromol. Chem.*, 182, 2007, 1981.
119. Haack, T. and Mutter, M., *Tetrahedron Lett.*, 33, 1589, 1992.
120. Haack, T., Zier, A., Nefzi, A., and Mutter, M., in *Peptides 1992. Proceedings of the 22nd European Peptides Symposium*, Schneider, C. H. and Eberle, A. N., Eds., ESCOM, Leiden, 1993, 595.
121. Haack, T., Nefzi, A., Dhanapal, D., and Mutter, M., in *Innovation and Perspectives in Solid Phase Synthesis. Peptides, Proteins and Nucleic Acids. Biological and Biomedical Applications*, Epton, R., Ed., Mayflower Worldwide Ltd, Birmingham, 1994, 521.
122. Wöhr, T. and Mutter, M., *Tetrahedron Lett.*, 36, 3847, 1995.
123. Weygand, F., Steglich, W., Bjarnason, J., Akhtar, R., and Khan, N. M., *Tetrahedron Lett.*, 3483, 1966.
124. Weygand, F., Steglich, W., and Bjarnason, J., *Chem. Ber.*, 101, 3642, 1968.
125. Isokawa, S., Tominaga, I., Asakura, T., and Narita, M., *Macromolecules*, 18, 878, 1985.
126. Narita, M., Ishikawa, K., Nakano, H., and Isokawa, S., *Int. J. Pept. Protein Res.*, 24, 14, 1984.
127. Blaakmeer, J., Tijsse-Keasen, T., and Tesser, G. I., *Int. J. Pept. Protein Res.*, 37, 556, 1991.
128. Bartl, R., Klöppel, K.-D., and Frank, R., in *Peptides. Chemistry and Biology. Proceedings of the 12th American Peptides Symposium*, Smith, J. A. and Rivier, J. E., Eds., ESCOM, Leiden, 1992, 505.
129. Bartl, R., Klöppel, K.-D., and Frank, R., in *Peptides 1992. Proceedings of the 22nd European Peptides Symposium*, Schneider, C. H. and Eberle, A. N., Eds., ESCOM, Leiden, 1993, 277.
130. Eckert, H. and Seidel, C., *Angew. Chem. Int. Ed. Engl.*, 25, 159, 1986.
131. Johnson, T., Quibell, M., Owen, D., and Sheppard, R. C., *J. Chem. Soc. Chem. Commun.*, 369, 1993.
132. Quibell, M., Turnell, W. G., and Johnson, T., in *Innovation and Perspectives in Solid Phase Synthesis. Peptides, Proteins and Nucleic Acids. Biological and Biomedical Applications*, Epton, R., Ed., Mayflower Worldwide Ltd, Birmingham, 1994, 653.
133. Quibell, M. and Johnson, T., in *Peptides 1994. Proceedings of the 23rd European Peptides Symposium*, Maia, H. L. S., Ed., ESCOM, Leiden, 1995, 173.
134. Macrae, R. and Young, G. T., *J. Chem. Soc. Perkin Trans. 1*, 1185, 1975.
135. Bayer, E. and Mutter, M., *Nature (London)*, 237, 512, 1972.
136. Merrifield, R. B. and Bach, A. E., *J. Org. Chem.*, 43, 4808, 1978.
137. Anzinger, H., Mutter, M., and Bayer, E., *Angew. Chem. Int. Ed. Engl.*, 18, 686, 1979.
138. Mutter, M., Oppliger, H., and Zier, A., *Makromol. Chem. Rapid Commun.*, 13, 151, 1992.
139. Ball, H. L., Kent, S. B. H., and Mascagni, P., *Int. J. Pept. Protein Res.*, 40, 370, 1992.
140. Funakoshi, S., Fukuda, H., and Fujii, N., *Proc. Natl. Acad. Sci. U.S.A.*, 88, 6981, 1991.
141. Ramage, R. and Raphy, G., *Tetrahedron Lett.*, 33, 385, 1992.

142. García-Echeverría, C., *J. Chem. Soc. Chem. Commun.*, 779, 1995.

143. Iselin, B., *Helv. Chim. Acta*, 44, 61, 1961.

144. Lloyd-Williams, P., Albericio, F., Gairí, M., and Giralt, E., in *Innovation and Perspectives in Solid Phase Synthesis*, Epton, R., Ed., Mayflower Worldwide, 1996, in press.

145. Rizo, J., Albericio, F., Romero, G., García-Echeverría, C., Claret, J., Muller, C., Giralt, E., and Pedroso, E., *J. Org. Chem.*, 53, 5386, 1988.

146. Suzuki, K., Sasaki, Y., and Endo, N., *Chem. Pharm. Bull.*, 24, 1, 1976.

147. Rizo, J., Albericio, F., Giralt, E., and Pedroso, E., *Tetrahedron Lett.*, 33, 397, 1992.

148. Narita, M., Honda, S., Umeyama, H., and Obana, S., *Bull. Chem. Soc. Jpn.*, 61, 281, 1988.

149. Narita, M., Honda, S., and Obana, S., *Bull. Chem. Soc. Jpn.*, 62, 342, 1989.

150. Narita, M., Umeyama, H., Isokawa, S., Honda, S., Sasaki, C., and Kakei, H., *Bull. Chem. Soc. Jpn.*, 62, 780, 1989.

151. Narita, M., Umeyama, H., and Yoshida, T., *Bull. Chem. Soc. Jpn.*, 62, 3582, 1989.

152. Kuroda, H., Chen, Y.-N., Kimura, T., and Sakakibara, S., *Int. J. Pept. Protein Res.*, 40, 294, 1992.

153. Zhang, L., Goldhammer, C., Henkel, B., Zühl, F., Panhaus, G., Jung, G., and Bayer, E., in *Innovation and Perspectives in Solid Phase Synthesis. Peptides, Proteins and Nucleic Acids. Biological and Biomedical Applications*, Epton, R., Ed., Mayflower Worldwide Ltd, Birmingham, 1994, 711.

154. Seebach, D., Thaler, A., and Beck, A. K., *Helv. Chim. Acta*, 72, 857, 1989.

155. Halverson, K., Fraser, P. E., Kirschner, D. A., and Lansbury, P. T., *Biochemistry*, 29, 2639, 1990.

156. Lloyd-Williams, P., Albericio, F., and Giralt, E., *Int. J. Pept. Protein Res.*, 37, 58, 1991.

157. Lindquist, B. and Storgards, T., *Nature (London)*, 175, 511, 1955.

158. Porath, J. and Flodin, P., *Nature (London)*, 183, 1657, 1959.

159. Porath, J., *Biochim. Biophys. Acta*, 39, 193, 1960.

160. Flodin, P., *J. Chromatogr.*, 5, 103, 1961.

161. Andrews, P., *Biochem. J.*, 91, 222, 1964.

162. Hearn, M. T. W., *Methods Enzymol.*, 104, 190, 1984.

163. Tarr, G. E. and Crabbe, J. W., *Anal. Biochem.*, 131, 99, 1983.

164. Welinder, B. S. and Sorenson, H. H., *J. Chromatogr.*, 537, 181, 1991.

165. Barber, M., Bordoli, R. S., Sedgwick, R. D., and Tyler, A. N., *J. Chem. Soc. Chem. Commun.*, 325, 1981.

166. Surman, D. J. and Vickerman, J. C., *J. Chem. Soc. Chem. Commun.*, 324, 1981.

167. Grandas, A., Pedroso, E., Figueras, A., Rivera, J., and Giralt, E., *Biomed. Environ. Mass Spectrom.*, 15, 681, 1988.

168. Celma, C. and Giralt, E., *Biomed. Environ. Mass Spectrom.*, 19, 235, 1990.

169. See James, T. L. and Oppenheimer, N. J., *Methods Enzymol.*, 239, 1994.

170. Albericio, F., Pons, M., Pedroso, E., and Giralt, E., *J. Org. Chem.*, 54, 360, 1989.

171. Felix, A. M. and Merrifield, R. B., *J. Am. Chem. Soc.*, 92, 1385, 1970.

172. Matsueda, R., Maruyama, H., Kitazawa, E., Takahagi, H., and Mukaiyama, T., *Bull. Chem. Soc. Jpn.*, 46, 3240, 1973.

173. Mukaiyama, T., Goto, K., Matsueda, R., and Ueki, M., *Tetrahedron Lett.*, 5293, 1970.

174. Matsueda, R., Maruyama, H., Kitazawa, E., Takahagi, H., and Mukaiyama, T., *J. Am. Chem. Soc.*, 97, 2573, 1975.

175. Maruyama, H., Matsueda, R., Kitasaura, E., Takahagi, H., and Mukaiyama, T., *Bull. Chem. Soc. Jpn.*, 49, 2259, 1976.

176. Rink, H., Born, W., and Fischer, J. A., in *Peptides. Chemistry, Structure and Biology. Proceedings of the 11th American Peptides Symposium*, Rivier, J. E. and Marshall, G. R., Eds., ESCOM, Leiden, 1990, 1041.

177. Carpino, L., El-Faham, A., and Albericio, F., *Tetrahedron Lett.*, 35, 2279, 1994.

178. Carpino, L. A., El-Fahan, A., Minor, C. A., and Albericio, F., *J. Chem. Soc. Chem. Commun.*, 201, 1994.

179. Dalcol, I., Rabanal, F., Ludevid, M.-D., Albericio, F., and Giralt, E., *J. Org. Chem.*, 60, 7575, 1995.

180. Tomalia, D. A., Naylor, A. M., and Goddard, W. A., *Angew. Chem. Int. Ed. Engl.*, 29, 138, 1990.

181. Bayer, E., Gil-Av, E., König, W. A., Nakapartksin, S., Oró, J., and Parr, W., *J. Am. Chem. Soc.*, 92, 1738, 1970.

182. Goodman, M., Keogh, P., and Anderson, H., *Bioorg. Chem.*, 6, 239, 1977.

183. Steinauer, R., Chen, F. M. F., and Benoiton, N. L., *J. Chromatogr.*, 325, 111, 1985.

184. Grandas, A., Albericio, F., Josa, J., Giralt, E., Pedroso, E., Sabatier, J. M., and van Rietschoten, J., *Tetrahedron*, 45, 4637, 1989.

185. Hendrix, J. C., Halverson, K. J., and Lansbury, P. T., *J. Am. Chem. Soc.*, 114, 7930, 1992.

186. Lelievre, D., Trudelle, Y., Heitz, F., and Spach, G., *Int. J. Pept. Protein Res.*, 33, 379, 1989.

187. Noda, K., Terada, S., Mitsuyasu, N., Waki, M., Kato, T., and Izumiya, N., *Mem. Fac. Sci. Kyushu Univ. Ser. B*, 7C, 189, 1970.

188. Lyle, T. A., Brady, S. F., Ciccarone, T. M., Colton, C. D., Paleveda, W. J., Veber, D. F., and Nutt, R. F., *J. Org. Chem.*, 52, 3752, 1987.

189. Barlos, K., Gatos, D., and Schäfer, W., *Angew. Chem. Int. Ed. Engl.*, 30, 590, 1991.

190. Quibell, M., Turnell, W. G., and Johnson, T., *J. Chem. Soc. Perkin Trans. 1*, 2019, 1995.

191. Hendrix, J. C. and Lansbury, P. T., *J. Org. Chem.*, 57, 3421, 1992.

192. Hendrix, J. C., Jarrett, J. T., Anisfield, S. T., and Lansbury, P. T., *J. Org. Chem.*, 57, 3414, 1992.

193. Atherton, E., Logan, C. J., and Sheppard, R. C., *J. Chem. Soc. Perkin Trans.*, 1, 538, 1981.

194. Abderhalden, E. and Nienburg, H., *Fermentforschung*, 13, 573, 1933.

195. Rothe, M. and Mazánek, J., *Angew. Chem. Int. Ed. Engl.*, 11, 293, 1972.

196. Rothe, M. and Mazánek, J., *Liebigs Ann. Chem.*, 439, 1974.

197. Rothe, M., Rott, M., and Mazánek, J., in *Peptides 1976. Proceedings of the 14th European Peptides Symposium*, Loffet, A., Ed., Editions de l' Université de Bruxelles, Bruxelles, 1976, 309.

198. Mergler, M., Nyfeler, R., Gosteli, J., and Tanner, R., *Tetrahedron Lett.*, 30, 6745, 1989.

199. Kamber, B. and Riniker, B., in *Peptides. Chemistry and Biology. Proceedings of the 12th American Peptides Symposium*, Smith, J. A. and Rivier, J. E., Eds., ESCOM, Leiden, 1992, 525.

200. Blake, J., *Int. J. Pept. Protein Res.*, 17, 273, 1981.

201. Blake, J. and Li, C. H., *Proc. Natl. Acad. Sci. U.S.A.*, 78, 4055, 1981.

202. Blake, J. and Li, C. H., *Proc. Natl. Acad. Sci. U.S.A.*, 80, 1556, 1983.

203. Blake, J., Westphal, M., and Li, C. H., *Int. J. Pept. Protein Res.*, 24, 498, 1984.

204. Blake, J., *Int. J. Pept. Protein Res.*, 27, 191, 1986.

205. Blake, J., Yamashiro, D., Ramasharma, K., and Li, C. H., *Int. J. Pept. Protein Res.*, 28, 468, 1986.

206. Yamashiro, D. and Li, C. H., *Int. J. Pept. Protein Res.*, 31, 322, 1988.

207. Cheng, H.-C. and Yamashiro, D., *Int. J. Pept. Protein Res.*, 38, 70, 1991.

208. Tesser, G. I. and Balvert-Geers, I. C., *Int. J. Pept. Protein Res.*, 7, 295, 1975.

209. Veber, D. F., Paleveda, W. J., Lee, Y. C., and Hirschmann, R., *J. Org. Chem.*, 42, 3286, 1977.

210. Dixon, H. B. F. and Perham, R. N., *Biochem. J.*, 109, 312, 1968.

211. Aimoto, S., Mizoguchi, N., Hojo, H., and Yoshimura, S., *Bull. Chem. Soc. Jpn.*, 62, 524, 1989.

212. Hojo, H. and Aimoto, S., *Bull. Chem. Soc. Jpn.*, 64, 111, 1991.

213. Hojo, H. and Aimoto, S., *Bull. Chem. Soc. Jpn.*, 65, 3055, 1992.

214. Hojo, H., Kwon, Y. D., and Aimoto, S., in *Peptide Chemistry 1991. Proceedings of the 29th Symposium on Peptide Chemistry*, Suzuki, A., Ed., Protein Research Foundation, Osaka, 1992, 115.

215. Kwon, Y., Zhang, R., Hojo, H., and Aimoto, S., in *Peptides 1992. Proceedings of the 2nd Japan Symposium on Peptide Chemistry*, Yanaihara, N., Ed., ESCOM, Leiden, 1993, 58.

216. Aimoto, S., Kwon, Y., and Hojo, H., in *Peptide Chemistry 1992. Proceedings of the 2nd Japan Symposium on Peptide Chemistry*, Yanaihara, N., Ed., ESCOM, Leiden, 1993, 54.

217. Aimoto, S. and Hojo, H., in *Peptides. Biology and Chemistry. Proceedings of the 1992 Chinese Peptide Symposium*, Du, Y.-C., Tam, J. P., and Zhang, Y.-S., Eds., ESCOM, Leiden, 1993, 273.

218. Muir, T. W. and Kent, S. B. H., *Curr. Opin. Biotechnol.*, 4, 420, 1993.

219. Drijfhout, J. W., Perdijk, E. W., Weijer, W. J., and Bloemhoff, W., *Int. J. Pept. Protein Res.*, 32, 161, 1988.

220. Albericio, F., Andreu, D., Giralt, E., Navalpotro, C., Pedroso, E., Ponsati, B., and Ruiz-Gayo, M., *Int. J. Pept. Protein Res.*, 34, 124, 1989.

221. van Regenmortel, M. H. V., Briand, J. P., Muller, S., and Plaué, S., in *Laboratory Techniques in Biochemistry and Molecular Biology*, Burdon, R. H. and van Knippenberg, P. H., Eds., Elsevier, Amsterdam, 1988, 95.

222. Schnölzer, M. and Kent, S. B. H., *Science*, 256, 221, 1992.

223. Schnölzer, M. and Kent, S. B. H., in *Peptides 1992. Proceedings of the 22nd European Peptides Symposium*, Schneider, C. H. and Eberle, A. N., Eds., ESCOM, Leiden, 1993, 237.
224. Lindner, W. and Robey, F. A., *Int. J. Pept. Protein Res.*, 30, 794, 1987.
225. Robey, F. A. and Fields, R. L., *Anal. Biochem.*, 177, 373, 1989.
226. Schneider, J. and Kent, S. B. H., *Cell*, 54, 363, 1988.
227. Wlodawer, A., Miller, M., Jaskólski, M., Sathyanarayana, B. K., Baldwin, E., Weber, I. T., Selk, L. M., Clawson, L., Schneider, J., and Kent, S. B. H., *Science*, 245, 616, 1989.
228. Miller, M., Schneider, J., Sathyanarayana, B. K., Toth, M. V., Marshall, G. R., Clawson, L., Selk, L., Kent, S. B. H., and Wlodawer, A., *Science*, 245, 1149, 1989.
229. Milton, R. C. D., Milton, S. C. F., Schnölzer, M., and Kent, S. B. H., in *Techniques in Protein Chemistry. IV. Proceedings of the Protein Society*, Angeletti, R., Ed., Academic Press, San Diego, 1993, 257.
230. Williams, M. J., Muir, T. W., Ginsberg, M. H., and Kent, S. B. H., *J. Am. Chem. Soc.*, 116, 10797, 1994.
231. Baca, M. and Kent, S. B. H., *Proc. Natl. Acad. Sci. U.S.A.*, 90, 11638, 1993.
232. Muir, T. W., Williams, M. J., Ginsberg, M. H., and Kent, S. B. H., *Biochemistry*, 33, 7701, 1994.
233. Baca, M., Muir, T. W., Schnölzer, M., and Kent, S. B. H., *J. Am. Chem. Soc.*, 117, 1881, 1995.
234. Englebretsen, D. R., Garnham, B. G., Bergman, D. A., and Alewood, P. F., *Tetrahedron Lett.*, 36, 8871, 1995.
235. Futaki, S., Ishikawa, T., Niwa, M., Kitagawa, K., and Yagami, T., *Tetrahedron Lett.*, 36, 5203, 1995.
236. McCafferty, D. G., Slate, C. A., Nakhle, B. M., Graham, H. D., Anstell, T. L., Vachet, R. W., Mullis, B. H., and Erickson, B. W., *Tetrahedron*, 51, 9859, 1995.
237. King, T. P., Zhao, S. W., and Lam, T., *Biochemistry*, 25, 5774, 1986.
238. Rose, K., Vilaseca, L. A., Werlen, R., Meunier, A., Fisch, I., Jones, R. M. L., and Offord, R. E., *Bioconjugate Chem.*, 2, 154, 1991.
239. Fisch, I., Künzi, G., Rose, K., and Offord, R. E., *Bioconjugate Chem.*, 3, 147, 1992.
240. Gaertner, H. F., Rose, K., Cotton, R., Timms, D., Camble, R., and Offord, R. E., *Bioconjugate Chem.*, 3, 262, 1992.
241. Gaertner, H. F., Offord, R. E., Cotton, R., Timms, D., Camble, R., and Rose, K., *J. Biol. Chem.*, 269, 7224, 1994.
242. Sahni, G., Cho, Y. J., Iyer, K. S., Khan, S. A., Seetharam, R., and Acharya, A. S., *Biochemistry*, 28, 5456, 1989.
243. Wallace, C. J. A., Guillemette, J. G., Hibiya, Y., and Smith, M., *J. Biol. Chem.*, 266, 21355, 1991.
244. Wallace, C. J. A. and Clark-Smith, I., *J. Biol. Chem.*, 267, 3852, 1992.
245. Jones, R. M. L. and Offord, R. E., *Biochem. J.*, 203, 125, 1982.
246. Rose, K., Herrero, C., Proudfoot, A. E. I., Offord, R., and Wallace, C. J. A., *Biochem. J.*, 249, 83, 1988.
247. Rose, K., Jones, M. L., Sundaram, G., and Offord, R. E., in *Peptides 1988. Proceedings of the 19th European Peptides Symposium*, Bayer, E. and Jung, G., Eds., Walter de Gruyter, Berlin, 1989, 274.
248. Geoghegan, K. F. and Stroh, J. G., *Bioconjugate Chem.*, 3, 138, 1992.
249. Pochon, S., Buchegger, F., Pélegrin, A., Mach, J.-P., Offord, R. E., Ryser, J. E., and Rose, K., *Int. J. Cancer*, 43, 1188, 1989.
250. Webb, R. R. and Kaneko, T., *Bioconjugate Chem.*, 1, 96, 1990.
251. Mikola, H. and Hanninen, E., *Bioconjugate Chem.*, 3, 182, 1992.
252. Rose, K., *J. Am. Chem. Soc.*, 116, 30, 1994.
253. Pallin, T. D. and Tam, J. P., *J. Chem. Soc. Chem. Commun.*, 2021, 1995.
254. Canne, L. E., Ferré-D'Amaré, A. R., Burley, S. K., and Kent, S. B. H., *J. Am. Chem. Soc.*, 117, 2998, 1995.
255. Brenner, M., Zimmermann, J. P., Wehrmüller, J., Quitt, P., Hartmann, A., Schneider, W., and Beglinger, U., *Helv. Chim. Acta*, 40, 1497, 1957.
256. Brenner, M. and Zimmermann, J. P., *Helv. Chim. Acta*, 40, 1933, 1957.
257. Brenner, M. and Wehrmüller, J., *Helv. Chim. Acta*, 40, 2374, 1957.
258. Brenner, M. and Hofer, W., *Helv. Chim. Acta*, 44, 1794, 1961.

259. Brenner, M. and Hofer, W., *Helv. Chim. Acta*, 44, 1798, 1961.
260. Brenner, M., in *Peptides. Proceedings of the 8th European Peptides Symposium*, Beyerman, H. C., van de Linde, A., and Maassen van den Brink, W., Eds., North-Holland Publishing Company, Amsterdam, 1967, 1.
261. Wieland, T., Bokelmann, E., Bauer, L., Lang, H. U., and Lau, H., *Liebigs Ann. Chem.*, 583, 129, 1953.
262. Russell, P. L., Topping, R. M., and Tutt, D.E., *J. Chem. Soc. (B)*, 657, 1971.
263. Sasaki, S., Shionoya, M., and Koga, K., *J. Am. Chem. Soc.*, 107, 3371, 1985.
264. Gennari, C., Molinari, F., and Piarulli, U., *Tetrahedron Lett.*, 31, 2929, 1990.
265. Wijkmans, J. C. H. M., van Boom, J. H., and Bloemhoff, W., *Tetrahedron Lett.*, 34, 7123, 1993.
266. Kemp, D. S., *Biochemistry*, 20, 1793, 1981.
267. Kemp, D. S. and Galakatos, N. G., *J. Org. Chem.*, 51, 1821, 1986.
268. Kemp, D. S., Galakatos, N. G., Bowen, B., and Tam, K., *J. Org. Chem.*, 51, 1829, 1986.
269. Kemp, D. S. and Carey, R. I., *Tetrahedron Lett.*, 32, 2845, 1991.
270. Fotouhi, N., Galakatos, N. G., and Kemp, D. S., *J. Org. Chem.*, 54, 2803, 1989.
271. Kemp, D. S. and Carey, R. I., *J. Org. Chem.*, 58, 2216, 1993.
272. Dawson, P. E., Muir, T. W., Clark-Lewis, I., and Kent, S. B. H., *Science*, 266, 776, 1994.
273. Tam, J. P., Cunningham-Rundles, W. F., Erickson, B. W., and Merrifield, R. B., *Tetrahedron Lett.*, 4001, 1977.
274. Tam, J. P., Lu, Y.-A., Liu, C.-F., and Shao, J., *Proc. Natl. Acad. Sci. U.S.A.*, 92, 12485, 1995.
275. Liu, C.-F., Rao, C., and Tam, J. P., *Tetrahedron Lett.*, 37, 933, 1996.
276. Liu, C.-F. and Tam, J. P., *Proc. Natl. Acad. Sci. U.S.A.*, 91, 6584, 1994.
277. Shao, J. and Tam, J. P., *J. Am. Chem. Soc.*, 117, 3893, 1995.
278. Spetzler, J. C. and Tam, J. P., *Int. J. Pept. Protein Res.*, 45, 78, 1995.
279. Tam, J. P., Rao, C., Liu, C.-F., and Shao, J., *Int. J. Pept. Protein Res.*, 45, 209, 1995.
280. Tam, J. P. and Spetzler, J. C., *Biomed. Pept. Proteins Nucleic Acids*, 1, 123, 1995.
281. Liu, C.-F., Rao, C., and Tam, J. P., *J. Am. Chem. Soc.*, 118, 307, 1996.
282. Tam, J. P., *Proc. Natl. Acad. Sci. U.S.A.*, 85, 5409, 1988.
283. Tam, J. P. and Lu, Y.-A., *Proc. Natl. Acad. Sci. U.S.A.*, 86, 9084, 1989.
284. Tam, J. P. and Zavala, F., *J. Immunol. Methods*, 124, 53, 1989.
285. Tam, J. P., Clavijo, P., Lu, Y.-A., Nussenzweig, V., Nussenzweig, R., and Zavala, F., *J. Exp. Med.*, 171, 299, 1990.
286. Tam, J. P., in *Peptides. Synthesis, Structures and Applications*, Gutte, B., Ed., Academic Press, New York, 1995, 456.
287. Drijfhout, J. W. and Bloemhoff, W., *Int. J. Pept. Protein Res.*, 37, 27, 1991.
288. Rao, C. and Tam, J. P., *J. Am. Chem. Soc.*, 116, 6975, 1994.
289. Mutter, M., Altmann, K.-H., Tuchscherer, G., and Vuilleumier, S., *Tetrahedron*, 44, 771, 1988.
290. Mutter, M., in *Peptides. Chemistry and Biology. Proceedings of the 10th American Peptides Symposium*, Marshall, G. R., Ed., ESCOM, Leiden, 1988, 349.
291. Mutter, M. and Vuilleumier, S., *Angew. Chem. Int. Ed. Engl.*, 28, 535, 1989.
292. Ernest, I., Vuilleumier, S., Fritz, H., and Mutter, M., *Tetrahedron Lett.*, 31, 4015, 1990.
293. Dörner, B., Carey, R. I., Mutter, M., Labhardt, A. M., Steiner, V., and Rink, H., in *Innovations and Perspectives in Solid-Phase Synthesis. Peptides, Polypeptides and Oligonucleotides*, Epton, R., Ed., Intercept Ltd, Andover, 1992, 163.
294. Arai, T., Ide, Y., Tanaka, Y., Fujimoto, T., and Nishino, N. *Chem. Lett.*, 381, 1995.
295. Nefzi, A., Sun, X., and Mutter, M., *Tetrahedron Lett.*, 36, 229, 1995.
296. Dawson, P. E. and Kent, S. B. H., *J. Am. Chem. Soc.*, 115, 7263, 1993.
297. Tuchscherer, G., *Tetrahedron Lett.*, 34, 8419, 1993.
298. Tuchscherer, G., Ernest, I., Rose, K., and Mutter, M., in *Peptides. Chemistry, Structure and Biology. Proceedings of the 13th American Peptides Symposium*, Hodges, R. S. and Smith, J. A., Eds., ESCOM, Leiden, 1994, 1067.

Chapter 5

Formation of Disulfide Bridges

Cysteine **1** is a special case among the DNA-encoded amino acids. The sulfhydryl group of its side chain can combine with that of a second cysteine residue to form the disulfide cystine **2**. Such disulfide bridges form part of the molecular framework of many naturally occurring peptides and proteins.[1-6] A given cystine residue may be formed from cysteines that are far apart in the amino acid sequence of a peptide chain, provided that their sulfhydryl groups are able to come sufficiently close together in space for disulfide formation to take place. Cysteine residues are sometimes referred to as half-cystines so that the formation of disulfide bridges may be considered to be brought about by the combination of two half-cystine residues.

1   2

Disulfide bridges reduce the flexibility of a peptide, making it more rigid and decreasing the number of conformations available to it in solution. Such conformational constraint is often necessary for biological activity, which may be lost if the disulfide is disrupted. Structural stabilization by disulfide bridge formation can also be introduced into nonnatural peptides. This allows the design and preparation of rigid artificial molecules, usually with enhanced thermal stability, that can be used to probe specific aspects of peptide structure or function.[7-14] Disulfide bridges are, therefore, of fundamental importance in peptide chemistry, and extensive investigations into methods for their formation have been carried out ever since the pioneering work of du Vigneaud.[15] Fundamental to the effective formation of disulfide bridges is the proper management of cysteine, and the protection of this residue together with the methods to be used for protecting group removal and for cysteine pairing must be very carefully considered.[16]

5.1 PEPTIDES WITH DISULFIDE BRIDGES

Disulfide bridges may be either intramolecular or intermolecular depending upon whether the cysteine residues involved are on the same chain or on different ones, respectively.

5.1.1 Intramolecular Disulfide Bridges

Many naturally occurring peptides and proteins have one or more intramolecular disulfide bridges. The peptide hormones oxytocin, vasopressin, somatostatin, and calcitonin, for example, all have one such bridge while the enzyme ribonuclease A has four. The ease with which intramolecular disulfide bridges can be formed depends predominantly upon conformational considerations.[17-24] If the peptide backbone can easily adopt a conformation in which the two cysteines are close together, then, assuming that the protecting group chemistry used is sufficiently optimized, disulfide formation can often be brought about efficiently. This can be easier in larger molecules that are inherently more flexible, although the exact sequence of the peptide backbone may play a crucial role. Smaller peptides, on the other hand, can pose more of a problem since favorable conformations for disulfide formation may be quite difficult to attain. The smallest possible case, that of an eight-membered ring incorporating a disulfide bridge is not commonly found in nature and can be difficult to synthesize.[25-31]

If the peptide chain has only two cysteine residues, then only one intramolecular disulfide bridge is possible, but for peptides with more than two the situation becomes appreciably more involved. Peptides with four, six, or eight cysteine residues can, in principle, give rise to 3, 15, and 105 different intramolecular isomers, respectively. It is obvious, therefore, that the synthesis of disulfide bridge-containing peptides can be very challenging and that strategies for the controlled, regioselective formation of disulfides are necessary if complex biologically active peptides are to be synthesized.

5.1.2 Intermolecular Disulfide Bridges

Intermolecular disulfide bridges are formed when cysteine residues located on different peptide chains are paired, and this structural characteristic is present in a variety of naturally occurring peptides. Among these, several are homodimers (dimers of peptide chains with identical amino acid sequences) linked by two disulfide bridges, either in an antiparallel (as in **3**) or parallel (as in **4**) fashion. The β-atrial natriuretic factor,[32-34] the F2 factor of *Locusta migratoria*,[35] uteroglobin,[36,37] and fibronectin[38] are all antiparallel homodimers of this type, while the hinge region of the immunoglobulins incorporates a parallel homodimer.[39]

Nonnatural dimers of the type **3** or **4** may also be useful in structural studies, and some display interesting binding properties.[39-44] An artificial parallel homodimer[45] of deamino oxytocin was found to act as a long-lasting prohormogen, presumably because of its slow disproportionation under physiological conditions.

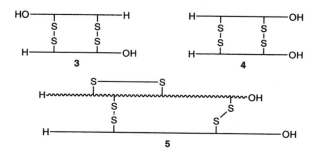

More-complicated examples of intermolecular disulfides are provided by insulin[46-49] and related peptides, such as the bombyxins.[50-52] These consist of two different chains linked together by two intermolecular bridges, with a third, intramolecular bridge also being present in one of the chains, as represented in **5**.

Although naturally occurring peptides consisting of two chains linked through a single intermolecular disulfide bridge are not common in nature, it may often be desirable to produce such structures in the laboratory. Immunological studies or the generation of enzyme active site models often requires that peptides be conjugated to a carrier protein or that they be dimerized to give homo- or heterodimers.[13,53-55] In these cases, assuming, of course, that the peptides in question incorporate cysteine residues, the formation of a single disulfide bridge is an obvious choice for linking them.

5.2 PROTECTION OF CYSTEINE IN PEPTIDE SYNTHESIS

The requirements for disulfide bridge formation place stringent demands on the groups used to protect cysteine.[56-61] A high degree of compatibility must be achieved between the protection used for the cysteine side chain and that used for the other functional groups in the molecule, if complex peptides with multiple disulfide bridges are to be synthesized successfully. The formation of disulfide bridges is often done toward the end of a synthesis so that cysteine protection, in common with that of other trifunctional amino acids, must be stable to all of the conditions used to effect chain elongation. The overall synthetic strategy, however, may additionally require either that sulfhydryl-protecting groups be removed in the presence of other side chain protection or, alternatively, that they be stable to the acidolytic removal of the majority of protecting groups from the peptide. Both of these scenarios require orthogonality between cysteine protection and all other protecting groups.

Among the most important factors to be taken into account when choosing a protecting group scheme for peptides with several cysteine residues is compatibility with the conditions required for repetitive removal of amino acid N^α protection. If peptide synthesis is to be carried out in solution, it must be remembered that the presence of cysteine residues is not compatible with standard catalytic hydrogenolysis of the Z group. Catalytic hydrogenolysis in liquid ammonia has been described in such situations,[62,63] but it is not clear that this represents a general approach.

In contemporary Boc/Bzl SPPS, the most widely used options for cysteine protection are the S-4-methylbenzyl[57,64-66] (S-Meb) **6**, S-4-methoxybenzyl[57,64,67-70]

(S-Mob) **7**, S-3-nitro-2-pyridinesulfenyl[71-77] (S-Npys) **8**, S-[(N'-methyl-N'-phenylcar-bamoyl)-sulfenyl][78] (S-S-Snm) **9**, S-9-fluorenylmethyl[20,79-81] (S-Fm) **10**, and S-2,4-dinitrophenylethyl[82,83] (S-Dnpe) **11** groups. All of these are stable to repetitive tri-fluoroacetic acid treatment, although some loss of the S-Mob **7** group may be observed in longer syntheses.[65]

Structures are drawn to include the cysteine sulfur atom.

Of these, **6** and **7** are removed by treatment with liquid hydrogen fluoride when the peptide is detached from the resin, **7** being somewhat more acid labile than **6**. Protecting groups **8**, **9**, **10**, and **11**, on the other hand, are stable to this acidolysis and must be removed in a separate step. In the case of **8** this can be done either by treatment with a thiol (which can be another cysteine sulfhydryl group) or a reducing agent such as tributylphosphine, whereas **9** is usually converted into an activated derivative for directed disulfide bond formation (see Section 5.3.3) on treatment with 2-thiopyridine. The base-labile groups **10** and **11** are removed on treatment with a solution of ammonia in methanol or with solutions of organic bases such as piperidine or 1,8-diazabicyclo[5.4.0]undec-7-ene (DBU). In addition, removal of protecting groups **6** and **7** can also be effected using metal salts, particularly those of mercury, thallium, and silver. This can be done in such a way as to generate the disulfide bridge directly (see Section 5.3.2), or, alternatively, the free thiol group is produced on treatment of the metal mercaptide with β-mercaptoethanol.

For Fmoc/tBu SPPS, protection with S-triphenylmethyl[47,58,64,84-89] (S-trityl, S-Trt) **12**, S-4,4',4''-trimethoxytriphenylmethyl[90] (S-TMTr) **13**, or S-2,4,6-trimethoxyben-zyl[90-92] (S-Tmob) **14** groups has been applied most commonly. These are all stable to the repetitive base treatments required for Fmoc group removal and can all be removed by different concentrations of trifluoroacetic acid. In the cases of **13** and **14** their acid lability is such that they can, if desired, be removed while the peptide is still bound to the resin, allowing disulfide bridge formation to take place on the solid support. The S-Trt group **12** requires stronger acid treatment so that it is removed when the peptide is detached from the resin. Suitable scavengers must be used when Trt- or Tmob-protected peptides are cleaved, since the stable cations produced can easily realkylate Cys residues.[86] These protecting groups are also susceptible to metal-assisted removal with mercury, thallium, and silver salts, either to give the disulfide directly (see Section 5.3.2) or the free thiol group on treatment with β-mercaptoethanol.

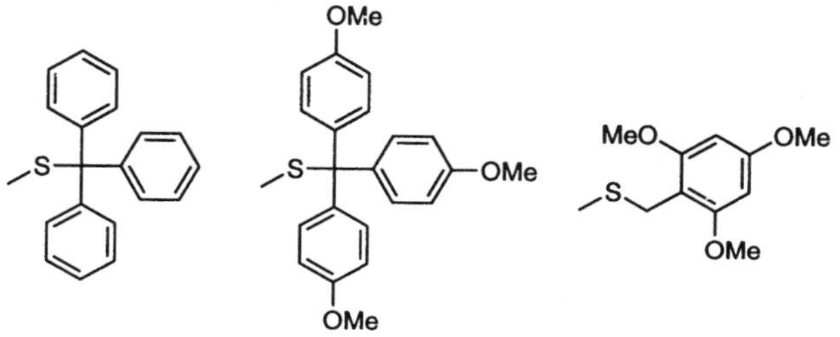

12   **13**   **14**

Structures are drawn to include the cysteine sulfur atom.

The S-acetamidomethyl[70,82,84,87,88,93-98] (S-Acm) group **15** and its derivatives, the S-trimethylacetamidomethyl[99] (S-Tacm) **16** and S-phenylacetamidomethyl[82,100-103] (S-Phacm) **17** groups, in addition to the S-tert-butylmercapto[94,104-108] (S-StBu) **18** and the S-tert-butyl[33,68,74,84,94,109-111] (S-tBu) **19** groups, are compatible with both Boc/Bzl and Fmoc/tBu SPPS.

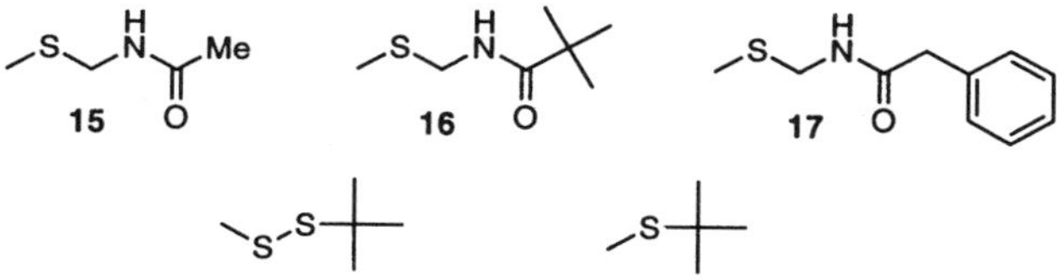

15   **16**   **17**

18   **19**

Structures are drawn to include the cysteine sulfur atom.

All of these, with the possible exception of the S-tBu group **18**, are stable to liquid hydrogen fluoride at 0°C although some, if not all, may be removed at higher temperatures. All are normally maintained in place after cleavage of the peptide from the resin (with concomitant removal of most other side chain-protecting groups) and are carried through to a later stage, where the partially protected peptide has been purified and the moment is ready for disulfide bridge formation. Removal of the S-Acm group **15** and its derivatives **16** and **17** can be achieved on treatment with silver, mercury, or thallium salts or, alternatively, with iodine to give the disulfide directly (see Section 5.3.2). In the case of metal-assisted removal, the free thiol group can be generated by treatment of the corresponding mercaptide with β-mercaptoethanol. In addition, the S-Acm group **15** can also be removed by electrophilic sulfur reagents of the general type RSCl. The S-Phacm group **17** can be removed enzymatically by treatment with penicillin amidohydrolase. The S-tBu group **18** must be removed by treatment with a thiol or a reducing agent such as tributyl phosphine while the S-tBu group **19** may be converted into the more reactive mixed disulfide **8**, upon treatment with 2-nitropyridenesulfenyl chloride (Npys-Cl). The free sulfhydryl group can then be obtained on treatment with reducing agents, or, alternatively, disulfide bridge formation can be brought about by reaction with the sulfhydryl group of another cysteine residue (see Section 5.3.3.3).

The range of protecting groups available and of methods for effecting their removal imparts a reasonable degree of flexibility to peptide synthesis involving multiple cysteine residues. However, given the intricacy of the chemistry involved, an accurate choice among the various options available is essential if the synthesis is to have a successful outcome.

5.3 CHEMICAL METHODS FOR THE FORMATION OF DISULFIDE BRIDGES

There are three basic chemical approaches that can be used for the formation of disulfide bridges in peptides, depending upon the way the various cysteine residues in the molecule are manipulated. These are considered in turn below.

5.3.1 From Precursors with Free Sulfhydryl Groups

The simplest method for the formation of disulfide bridges is represented in Scheme 5.1. In protected peptide **20**, which can be elaborated either by solid-phase methods or in solution, both of the cysteine residues to be paired are protected with the same protecting group. This is removed along with all other side chain protection in the final acidolysis, generating a linear precursor with free sulfhydryl groups **21**. Oxidation of this precursor then gives the desired disulfide bridge-containing peptide **22**.

The advantages of the strategy are that a single protecting group suffices for all cysteine residues and that the number of chemical operations necessary after chain assembly is reduced to a minimum. The choice of cysteine protection is dictated largely by its compatibility with the chemistry adopted for chain elongation. For Boc/Bzl synthesis p-methylbenzyl **6**, p-methoxybenzyl **7**, or even *tert*-butyl **19** thioethers are good choices, whereas in the Fmoc/tBu approach, S-trityl **12**, S-methoxytrityl **13**, or S-trimethoxybenzyl **14** groups would be indicated. For peptide synthesis in solution, the choice of cysteine protection would again depend upon the N^α-protecting group used.

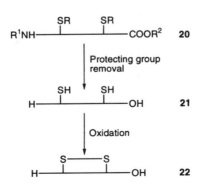

SCHEME 5.1   Formation of an intramolecular disulfide bridge from a precursor with free sulfhydryl groups.

Experimental conditions must be carefully controlled if acceptable yields of the desired product are to be obtained. After the completion of chain elongation and removal of side chain protection (with concomitant cleavage from the resin in the case of SPPS), the crude peptide is treated with reducing agents such as dithiothreitol or mercaptoethanol. This is to ensure that it is a discrete linear monomeric species, free of oligomers or of intra- or intermolecular disulfide bridges caused by premature disulfide formation between the free sulfhydryl groups.[112-114] The linear precursor is then normally purified at acidic pH, again to minimize unwanted premature inter- or intramolecular disulfide formation. Controlled oxidation to the desired disulfide is carried out in dilute solution, to reduce the formation of oligomers, using air or other suitable oxidizing agents. Molecular oxygen, which promotes disulfide formation at slightly alkaline pH by what is probably a radical mechanism,[115] may sometimes give low yields and can be too slow. In these cases, the use of other mild oxidizing agents such as potassium ferricyanide[116-119] or dimethylsulfoxide[91,93,120-124] can give better results.

In more complex cases, where more than one intramolecular disulfide bridge is to be formed, the use of a single protecting group for cysteine, following the strategy outlined in Scheme 5.1, can still give good results. Its application to the formation of the four intramolecular disulfide bridges in ribonuclease A is discussed in Section 3.6.3. The generation of free sulfhydryl groups for disulfide bridge formation need not necessarily, however, be brought about in the final acidolysis. If desired, it can be done at some other convenient point in the synthesis of a peptide, depending upon the groups chosen for the protection of cysteine. Sulhydryl groups can be liberated when all other side chain-protecting groups are still in place, or, alternatively, they can be retained in their protected form after the final acidolysis and liberated after purification of the partially protected peptide intermediate. The decision as to when to remove them and when to form the disulfide bridges will be dictated by the characteristics of the peptide to be synthesized and by the strategy adopted. Maintenance of Cys protection after the final acidolysis has the advantage of simplifying the purification of the linear precursor, before disulfide formation is brought about, something that is especially important in multiple Cys-containing peptides.

Irrespective of the point at which Cys thiols are liberated, the production of the native arrangement of disulfide bridges can only occur if the free sulhydryl group precursor can fold into the correct conformation for native half-cystine pairing. In complex molecules containing several disulfide bridges, this can be a very delicate undertaking and can be aided by the use of redox buffers that mimic physiological conditions.[125-128] One of the most useful methods is to use oxidized and reduced glutathione, respectively.[123,126] In such systems the polythiol peptide is allowed to equilibrate in the redox buffer and even if nonnative structures are favored kinetically they can equilibrate ultimately to the native arrangement by thiol–disulfide exchange mechanisms. Another possibility, which has as yet not been applied to synthetic peptides, is the use of protein disulfide isomerase.[129-132] The five disulfide bridges present in the 121-residue human midkine[133,134] were formed after purification of a partially protected 10 S-Acm-containing linear precursor. Liberation of the thiol

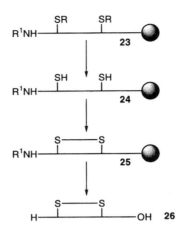

SCHEME 5.2 Formation of an intramolecular disulfide bridge from a precursor with free thiol groups, on the solid support.

groups followed by slow oxidative folding in the presence of glutathione gave good yields of the synthetic protein.

Disulfide bridge formation from free sulfhydryl precursors may also be carried out on the solid support, prior to cleavage of the peptide from the resin, as shown in Scheme 5.2. This has several advantages, including the elimination of problems of peptide solubility and the facile removal of oxidizing agents and solvents by filtration. The pseudo-dilution phenomenon[20,135,136] that applies to solid-phase reactions favors intramolecular disulfide bridge formation over intermolecular reaction and yields can be high in favorable cases.

The solid-phase formation of disulfide bridges from precursors with free sulfhydryl groups normally requires that cysteine protection in peptide–resin **23** be orthogonal to that of other side chain-protecting groups. For Boc/Bzl SPPS S-Fm **10** or S-Dnpe **11** protection can be removed selectively from the peptide on the resin to give the necessary free sulfhydryl group precursors. In Fmoc/tBu synthesis the S-StBu group **18** has been used,[108] treatment with β-mercaptoethanol being necessary for generation of the free thiol precursor. Other options include the S-TMTr **13** or S-Tmob **14** groups, but care must be taken that the acidolysis required for their removal does not cleave the peptide from the resin. Simultaneous removal of the protecting groups from the cysteine residues to be paired gives the free thiol peptide–resin **24**. Oxidation using air or some other suitable reagent[90-92,137-139] then gives the required disulfide bridge in the resin-bound peptide **25**. Cleavage from the solid support with concomitant removal of all other side chain-protecting groups affords the crude disulfide bridge peptide **26**. This procedure can, in principle, be applied to the synthesis of multiple disulfide bridge-containing peptides.

The preparation of intermolecular disulfides from two precursors with free sulfhydryl groups, such as **27** and **28**, is normally not a practical proposition. The formation of the two homodimeric disulfides **29** and **31** will compete to a significant extent with the desired heterodimer **30**, which may itself only be one of the minor products of the reaction, as shown in Scheme 5.3.

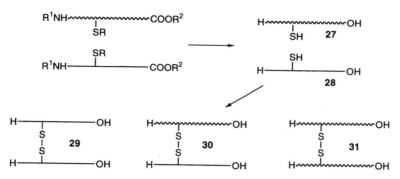

SCHEME 5.3 Uncontrolled formation of intermolecular disulfides from two different peptide chains each having a free thiol group.

In general, the formation of intermolecular disulfide bridges requires a greater degree of chemical sophistication, and, where possible, directed methods (see Section 5.3.3) are used. However, for the simple case of the synthesis of homodimeric disulfides (Scheme 5.3, where **27** and **28** are identical), it is only necessary to synthesize one single peptide chain, and, since the formation of other unwanted disulfide side products is not possible, free thiol precursors can be used to form them.[13,53-55] The cysteine-protecting group is removed at some opportune moment, liberating the sulfhydryl groups, and disulfide bridge formation is brought about by oxidation with air or with some other suitable reagent.

In principle, the synthesis of homodimers linked by two or more intermolecular disulfide bridges should, at least in certain cases, be possible from precursors having free sulfhydryl groups. The two chains would have to orientate themselves correctly for the formation of the desired disulfide arrangement, and thermodynamic considerations would be of paramount importance. However, in the case of the synthesis of naturally occurring homodimers, where the monomers might have a tendency to adopt native conformations, intermolecular disulfide bridge formation to give the thermodynamically more stable natural product might reasonably be expected to take place.

5.3.2 From Precursors with Protected Sulfhydryl Groups

Disulfide bridges can also be formed directly from protected cysteine residues, without having to generate the free thiol intermediates as precursors. This has the advantage of suppressing the unwanted premature inter- or intramolecular disulfide formation that can occur when free sulfhydryl group-containing peptides are manipulated. Both of the cysteine residues to be paired are again protected with the same group, but this is now removed oxidatively, forming the disulfide directly.

The most widely applied variation of the approach is the iodine-mediated oxidation of cysteine residues that are S-trityl **12**, S-Tmob **14**, or S-Acm **15** protected.[87,92,140,141] Such oxidation can be carried out when the peptide precursors **32** are still fully protected, so that, if desired, further synthetic elaboration can be performed on peptide **33**, after disulfide bridge formation has occurred. The mechanism

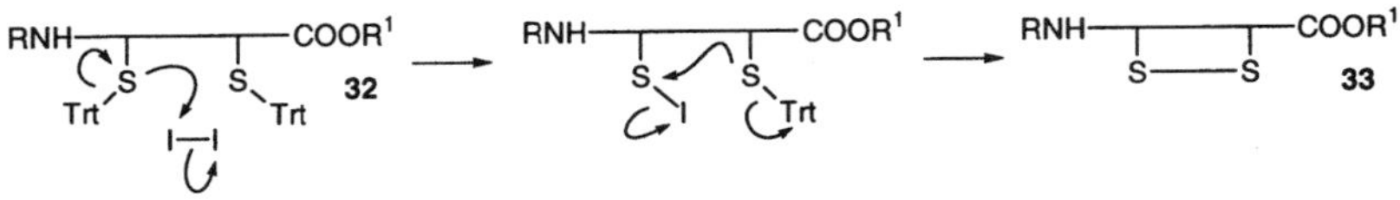

SCHEME 5.4 Disulfide bridge formation by iodine-mediated oxidative removal of Cys(Trt) groups.

is thought to be that outlined in Scheme 5.4, illustrated for cysteines protected with the trityl group **12**. Similar mechanisms are thought to apply in the case of *S*-Tmob **14** or *S*-Acm **15** protection.

The rates of oxidative removal are very sensitive to changes in solvent so that for peptides having combinations of the *S*-protecting groups **12** and **15**, selective pairing of one type of protected cysteine residue in the presence of another is possible. *S*-Acm is selectively oxidized in the presence of *S*-Trt in dimethylformamide or dimethylformamide–water mixtures, whereas *S*-Trt is selectively oxidized in the presence of *S*-Acm in chloroform or dichloromethane.[87]

Other oxidizing agents can also be used, providing even more flexibility. Cyanogen iodide[142] or *N*-iodosuccinimide[143] both act as an alternative source of the iodonium ion, but a milder alternative reagent is thallium(III) trifluoroacetate.[88] Its use reduces the risk of the unwanted oxidation of sensitive residues such as Trp or Met; ideally, however, they should be protected.[87,119,144-146] Mechanistically, this oxidation is thought to proceed as outlined in Scheme 5.5 for protected peptide **33**.

The method is applicable to peptides having Cys residues that are Mob **7**, Trt **12**, Tmob **14**, Acm **15**, or *t*Bu **19** protected. Another useful reagent is constituted by a mixture of alkyltrichlorosilanes and sulfoxides, for example, methyltrichlorosilane and diphenyl sulfoxide.[33,147] The reaction is thought to proceed as outlined in Scheme 5.6, for a protected peptide **34** having Cys residues that are Meb **6**, Mob **7**, Acm **15**, Tacm **16**, or *t*Bu **19** protected.

The oxidative removal of cysteine-protecting groups to form disulfide bridges directly can also be carried out on solid supports by application of the proper oxidizing agent to the peptide–resin in question,[20,82,91,92,148-151] and yields can be high in favorable cases. The general reaction scheme is similar to that shown in Scheme 5.2, except that the free sulfhydryl group intermediates are not generated. The method has also been used to prepare large (56-atom and 62-atom) disulfide peptides on the

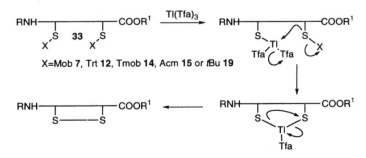

SCHEME 5.5 Disulfide bridge formation by thallium(III) trifluoroacetate-mediated removal of Cys-protecting groups.

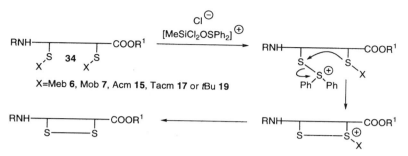

SCHEME 5.6 Disulfide bridge formation by oxidative removal of Cys-protecting groups using the sulfoxide-alkyltrichlorosilane method.

solid support,[152,153] and its application in the synthesis of α-conotoxin is discussed in more detail in Section 5.4.

As is the case for the formation of disulfide bridges from free sulfhydryl group-containing precursors, approaches based on the oxidative removal of cysteine-protecting groups are, generally speaking, only applicable to the formation of intramolecular or homodimeric disulfide bridges. The formation of intermolecular disulfides is subject to the same inconveniences already discussed in Section 5.3.1.

5.3.3 Directed Disulfide Bridge Formation

A more-sophisticated approach to disulfide bridge formation involves the protection of each of the Cys residues to be paired with a different group. These must be chosen carefully, since it must be possible to activate one cysteine residue selectively toward nucleophilic attack by the other, which itself either may be in its protected form or which may have a free sulfhydryl group. Directed disulfide formation then takes place by a displacement reaction. If the attacking cysteine residue is required with a free sulfhydryl group then, additionally, it may be necessary to remove one of the cysteine-protecting groups in the presence of the other.

Methods for directed disulfide formation are necessarily more intricate and require a greater number of chemical operations for their realization than nondirected ones. They do, however, provide greater synthetic control and are invaluable in the synthesis of complex peptides containing multiple disulfide bridges. They are particularly useful for the synthesis of intermolecular heterodisulfides, which are difficult to synthesize from free thiol precursors or by co-oxidation methods,[87] since the competitive formation of unwanted subproducts usually means that yields are low. The differentiation of cysteine residues for directed disulfide formation can be brought about in several different ways.

5.3.3.1 Selective Activation of Cysteine Residues *In Situ*

Activation of the sulfhydryl group of a cysteine residue toward nucleophilic attack may be achieved by treatment of the cysteine-containing peptide, either as the protected or unprotected derivative **35**, with thiocyanogen.[58] The active intermediate

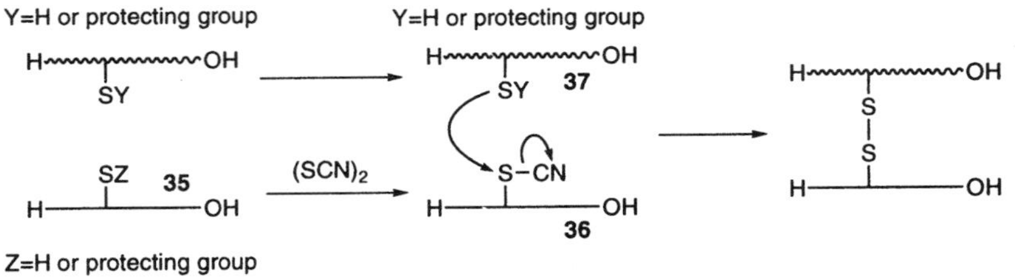

SCHEME 5.7 Directed disulfide bridge formation by *in situ* activation of a Cys residue with thiocyanogen.

36 is not isolated but is allowed to react with the free (or protected) sulfhydryl group of the second peptide chain **37**. This, by now classical, approach is outlined in Scheme 5.7.

Although the method can give satisfactory results, there are two main drawbacks associated with its use. First, the rather high electrophilic reactivity of the reagent is cause for concern especially when sensitive residues or protecting groups are present and may provoke several undesired side reactions. A second problem occurs if the activation step is slow or does not go to completion for any reason, since substantial amounts of the homodimeric disulfides can then be formed.

5.3.3.2 Selective Activation that Gives a Stable, Isolable Intermediate

A milder alternative is to activate the cysteine of the peptide in question with groups that give rise to a stable, isolable derivative. After purification this is then made to react with the peptide containing the second sulfhydryl group, whether in its free or protected form. One of the most popular alternatives is to form a mixed disulfide of the sulfhydryl group of the cysteine to be activated, and several possibilities are available. Among the most frequently used[78,154-158] are the *S*-methoxycarbonylsulfenyl (*S*-Scm) **38**, the 2-nitrophenylsulfenyl (*S*-Nps) **39**, the 2-pyridinesulfenyl (*S*-Pyr) **40** derivatives together with that formed from Ellman's reagent 5,5′-dithiobis(2-nitrobenzoic acid), **41**. A promising new alternative[159] is the formation of mixed disulfides with 2,2′-dithiobis(5-nitropyridine).

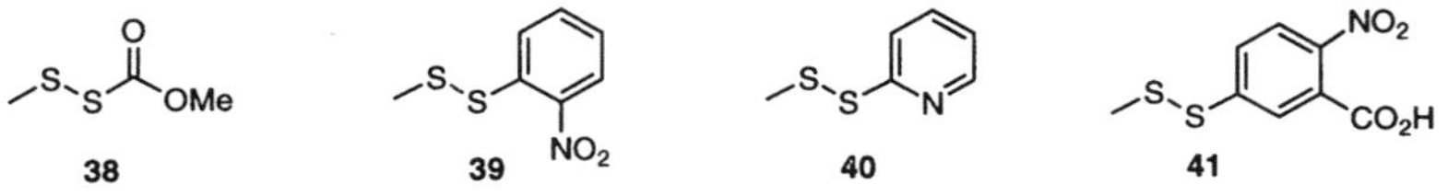

Structures are drawn to include the sulfur atom of Cys.

A peptide such as **42**, which has the Cys residue either as the free thiol or protected with the *S*-Trt **12** or *S*-Acm **15** groups, can be activated as the *S*-Scm disulfide **38** on treatment with methoxycarbonylsulfenyl chloride (ScmCl). The activated *S*-Scm peptide **43** can be isolated under neutral or slightly acidic conditions and disulfide bridge formation brought about by adding the second peptide chain

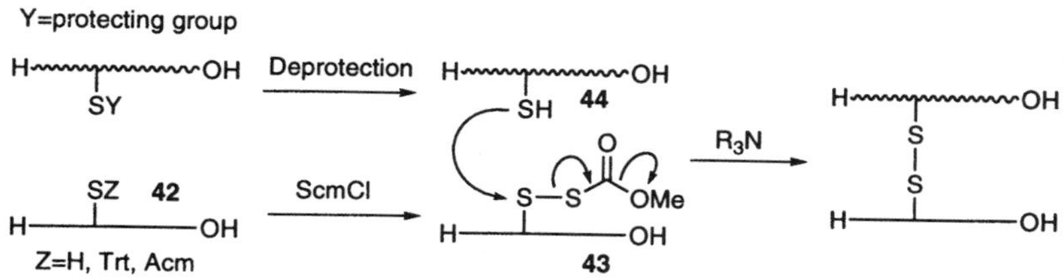

SCHEME 5.8 Directed disulfide formation by S-Scm activation of a Cys residue.

44 under slightly basic conditions. The displaced S-Scm group then decomposes to carbonyl sulfide and methanol, as outlined in Scheme 5.8.

Both free or protected sulfhydryl groups can also be activated by reaction with aromatic sulfenyl halides, such as 2-nitrophenylsulfenyl chloride or 2-pyridenesulfenyl chloride,[111,160,161] giving the mixed disulfides **39** or **40**. A drawback, however, is that the aromatic sulfenyl halides required can be rather difficult to manipulate. For free sulfhydryl group–containing peptides, derivatives of type **40** can, as an alternative, be formed on treatment with 2,2'-dithiopyridine[33,43,44,50] and derivatives of type **41** by reaction of the free sulfhydryl group with Ellman's reagent.[162] A further possibility for the formation of **40** is treatment of the S-[(N'-methyl-N'-phenylcarbamoyl)-sulfenyl] (S-Snm) protected derivative **9** with 2-thiopyridine, a highly specific transformation that often proceeds in good yield.[78]

Any of these methods can be used to transform a peptide such as **45** into the activated arylsulfenyl derivative **46**. Disulfide formation can then take place on addition of the second peptide thiol **47**, as outlined in Scheme 5.9.

An alternative to the formation of activated mixed disulfides is the use of bis(*tert*-butyl)azodicarboxylate,[39,163-165] which reacts with free cysteine sulfhydryl groups giving rise to stable, isolable sulfenylhydrazides of the general type **48**. Treatment with a second peptide **49** containing a free sulfhydryl group then gives rise to intermolecular disulfides, as shown in Scheme 5.10.

Another different but still related approach allows unsymmetrical disulfides to be formed under strong-acid conditions.[64,166,167] One of the residues to be paired is both protected and oxidized to the sulfoxide. The cysteine residue in question can in fact be incorporated into the growing peptide chain in this form from the outset. Combination of the peptide **50** incorporating such a cysteine residue with a second peptide chain **51** in which the sulfhydryl group is free, under strongly acidic

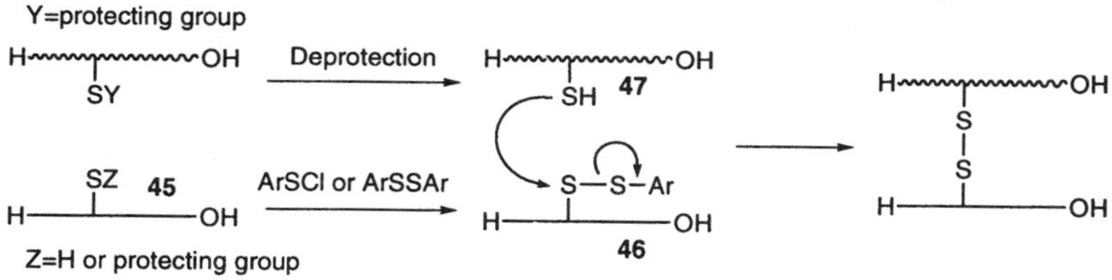

SCHEME 5.9 Directed disulfide bridge formation by arylsulfenyl activation of a Cys residue.

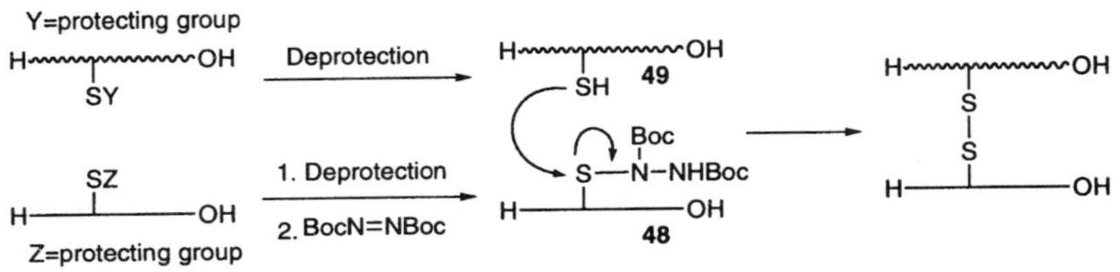

SCHEME 5.10 Directed disulfide bridge formation by bis(*tert*-butyl)azodicarboxylate activation of a Cys residue.

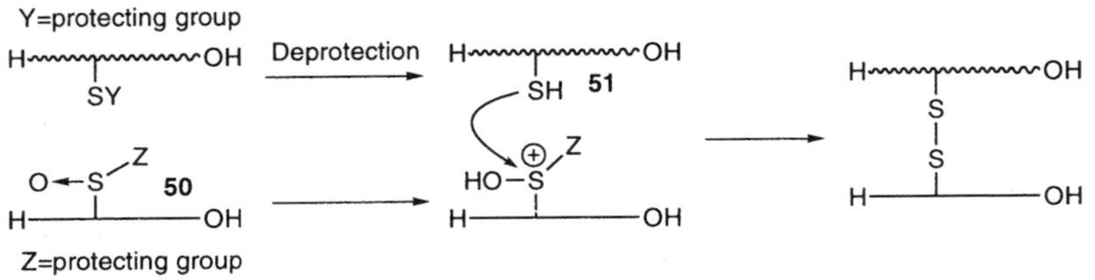

SCHEME 5.11 Directed disulfide bridge formation from a precursor having a Cys(O) residue.

conditions, then leads to the formation of the required unsymmetrical disulfide, with the elimination of water, as shown in Scheme 5.11.

All these directed methods have been used in peptide synthesis in solution although, as yet, not all have been applied to disulfide bridge formation on solid supports. They have been used for the synthesis of a range of target molecules having both intra- and intermolecular disulfides.

5.3.3.3 Use of the *S*-3-Nitro-2-Pyridinesulfenyl Group

Although the methods for the production of the stable, activated intermediates above normally proceed cleanly and in good yield, the extra synthetic step can complicate matters. An alternative is to take advantage of the dual nature of *S*-Npys protection **8**. This group is stable to the chain elongation conditions of Boc/Bzl SPPS and is also stable to the hydrogen fluoride-mediated cleavage of the peptide from the solid support. The Npys group serves both to protect the sulfhydryl of the cysteine residue and, at the same time, acts as an activating group so that directed disulfide bridge formation can be brought about by attack of the free sulfhydryl group of a second cysteine residue.[41,43,44,71,74,76,168-170] Disulfide formation occurs over quite a wide pH range and can be monitored spectrophotometrically by titration of the released 3-nitro-2-pyridinethiol.[171]

This is a versatile approach to disulfide bond formation and has been used both in intramolecular and intermolecular disulfide bridges both in solution and on solid supports. Unfortunately, the Npys group cannot be used in Fmoc/*t*Bu SPPS because it is not stable to the repetitive base treatments required for Fmoc group removal. Boc-Cys(Npys)-OH can, however, be incorporated as the *N*-terminal residue into peptides synthesized using the Fmoc/*t*Bu approach. Alternatively, cysteine residues protected with *S*-Trt **12**, *S*-Acm **15**, or *S*-*t*Bu **19** groups can be converted into the corresponding Cys(Npys)-protected peptides by treatment with NpysCl, at a later stage using methods similar to those described in Section 5.3.2.

5.3.4 From Cystine

A completely different approach to the synthesis of peptides with disulfide bridges, whether intramolecular or intermolecular, is to start from cystine itself and to elongate each of the chains in the required manner. Although such a strategy has been used in the synthesis of insulin partial sequences,[172-174] its general application is rather impractical. The construction of different peptide chains on the same molecule requires several levels of orthogonality between protecting groups, making the overall protection scheme very intricate. Furthermore, although disulfide bridges are reasonably resistant to the chemistry involved in chain elongation, their survival throughout the conditions necessary to effect the construction of large complex peptides is doubtful.

5.4 REGIOSELECTIVE DISULFIDE BRIDGE FORMATION

The methods described above have been widely used for the formation of disulfide bridges, but the sternest test of their efficiency is the synthesis of peptides containing multiple disulfide bridges.[175,176] The strategy adopted for the synthesis of such complex peptides will depend upon the various structural characteristics of the target molecule. Of primary importance among these will be the type and number of disulfide bridges present, whether they are intramolecular, intermolecular, or both.

5.4.1 Intramolecular Disulfide Bridges

A common general approach to the regioselective[175] formation of intramolecular disulfide bridges in multiple disulfide-containing peptides is to protect each of the various pairs of cysteine residues with the same group, but to use different protecting groups for different pairs. Sulfhydryl functions can then be liberated pairwise in turn and disulfide bridge formation allowed to take place sequentially. This strategy is preferable to nonregioselective disulfide bridge formation either from free thiol precursor or from precursors with protected Cys residues, for multiple disulfide bridge-containing synthetic peptides. It does, however, require an accurate choice of cysteine-protecting groups as well as of the order of deprotection. The strategy can also be applied in multiple disulfide bridge formation on solid supports, by an extension of the procedure outlined in Scheme 5.2 (see Scheme 5.16, for an example).

If the peptide in question has been synthesized by SPPS, one of the classes of Cys protection might be a group that is removed acidolytically in the cleavage reaction, while the others would be stable to such acidolysis. A widely used combination in Boc/Bzl synthesis is hydrogen fluoride-labile *S*-Meb **6** protection with hydrogen fluoride-stable *S*-Acm **15** protection. This is exemplified in a synthesis[177] of one of the conotoxin peptides, specifically conotoxin GI, carried out in solution

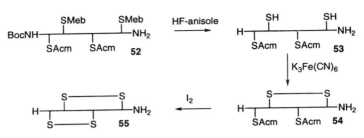

SCHEME 5.12 Synthesis of conotoxin GI in solution.

and outlined in Scheme 5.12. Acidolysis of fully protected peptide **52** gave the bis *S*-Acm peptide **53**, which after purification and air oxidation gave **54**, containing the first disulfide bridge. The second was then formed oxidatively, by treatment of **54** with iodine, giving conotoxin GI **55**.

Another combination of Cys-protecting groups that is useful in Boc/Bzl synthesis is *S*-Fm **10** protection in combination with the *S*-Meb group **6** or the *S*-Acm group **15**. The *S*-Fm group **10** is stable to hydrogen fluoride treatment so that it is retained on cleavage. It can, however, be removed selectively on treatment of the peptide with piperidine. The orthogonal combination of *S*-Fm **10** and *S*-Meb **6** protection was used in a synthesis of a fragment of a bovine pituitary peptide,[178] outlined in Scheme 5.13. Here, the first disulfide bridge was formed on the solid support, a phenylacetamidomethyl resin (see Section 2.3). After elaboration of the desired peptide–resin **56**, treatment with piperidine in dimethylformamide removed both *S*-Fm protecting groups together with the formyl group used to protect Trp. Air oxidation then brought about the efficient formation of the first disulfide bridge while the peptide was still attached to the resin. Treatment of peptide–resin **57** with hydrogen fluoride led to cleavage of bis thiol peptide **58**, without affecting the disulfide bridge. Purification of **58**, followed by air oxidation then gave the desired fragment **59**.

The combination of *S*-Meb **6** and *S*-Acm **15** Cys protection has also been used for the synthesis of a 13-residue enterotoxin fragment containing three disulfide bridges.[179] Chain elongation was carried out by Boc/Bzl SPPS, giving peptide–resin **60**. Acidolysis with liquid hydrogen fluoride detached the peptide from the resin and removed all Meb groups, giving bis Acm peptide **61**. Air oxidation of this then led to the formation of peptide **62**, having two disulfide bridges with the correct half-cystine pairings. In such favorable cases, where the peptide backbone can easily adopt the conformation required for the native disulfide arrangement, the overall protection scheme can be significantly simplified. The third disulfide bridge was subsequently formed by iodine-mediated oxidative removal of the *S*-Acm groups, giving the desired target molecule **63**, as shown below in Scheme 5.14.

For Fmoc/*t*Bu SPPS, the same strategy can be used for the regioselective formation of disulfide bridges. This was first demonstrated by Sheppard[180] in a synthesis of conotoxin GI. A combination of *S*-Acm and *S*-*t*Bu Cys protection was used to ensure regioselectivity, as outlined in Scheme 5.15. Chain elongation was carried out using the Fmoc/*t*Bu approach to give the peptide resin **64**. After Fmoc

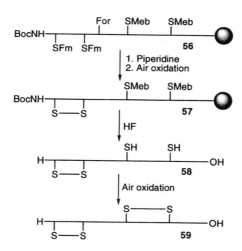

SCHEME 5.13 Synthesis of a bovine pituitary peptide fragment by disulfide bridge formation on the solid support.

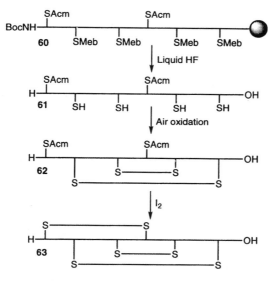

SCHEME 5.14 Synthesis of an enterotoxin fragment.

group removal, treatment with trifluoroacetic acid removed most of the side chain-protecting groups and ammonolysis detached the peptide from the resin as the amide **65**. Treatment of this with tributylphosphine, followed by air oxidation formed the first disulfide bridge in peptide **66**. The second was produced on iodine-mediated oxidative removal of the S-Acm groups, giving the desired conotoxin **67**.

Other combinations of Cys-protecting groups are also useful for the regioselective formation of disulfide bridges in Fmoc/tBu SPPS. Another conotoxin synthesis[91] used

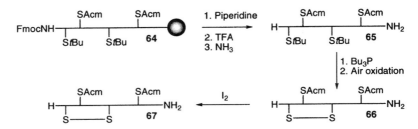

SCHEME 5.15 Solid-phase synthesis of conotoxin GI.

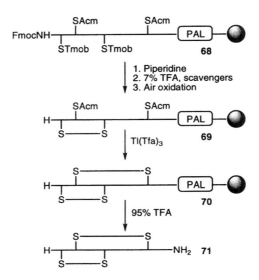

SCHEME 5.16 Solid-phase synthesis of conotoxin, with formation of both disulfide bridges on the solid support.

S-Acm **15** and S-Tmob **14** protection to differentiate the Cys residues. Here, both of the disulfide bridges were formed on the solid support, as outlined in Scheme 5.16.

The peptide chain was synthesized using Fmoc/tBu SPPS, on a tris(alkoxy)benzylamide (PAL) support (see Section 2.3.2), giving peptide–resin **68**. After completion of chain assembly, the N-terminal Fmoc group was removed by treatment with piperidine. The Tmob groups were removed selectively by mild acidolysis with dilute (7%) trifluoroacetic acid containing triethylsilane and water as scavengers. The resin-bound bis(acetamidomethyl) peptide was then oxidized under mild conditions giving the peptide–resin **69**, in which the first disulfide bridge had been formed. The second was again formed on-resin, by oxidative removal of S-Acm protection using thallium(III) trifluoroacetate, giving the double disulfide bridge-containing peptide–resin **70**. Cleavage of the bicyclic peptide from the solid support using concentrated trifluoroacetic acid, followed by purification afforded the desired conotoxin **71**. The advantage of the formation of disulfide bridges on the solid

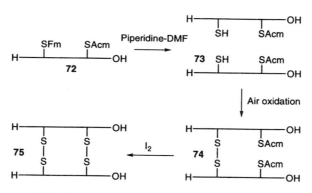

SCHEME 5.17 Synthesis of parallel bis-cystine dimers.

support is its operational simplicity, since all excess reagents and soluble by-products are removed from the system by simple filtration.

5.4.2 Intermolecular Disulfide Bridges

5.4.2.1 Preparation of Parallel and Antiparallel Homodimers Linked by Two Disulfide Bridges

Although unwanted dimeric products often arise during the formation of disulfide bridges, the controlled formation of dimeric peptides linked by disulfide bridges is often challenging. When the objective is the regioselective synthesis of dimers linked by two disulfide bridges, then the strategy adopted will depend upon whether the parallel or antiparallel dimer is to be formed.[33,34,39,41-44] Parallel bis-cystine dimers can be prepared from a single peptide having two orthogonally removable protecting groups for the cysteine residues. The strategy is exemplified by the synthesis[43,44] of a parallel uteroglobin-like dimer, prepared from peptide **72**, synthesized by Boc/Bzl SPPS, in which the Cys residues are protected with S-Fm **10** and S-Acm **15** groups, as outlined in Scheme 5.17.

Removal of the S-Fm group by treatment with piperidine gave the free sulfhydryl group precursor **73**, from which homodimer **74** was produced by air oxidation. The formation of the second disulfide bridge was brought about under high-dilution conditions, to minimize the formation of unwanted oligomers, by oxidative removal of the S-Acm group, giving the desired parallel dimer **75**. A similar strategy was applied in an earlier synthesis of peptide bis-cystine dimers,[32] where a combination of S-Meb and S-Acm Cys protection was used. The synthesis of a parallel dimer of deaminooxytocin was carried out using the same overall strategy, but in this case both disulfide bridges were formed on the solid support, using a combination of S-Tmob with S-Acm Cys protection.[45]

For antiparallel dimers, on the other hand, the situation is rather more involved, and two versions of the same monomeric precursor must be prepared. In each of these, one of the cysteine residues must have a protecting group in common, although it must be on a different cysteine residue in each monomer. The second cysteine

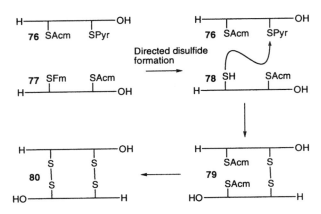

SCHEME 5.18 Synthesis of antiparallel bis-cystine dimers.

residue must then be protected in such a way that one of them is activated toward nucleophilic attack and the other can furnish the corresponding thiol. This strategy is exemplified by the synthesis of an antiparallel uteroglobin-like dimer,[43,44] as outlined in Scheme 5.18.

Of the two peptides required to form the antiparallel dimer, the first, **76**, had one of its Cys protected with the S-Acm group **15** and the other was activated as the S-Pyr mixed disulfide **40**. The second peptide **77** had its Cys residues protected with the S-Fm **10** and S-Acm **15** groups. Directed formation of the first disulfide bridge was brought about by attack of the free thiol of **78** at the activated Cys residue. Directed formation was necessary; otherwise the two possible homodimers that would ultimately lead to the parallel dimer would be formed preferentially. After the first disulfide linkage had been achieved, the second was then formed by treatment of **79** with iodine, under high-dilution conditions, giving the antiparallel dimer **80**.

Syntheses of the antiparallel dimer of human natriuretic peptide have employed a very similar protecting group combination,[33] or one based upon S-Acm protection in combination with directed disulfide formation using Cys(Npys) activation.[32]

5.4.2.2 Insulin and Related Peptides

The classic synthesis of insulin carried out by the Ciba–Geigy group[47,48,181,182] is a model of elegance and economy. Although not the first synthesis of this molecule, it was the first in which the formation of the two intermolecular and one intramolecular disulfide bridges was achieved regioselectively.

Chain elongation was carried out in solution by a convergent, segment-coupling strategy using trityl (Trt), 2-(4-biphenylyl)isopropoxycarbonyl (Bpoc), and Boc groups for the protection of the N^α-amino group at various points in the synthesis. Side chain protection was based on the *tert*-butyl group for all amino acids, except His, Arg, Asn, and Gln, whose side chains were left unprotected; the *tert*-butyl ester was used for C-terminal protection (see Chapter 3). The synthesis is outlined in

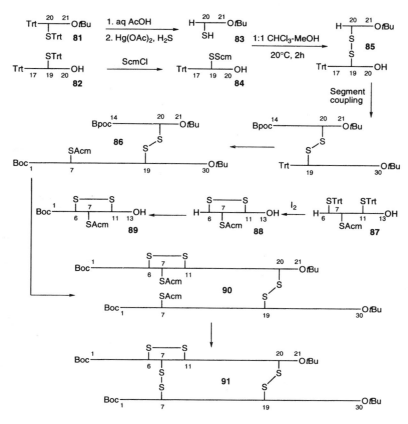

SCHEME 5.19   Synthesis of insulin in solution.

Scheme 5.19. Initially, directed intermolecular disulfide bridge formation was brought about between two peptide segments **81** and **82**, the precursors of the A and B chains, respectively. The *S*-Trt group of **81** was deprotected by mild acidolysis while the *S*-Trt-protected cysteine residue of **82** was converted into the *S*-Scm derivative, by treatment with ScmCl. Directed disulfide formation between **83** and **84** then gave cystine peptide **85**. The A and B chains of **85** were elongated by segment-coupling reactions mediated by dicyclohexylcarbodiimide in the presence of 1-hydroxybenzotriazole, ultimately affording protected peptide **86**. The key A chain protected peptide segment **87** contained three cysteine residues, two protected with the Trt group and the other with the Acm group. Intramolecular disulfide formation was brought about by selective iodine-mediated oxidative removal of the *S*-Trt groups in the presence of the *S*-Acm group, giving **88** (see Section 5.3.2). After further chain elongation, the intramolecular disulfide bridge-containing protected peptide **89** was coupled to cystine peptide **86** forming an intermediate **90** in which two of the three disulfide bridges were now formed. The remaining cysteine residues to be paired were both *S*-Acm protected so that oxidation with iodine, under carefully controlled conditions, brought about formation of the third disulfide bridge

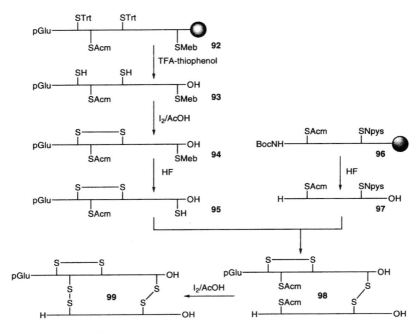

SCHEME 5.20   Solid-phase synthesis of human relaxin.

in **91**. Final acidolytic deprotection then gave human insulin, which after purification was found to exhibit full biological activity.

A modern synthesis of another insulin-like peptide, human relaxin, was carried out in an equally elegant manner by Büllesbach and Schwabe,[183] using solid-phase methods for accomplishing chain elongation. The strategy adopted is outlined in Scheme 5.20.

The A chain precursor **92** was elaborated on a *p*-alkoxybenzyl alcohol-based resin using the Fmoc/*t*Bu approach. Of the four Cys residues that it contained, two were protected as the *S*-Trt derivatives while the others had *S*-Acm and *S*-Meb protection, respectively. Cleavage from the resin by treatment with trifluoroacetic acid gave the bis thiol peptide **93**, which was oxidized with iodine to give the intramolecular disulfide **94**. This was then treated with liquid hydrogen fluoride to remove the *S*-Meb group selectively, affording thiol peptide **95**. The B chain precursor **96** was synthesized on a Pam resin (see Section 2.3.2) by Boc/Bzl synthesis. Hydrogen fluoride–mediated cleavage from the solid support gave peptide **97**, having *S*-Acm- and *S*-Npys-protected Cys residues. Directed intermolecular heterodisulfide formation was brought about between the free thiol group of **95** and the activated Cys(Npys) residue of **97**, to give bis *S*-Acm peptide **98**. Oxidative removal of the *S*-Acm groups with iodine then furnished the peptide **99**. Final deprotection of the still-protected Trp and Met residues (*N*-formyl and sulfoxide derivatives, respectively) followed by purification gave human relaxin with full biological activity.

REFERENCES

1. Creighton, T. E., *Methods Enzymol.*, 131, 83, 1986.
2. Creighton, T. E., *BioEssays*, 8, 57, 1988.
3. Srinivasan, N., Sowdhamini, R., Ramakrishnan, C., and Balaram, P., *Int. J. Pept. Protein Res.*, 36, 147, 1990.
4. Kikuchi, T., Némethy, G., and Scheraga, H. A., *J. Comput. Chem.*, 7, 67, 1986.
5. Mao, B., *J. Am. Chem. Soc.*, 111, 6132, 1989.
6. Warne, N. W. and Laskowski, M., *Biochem. Biophys. Res. Commun.*, 172, 1364, 1990.
7. Rizo, J., Albericio, F., Romero, G., García-Echeverría, C., Claret, J., Muller, C., Giralt, E., and Pedroso, E., *J. Org. Chem.*, 53, 5386, 1988.
8. Hruby, V. J., *Life Sci.*, 31, 189, 1982.
9. Hruby, V. J., Al-Obeidi, F., and Kazmierski, W., *Biochem. J.*, 268, 249, 1990.
10. Bolin, D. R., Cottrell, J., Garippa, R., O' Neill, N., Simko, B., and O'Donell, M., *Int. J. Pept. Protein Res.*, 41, 124, 1993.
11. Pohl, M., Ambrosius, D., Grötzinger, J., Kretzschmar, T., Saunders, D., Wollmer, A., Brandenburg, D., Bitter-Suermann, D., and Höcker, H. *Int. J. Pept. Protein Res.*, 41, 362, 1993.
12. Wetzel, R., *Trends Biochem. Sci.*, 12, 478, 1987.
13. Hodges, R. S., Zhou, N. E., Kay, C. M., and Semchuk, P. D., *Pept. Res.*, 3, 123, 1990.
14. Kanaya, S., Katsuda, C., Kimura, S., Nakai, T., Kitakuni, E., Nakamura, H., Kitayanagi, K., Morikawa, K., and Ikehara, M., *J. Biol. Chem.*, 266, 6038, 1991.
15. du Vigneaud, V., Ressler, C., Swain, J. M., Roberts, C. W., Katsoyannis, P. G., and Gordon, S., *J. Am. Chem. Soc.*, 75, 4879, 1953.
16. Andreu, D., Albericio, F., Solé, N., Munson, M., Ferrer, M., and Barany, G., in *Methods in Molecular Biology, Vol. 35: Peptide Synthesis Protocols*, Pennington, M. W. and Dunn, B. M., Eds., Humana Press, Totowa, NJ, 1994, 91.
17. Creighton, T. E. and Goldenberg, D. P., *J. Mol. Biol.*, 179, 497, 1984.
18. Zhang, R. and Snyder, G. H., *J. Biol. Chem.*, 264, 18472, 1989.
19. Zhang, R. and Snyder, G. H., *Biochemistry*, 30, 11343, 1991.
20. Albericio, F., Hammer, R. P., García-Echeverría, C., Molins, M. A., Chang, J. L., Munson, M., Pons, M., Giralt, E., and Barany, G., *Int. J. Pept. Protein Res.*, 37, 402, 1991.
21. Falcomer, C. M., Meinwald, Y. C., Choudhary, I., Talluri, S., Milburn, P. J., Clardy, J., and Scheraga, H. A., *J. Am. Chem. Soc.*, 114, 4036, 1992.
22. Altmann, K.-H. and Scheraga, H. A., *J. Am. Chem. Soc.*, 112, 4926, 1990.
23. Noszál, B., Guo, W., and Rabenstein, D. L., *J. Org. Chem.*, 57, 2327, 1992.
24. Fiser, A., Cserzö, M., Tüdös, E., and Simon, I., *FEBS Lett.*, 302, 117, 1992.
25. Bodanszky, M. and Stahl, G. L., *Proc. Natl. Acad. Sci. U.S.A.*, 71, 2791, 1974.
26. Kao, P. N. and Karlin, A., *J. Biol. Chem.*, 261, 8085, 1986.
27. Marti, T., Rösselet, S. J., and Walsh, K. A., *Biochemistry*, 26, 8099, 1987.
28. Mosckovitz, R. and Gershoni, J. M., *J. Biol. Chem.*, 263, 1017, 1988.
29. Prorok, M. and Lawrence, D. S., *J. Am. Chem. Soc.*, 112, 8626, 1990.
30. Sukumaran, D. K., Prorok, M., and Lawrence, D. S., *J. Am. Chem. Soc.*, 113, 706, 1991.
31. Brady, S. F., Paleveda, W. J., Arison, B. H., Saperstein, R., Brady, E. J., Raynor, K., Reisine, T., Veber, D. F., and Freidinger, R. M., *Tetrahedron*, 49, 3449, 1993.
32. Chino, N., Yoshizawa, K., Kumagaye, K. Y., Noda, Y., Watanabe, T. X., Kimura, T., and Sakakibara, S., *Biochem. Biophys. Res. Commun.*, 141, 665, 1986.
33. Akaji, K., Fujino, K., Tatsumi, T., and Kiso, Y., *Tetrahedron Lett.*, 33, 1073, 1992.
34. Kangawa, K., Fukuda, A., and Matsuo, H., *Nature (London)*, 313, 397, 1985.
35. Kellenberger, C., Hietter, H., Trifilieff, E., and Luu, B., in *Innovation and Perspectives in Solid Phase Synthesis. Peptides, Proteins and Nucleic Acids. Biological and Biomedical Applications*, Epton, R., Ed., Mayflower Worldwide Ltd, Birmingham, 1994, 567.
36. Nieto, A., Ponstingl, H., and Beato, M., *Arch. Biochem. Biophys.*, 180, 82, 1977.
37. Morize, I., Surcouf, E., Vaney, M. C., Epelboin, Y., Buehner, M., Fridlansky, F., Milgrom, E., and Mornon, J. P., *J. Mol. Biol.*, 194, 725, 1987.

38. An, S. S. A., Jiménez-Barbero, J., Petersen, T. E., and Llinás, M., *Biochemistry*, 31, 9927, 1992.
39. Wünsch, E., Moroder, L., Göhring-Romani, S., Musiol, H.-J., Göhring, W., and Bovermann, G., *Int. J. Pept. Protein Res.*, 32, 368, 1988.
40. Hiskey, R. G., Davis, G. W., Safdy, M. F., Inui, T., Upham, R. A., and Jones, W. C., *J. Org. Chem.*, 35, 4148, 1970.
41. Royo, M., Albericio, F., Giralt, E., and Pons, M., in *Peptides 1992. Proceedings of the 22nd European Peptides Symposium*, Schneider, C. H. and Eberle, A. N., Ed., ESCOM, Leiden, 1993, 487.
42. García-Echeverría, C., Albericio, F., Giralt, E., and Pons, M., *J. Am. Chem. Soc.*, 115, 11663, 1993.
43. Ruiz-Gayo, M., Albericio, F., Pons, M., Royo, M., Pedroso, E., and Giralt, E., *Tetrahedron Lett.*, 29, 3845, 1988.
44. Ruiz-Gayo, M., Royo, M., Fernández, I., Albericio, F., Giralt, E., and Pons, M., *J. Org. Chem.*, 58, 6319, 1993.
45. Munson, M. C., Lebl, M., Slaninová, J., and Barany, G., *Pept. Res.*, 6, 155, 1993.
46. Ryle, A. P., Sanger, F., Smith, L. F., and Kitai, R., *Biochem. J.*, 60, 541, 1955.
47. Sieber, P., Kamber, B., Hartmann, A., Jöhl, A., Riniker, B., and Rittel, W., *Helv. Chim. Acta*, 57, 2617, 1974.
48. Sieber, P., Kamber, B., Hartmann, A., Jöhl, A., Riniker, B., and Rittel, W., *Helv. Chim. Acta*, 60, 27, 1977.
49. Sieber, P., Eisler, K., Kamber, B., Riniker, B., Rittel, W., Märki, F., and de Gasparo, M., *Hoppe-Seyler's Z. Physiol. Chem.*, 359, 113, 1978.
50. Maruyama, K., Nagata, K., Tanaka, M., Nagasawa, H., Isogai, A., Ishizaki, H., and Suzuki, A., *J. Protein Chem.*, 11, 1, 1992.
51. Maruyama, K., Nagasawa, H., Isogai, A., Ishikazi, H., and Suzuki, A., *J. Protein Chem.*, 11, 13, 1992.
52. Nagata, K., Maruyama, K., Nagasawa, H., Urushibata, I., Isogai, A., Ishizaki, H., and Suzuki, A., *Peptides*, 13, 653, 1992.
53. Hodges, R. S., Semchuk, P. D., Taneja, A. K., Kay, C. M., Parker, J. M. R., and Mant, C. T., *Pept. Res.*, 1, 19, 1988.
54. Zhou, N. E., Kay, C. M., and Hodges, R. S., *J. Biol. Chem.*, 267, 2664, 1992.
55. Monera, O. D., Zhou, N. E., Kay, C. M., and Hodges, R. S., *J. Biol. Chem.*, 268, 19218, 1993.
56. Wünsch, E., in *Methoden der Organischen Chemie (Houben-Weyl)*, Vol. 15/1, Müller, E., Ed., Georg Thieme Verlag, Stuttgart, 1974, 469.
57. Barany, G. and Merrifield, R. B., in *The Peptides. Analysis, Synthesis, Biology*, Vol. 2, *Special Methods in Peptide Synthesis, Part A*, Gross, E. and Meienhofer, J., Eds., Academic Press, New York, 1979, 1.
58. Hiskey, R. G., in *The Peptides. Analysis, Synthesis, Biology*, Vol. 3, *Protection of Functional Groups in Peptide Synthesis*, Gross, E. and Meienhofer, J., Eds., Academic Press, New York, 1981, 137.
59. Cavelier, F., Daunis, J., and Jacquier, R., *Bull. Soc. Chim.*, 788, 1989.
60. Cavelier, F., Daunis, J., and Jacquier, R., *Bull. Soc. Chim.*, 210, 1990.
61. Fields, G. B., Tian, Z., and Barany, G., in *Synthetic Peptides. A User's Guide*, Grant, G. A., Ed., W. H. Freeman, New York, 1992, 77.
62. Kuromizu, K. and Meienhofer, J., *J. Am. Chem. Soc.*, 96, 4978, 1974.
63. Felix, A. M., Jimenez, M. H., Mowles, T., and Meienhofer, J., *Int. J. Pept. Protein Res.*, 11, 329, 1978.
64. Yajima, H., Fujii, N., Funakoshi, S., Watanabe, T., Murayama, E., and Otaka, A., *Tetrahedron*, 44, 805, 1988.
65. Erickson, B. W. and Merrifield, R. B., *J. Am. Chem. Soc.*, 95, 3750, 1973.
66. Heath, W. F., Tam, J. P., and Merrifield, R. B., *Int. J. Pept. Protein Res.*, 28, 498, 1986.
67. Akabori, S., Sakakibara, S., Shimonishi, Y., and Nobuhara, Y., *Bull. Chem. Soc. Jpn.*, 37, 433, 1964.
68. Nishimura, O., Kitada, C., and Fujino, M., *Chem. Pharm. Bull.*, 26, 1576, 1978.
69. Platen, M. and Steckhan, E., *Liebigs Ann. Chem.*, 1563, 1984.
70. Fujii, N., Otaka, A., Watanabe, T., Okamachi, A., Tamamura, H., Yajima, H., Inagaki, Y., Nomizu, M., and Asano, K., *J. Chem. Soc. Chem. Commun.*, 283, 1989.

71. Bernatowicz, M. S., Matsueda, R., and Matsueda, G. R., *Int. J. Pept. Protein Res.*, 28, 107, 1986.
72. Albericio, F., Andreu, D., Giralt, E., Navalpotro, C., Pedroso, E., Ponsati, B., and Ruiz-Gayo, M., *Int. J. Pept. Protein Res.*, 34, 124, 1989.
73. Simmonds, R. G., Tupper, D. E., and Harris, J. R., *Int. J. Pept. Protein Res.*, 43, 363, 1994.
74. Ploux, O., Chassaing, G., and Marquet, A., *Int. J. Pept. Protein Res.*, 29, 162, 1987.
75. Matsueda, R., Kimura, T., Kaiser, E. T., and Matsueda, G. R., *Chem. Lett.*, 737, 1981.
76. Ridge, R. J., Matsueda, G. R., Haber, E., and Matsueda, R., *Int. J. Pept. Protein Res.*, 19, 490, 1982.
77. Matsueda, R., Higashida, S., Albericio, F., and Andreu, D., *Pept. Res.*, 5, 262, 1992.
78. Schroll, A. L. and Barany, G., *J. Org. Chem.*, 54, 244, 1989.
79. Bodanszky, M. and Bednarek, M. A., *Int. J. Pept. Protein Res.*, 20, 434, 1982.
80. Ruiz-Gayo, M., Albericio, F., Pedroso, E., and Giralt, E., *J. Chem. Soc. Chem. Commun.*, 1501, 1986.
81. Albericio, F., Nicolás, E., Rizo, J., Ruiz-Gayo, M., Pedroso, E., and Giralt, E., *Synthesis*, 119, 1990.
82. Albericio, F., Royo, M., Munson, M. C., Solé, N. A., van Abel, R. J., Alsina, J., García-Echeverría, C., Slomczynska, U., Eritja, R., Giralt, E., and Barany, G., in *Peptide Chemistry 1992. Proceedings of the 2nd Japan Symposium on Peptide Chemistry*, Yanaihara, N., Ed., ESCOM, Leiden, 1993, 19.
83. Royo, M., García-Echeverría, C., Giralt, E., Eritja, R., and Albericio, F., *Tetrahedron Lett.*, 33, 2391, 1992.
84. McCurdy, S. N., *Pept. Res.*, 2, 147, 1989.
85. Kamber, B. and Rittel, W., *Helv. Chim. Acta*, 51, 2061, 1968.
86. Photaki, I., Taylor-Papadimitriou, J., Sakarellas, C., Mazarakis, P., and Zervas, L., *J. Chem. Soc. (C)*, 2683, 1970.
87. Kamber, B., Hartmann, A., Eisler, K., Riniker, B., Rink, H., Sieber, P., and Rittel, W., *Helv. Chim. Acta*, 63, 899, 1980.
88. Fujii, N., Okada, A., Funakoshi, S., Bessho, K., Watanabe, T., Akaji, K., and Yajima, H., *Chem. Pharm. Bull.*, 35, 2339, 1987.
89. Frankel, A. D., Biancalana, S., and Hudson, D., *Proc. Natl. Acad. Sci. U.S.A.*, 86, 7397, 1989.
90. Munson, M. C., García-Echeverría, C., Albericio, F., and Barany, G., in *Peptides. Chemistry and Biology. Proceedings of the 12th American Peptides Symposium*, Smith, J. A. and Rivier, J. E., Eds., ESCOM, Leiden, 1992, 605.
91. Munson, M. and Barany, G., *J. Am. Chem. Soc.*, 115, 10203, 1993.
92. Munson, M. C., García-Echeverría, C., Albericio, F., and Barany, G., *J. Org. Chem.*, 57, 3013, 1992.
93. Tam, J. P. and Shan, Z.-Y., *Int. J. Pept. Protein Res.*, 39, 464, 1992.
94. Atherton, E., Pinori, M., and Sheppard, R. C., *J. Chem. Soc. Perkin Trans. 1*, 2057, 1985.
95. Veber, D. F., Milkowski, J. D., Denkewalter, R. G., and Hirschmann, R., *Tetrahedron Lett.*, 3057, 1968.
96. Veber, D. F., Milkowski, J. D., Varga, S. L., Denkewalter, R. G., and Hirschmann, R., *J. Am. Chem. Soc.*, 94, 5456, 1972.
97. Kamber, B., *Helv. Chim. Acta*, 54, 927, 1971.
98. Liu, W., Shiue, G. H., and Tam, J. P., in *Peptides. Chemistry, Structure and Biology. Proceedings of the 11th American Peptides Symposium*, Rivier, J. E. and Marshall, G. R., Eds., ESCOM, Leiden, 1990, 271.
99. Kiso, Y., Yoshida, M., Kimura, T., Fujiwara, Y., and Shimokura, M., *Tetrahedron Lett.*, 30, 1979, 1989.
100. Greiner, G. and Hermann, P., in *Peptides 1990. Proceedings of the 21st European Peptides Symposium*, Giralt, E. and Andreu, D., Eds., ESCOM, Leiden, 1991, 277.
101. Hermann, P. and Schillings, T., in *Peptides 1992. Proceedings of the 22nd European Peptides Symposium*, Schneider, C. H. and Eberle, A. N., Eds., ESCOM, Leiden, 1993, 411.
102. Royo, M., Alsina, J., Giralt, E., Slomczynska, U., and Albericio, F., in *Peptides. Chemistry, Structure and Biology. Proceedings of the 13th American Peptides Symposium*, Hodges, R. S. and Smith, J. A., Eds., ESCOM, Leiden, 1994, 116.
103. Royo, M., Alsina, J., Giralt, E., Slomczynska, U., and Albericio, F., *J. Chem. Soc. Perkin Trans. 1*, 1095, 1995.
104. Moroder, L., Gemeiner, M., Goehring, W., Jaeger, E., Thamm, P., and Wünsch, E., *Biopolymers*, 20, 17, 1981.

105. van Rietschoten, J., Grabier, C., Rochat, H., Lissitzky, S., and Miranda, F., *Eur. J. Biochem.*, 56, 35, 1975.

106. van Rietschoten, J., Pedroso Muller, E., and Granier, C., in *Peptides. Proceedings of the 5th American Peptides Symposium*, Goodman, M. and Meienhofer, J., Eds., John Wiley & Sons, New York, 1977, 522.

107. Weber, U. and Hartter, P., *Hoppe-Seyler's Z. Physiol. Chem.*, 351, 1384, 1970.

108. Eritja, R., Zieher-Martin, J. P., Walker, P. A., Lee, T. D., Legesse, K., Albericio, F., and Kaplan, B. D., *Tetrahedron*, 43, 2675, 1987.

109. Chang, C.-D., Felix, A. M., Jimenez, M. H., and Meienhofer, J., *Int. J. Pept. Protein Res.*, 15, 485, 1980.

110. Sakakibara, S., Shimonishi, Y., Kishida, Y., Okada, M., and Sugihara, H., *Bull. Chem. Soc. Jpn.*, 40, 2164, 1967.

111. Pastuszak, J. J. and Chimiak, A., *J. Org. Chem.*, 46, 1868, 1981.

112. Singh, R. and Whitesides, G. M., *J. Org. Chem.*, 56, 2332, 1991.

113. Ranganathan, S. and Jayaraman, N., *J. Chem. Soc. Chem. Commun.*, 934, 1991.

114. Lamoureux, G. V. and Whitesides, G. M., *J. Org. Chem.*, 58, 633, 1993.

115. Tarbell, D. S., in *Organic Sulfur Compounds*, Vol. 1, Kharasch, N., Ed., Pergamon Press, Oxford, 1961, 97.

116. Hope, D. B., Murti, V. V. S., and du Vigneaud, V., *J. Biol. Chem.*, 237, 1563, 1962.

117. Live, D. H., Agosta, W. C., and Cowburn, D., *J. Org. Chem.*, 42, 3556, 1977.

118. Rivier, J. E., Kaiser, R., and Galyean, R., *Biopolymers*, 17, 1927, 1978.

119. Zonta, A. and Adermann, K., in *Peptides 1992. Proceedings of the 22nd European Peptides Symposium*, Schneider, C. H. and Eberle, A. N., Eds., ESCOM, Leiden, 1993, 397.

120. Wallace, T. J. and Mahon, J. J., *J. Org. Chem.*, 30, 1502, 1965.

121. Tam, J. P., Cunningham-Rundles, W. F., Erickson, B. W., and Merrifield, R. B., *Tetrahedron Lett.*, 4001, 1977.

122. Tam, J. P., Wu, C.-R., Liu, W., and Zhang, J.-W., *J. Am. Chem. Soc.*, 113, 6657, 1991.

123. Ferrer, M., Woodward, C., and Barany, G., *Int. J. Pept. Protein Res.*, 40, 194, 1992.

124. Fujii, N., Otaka, A., Okamachi, A., Watanabe, T., Arai, H., Tamamura, H., Funakoshi, S., and Yajima, H., in *Peptides 1988. Proceedings of the 20th European Peptides Symposium*, Jung, G. and Bayer, E., Eds., Walter de Gruyter, Berlin, 1989, 58.

125. Hantgan, R. R., Hammes, G. G., and Scheraga, H., *Biochemistry*, 13, 3421, 1974.

126. Ahmed, A. K., Schaffer, S. W., and Wetlaufer, D. B., *J. Biol. Chem.*, 250, 8477, 1975.

127. Rothwarf, D. M. and Scheraga, H. A., *Biochemistry*, 32, 2671, 1993.

128. Pigiet, V. P. and Schuster, B. J., *Proc. Natl. Acad. Sci. U.S.A.*, 83, 7643, 1986.

129. Hillson, D. A., Lambert, N., and Freedman, R. B., *Methods Enzymol.*, 107, 281, 1984.

130. Edman, J. C., Ellis, L., Blacher, R. W., Roth, R. A., and Rutter, W. J., *Nature (London)*, 317, 267, 1985.

131. Lyles, M. M. and Gilbert, H. F., *Biochemistry*, 30, 613, 1991.

132. Lyles, M. M. and Gilbert, H. F., *Biochemistry*, 30, 619, 1991.

133. Inui, T., Kubo, S., Bodí, J., Kimura, T., and Sakakibara, S., in *Peptides 1994. Proceedings of the 23rd European Peptides Symposium*, Maia, H. L. S., Ed., ESCOM, Leiden, 1995, 36.

134. Inui, T., Bódi, J., Kubo, S., Nishio, H., Kimura, T., Kojima, S., Maruta, H., Muramatsu, T., and Sakakibara, S., *J. Pept. Sci.*, 2, 28, 1996.

135. Mazur, S. and Jayalekshmy, P., *J. Am. Chem. Soc.*, 101, 677, 1979.

136. García-Echeverría, C., Albericio, F., Pons, M., Barany, G., and Giralt, E., *Tetrahedron Lett.*, 30, 2441, 1989.

137. Gray, W. R., Rivier, J. E., Galyean, R., Cruz, L. J., and Olivera, B. M., *J. Biol. Chem.*, 258, 12247, 1983.

138. Inukai, N., Nakano, K., and Murakami, M., *Bull. Chem. Soc. Jpn.*, 41, 182, 1968.

139. Buchta, R., Bondi, E., and Fridkin, M., *Int. J. Pept. Protein Res.*, 28, 289, 1986.

140. Brady, S. F., Freidinger, R. M., Paleveda, W. J., Colton, C. D., Homnick, C. F., Whitter, W. L., Curley, P., Nutt, R. F., and Veber, D. F., *J. Org. Chem.*, 52, 764, 1987.

141. Ruiz-Gayo, M., Albericio, F., Royo, M., García-Echeverría, C., Pedroso, E., Pons, M., and Giralt, E., *An. Quim.*, 85C, 116, 1989.

142. Bishop, P. and Chmielewski, J., *Tetrahedron Lett.*, 33, 6263, 1992.
143. Shih, H., *J. Org. Chem.*, 58, 3003, 1993.
144. Casaretto, R. and Nyfeler, R., in *Peptides 1990. Proceedings of the 21st European Peptides Symposium*, Giralt, E. and Andreu, D., Eds., ESCOM, Leiden, 1991, 181.
145. Cotton, R., Dutta, A. S., Giles, M. B., and Hayward, C. F., in *Peptides. Chemistry and Biology. Proceedings of the 12th American Peptides Symposium*, Smith, J. A. and Rivier, J. E., Eds., ESCOM, Leiden, 1992, 639.
146. Lamthanh, H., Romestand, C., Deprun, C., and Ménez, A., *Int. J. Pept. Protein Res.*, 41, 85, 1993.
147. Koide, T., Otaka, A., Suzuki, H., and Fujii, N., *Synlett*, 345, 1991.
148. Funakoshi, S., Murayama, E., Guo, L., Fujii, N., and Yajima, H., *J. Chem. Soc. Chem. Commun.*, 382, 1988.
149. Seidel, C., Klein, C., Empl, B., Bayer, E., Lin, M., and Batz, H.-G., in *Peptides 1990. Proceedings of the 21st European Peptides Symposium*, Giralt, E. and Andreu, D., Eds., ESCOM, Leiden, 1991, 236.
150. Albrecht, E., Harada, Y., Cooper, G. J. S., Jones, H., and Lehman de Gaeta, L., in *Peptides. Chemistry and Biology. Proceedings of the 12th American Peptides Symposium*, Smith, J. A. and Rivier, J. E., Eds., ESCOM, Leiden, 1992, 441.
151. Canas, M., Jodas, G., Albericio, F., Andreu, D., García-Anton, J. M., Parente, A., and Ponsati, B., in *Peptides 1992. Proceedings of the 22nd European Peptides Symposium*, Schneider, C. H. and Eberle, A. N., Eds., ESCOM, Leiden, 1993, 401.
152. Camarero, J. A., Giralt, E., and Andreu, D., *Tetrahedron Lett.*, 36, 1137, 1995.
153. Adeva, A., Camarero, J., Giralt, E., and Andreu, D., *Tetrahedron Lett.*, 36, 3885, 1995.
154. Kullman, W. and Gutte, B., *Int. J. Pept. Protein Res.*, 12, 17, 1978.
155. Kamber, B., *Helv. Chim. Acta*, 56, 1371, 1973.
156. Nokihara, K. and Berndt, H., *J. Org. Chem.*, 43, 4893, 1978.
157. Le-Nguyen, D. and Rivier, J., *Int. J. Pept. Protein Res.*, 27, 285, 1986.
158. Brois, S. J., Pilot, J. F., and Barnum, H. W., *J. Am. Chem. Soc.*, 92, 7629, 1970.
159. Rabanal, F., DeGrado, W. F., and Dutton, P. L., *Tetrahedron Lett.*, 37, 1347, 1996.
160. Moroder, L., Marchiori, F., Borin, G., and Scoffone, E., *Biopolymers*, 12, 493, 1973.
161. Castell, J. V. and Tung-Kyi, A., *Helv. Chim. Acta*, 62, 2507, 1979.
162. Crimmins, D. L., *Pept. Res.*, 2, 395, 1989.
163. Romani, S., Moroder, L., Göhring, W., Scharf, R., Wünsch, E., Barde, Y. A., and Thoenen, H., *Int. J. Pept. Protein Res.*, 29, 107, 1987.
164. Mukaiyama, T. and Takahashi, K., *Tetrahedron Lett.*, 5907, 1968.
165. Wünsch, E. and Romani, S., *Hoppe-Seyler's Z. Physiol. Chem.*, 363, 449, 1982.
166. Fujii, N., Otaka, A., Watanabe, T., Arai, H., Funakoshi, S., Bessho, K., and Yajima, H., *J. Chem. Soc. Chem. Commun.*, 1676, 1987.
167. Fujii, N., Watanabe, T., Aotake, T., Otaka, A., Yamamoto, I., Konishi, J., and Yajima, H., *Chem. Pharm. Bull.*, 36, 3304, 1988.
168. Drijfhout, J. W., Perdijk, E. W., Weijer, W. J., and Bloemhoff, W., *Int. J. Pept. Protein Res.*, 32, 161, 1988.
169. Ponsati, B., Ruiz-Gayo, M., Giralt, E., Albericio, F., and Andreu, D., *J. Am. Chem. Soc.*, 112, 5345, 1990.
170. Baleux, F. and Dubois, P., *Int. J. Pept. Protein Res.*, 40, 7, 1992.
171. Kimura, T., Matsueda, R., Nakagawa, Y., and Kaiser, E. T., *Anal. Biochem.*, 122, 274, 1982.
172. Zahn, H. and Schmidt, G., *Tetrahedron Lett.*, 5095, 1967.
173. Zahn, H. and Schmidt, G., *Liebigs Ann. Chem.*, 731, 101, 1970.
174. Zahn, H. and Schmidt, G., *Liebigs Ann. Chem.*, 731, 91, 1970.
175. Büllesbach, E. E., *Kontakte (Darmstadt)*, 1, 21, 1992.
176. Moroder, L., Besse, D., Musiol, H.-J., Rudolph-Böhner, S., and Siedler, F., *Biopolymers (Pept. Sci.)*, 40, 207, 1996.
177. Nishiuchi, N. and Sakakibara, S., *FEBS Lett.*, 148, 260, 1982.
178. Ponsati, B., Giralt, E., and Andreu, D., *Tetrahedron*, 46, 8255, 1990.
179. Shimonishi, Y., Hudaka, Y., Koizumi, M., Hane, M., Aimoto, S., Takeda, T., Miwatani, T., and Takeda, Y., *FEBS Lett.*, 215, 165, 1987.

180. Atherton, E., Sheppard, R. C., and Ward, P., *J. Chem. Soc. Perkin Trans. 1*, 2065, 1985.
181. Sieber, P., Kamber, B., Eisler, K., Hartmann, A., Riniker, B., and Rittel, W., *Helv. Chim. Acta*, 59, 1489, 1976.
182. Kamber, B., Riniker, B., Sieber, P., and Rittel, W., *Helv. Chim. Acta*, 59, 2830, 1976.
183. Büllesbach, E. E. and Schwabe, C., *J. Biol. Chem.*, 266, 10754, 1991.

Chapter 6

Peptide Libraries

Over the last decade or so there has been an escalation in the demand for synthetic peptides for a variety of purposes. The search for the most-promising biologically active peptide analogues, for example, and the development of vaccines and the elucidation of the factors influencing the overall three-dimensional structure of proteins — all require the synthesis and evaluation of many different peptide molecules. The limiting factor in such studies has often been the time and effort required for peptide synthesis, since the traditional methods that involve the elaboration of one molecule at a time are simply too slow. As a consequence, attention has focused more and more on the problem of how large numbers of peptides can be produced quickly, reliably, and economically. Biochemical and chemical approaches to achieving this have evolved in parallel, and the display of vast numbers of peptides on bacteriophage particles is one of the techniques that have been developed.[1] In this chapter, however, only chemical methods for the generation of large numbers of peptides are considered.

Since SPPS lends itself easily to automation, the first steps taken were to design and produce machines with increased numbers of reaction vessels, so that different peptides could be synthesized in parallel.[2-8] Automated synthesizers that can elaborate up to 50 different molecules simultaneously are now available, and this greatly increases the availability of individual peptides. However, the number of possible analogues of even quite short peptides, even if only the proteinogenic amino acids are considered, is immense. There are, for example, no less than 64 million (20^6) possible permutations of the 20 DNA-encoded amino acids in a hexapeptide. If other amino acids such as the D-enantiomers or nonproteinogenic amino acids are also taken into consideration, then the number of possible permutations becomes truly astronomical. Clearly, conventional synthetic methods (even those allowing the parallel synthesis of hundreds of peptides) are not up to the task.

A way around the problem is to synthesize peptide mixtures (known as peptide libraries) in a controlled manner, rather than attempting to synthesize each peptide separately. The controlled production of mixtures of large numbers of compounds is often referred to as the generation of molecular (or chemical) diversity, and the chemical methods used to do this have come to be known as combinatorial synthesis. These methods stand in sharp contrast to the traditional ones of organic chemistry, which are directed toward the preparation of a single, target molecule in as pure a state as possible. Combinatorial synthesis is not, however, simply a question of quantity rather than quality. If it is to provide meaningful results, a peptide library must be very carefully designed and must be synthesized using highly optimized chemistry.

6.1 MULTIPLE-PEPTIDE SYNTHESIS IN PARALLEL

Two different peptides can be been synthesized in parallel by taking advantage of the differing physical properties of polystyrene-1% divinylbenzene and Kel-F-g-styrene.[9] For all common amino acid–coupling and N^α-deprotection steps, peptide synthesis takes place on a mixture of these solid supports. When the peptides are to be differentiated, however, they can be separated by flotation since the resins have different densities. This allows each chain to be elaborated individually, when necessary. The resins can then be recombined for further common synthetic operations.

Conceptually, however, the most straightforward method for multiple peptide synthesis is to elaborate the various chains in parallel, each one on its own solid support in a separate reaction vessel. This being so, both the structure and location of each of the target peptides is known and can be confirmed by standard analytical techniques. Multiple-peptide synthesis is said to be spatially addressable if the location of a given peptide structure can be known precisely. Most automated synthesizers function in this way, the different peptides having their own batches of resin in separate reaction vessels. Reagents and wash solvents are delivered to each one, as necessary, and the whole process is controlled by a dedicated microcomputer.

However, since the majority of steps in simultaneous peptide synthesis, such as washing, N^α deprotection, and even many of the amino acid–coupling reactions, are often common to all of the different peptides, a separate reaction vessel for each peptide is not always necessary. As long as any given portion of the solid support carrying a given individual peptide can be separated from all other portions when necessary, then all other operations can be performed in the same reaction vessel. Solid support portions must be separated when a variation in procedure unique to a given peptide is carried out. This might be, for example, the incorporation or omission of a specific amino acid, among other possibilities. As long as the various batches of the solid support, each carrying its own different molecule, are adequately labeled, then the structure and location of any given peptide can still be known.

6.1.1 Peptide Synthesis on Polyethylene Pins

The origins of peptide library synthesis can be traced back to Geysen[10-13] who, in an effort to develop a more efficient method for epitope mapping, designed an apparatus for the parallel synthesis of 96 peptides. Epitopes, also known as determinants, are regions of a protein antigen that are recognized by antibodies. Antigens are foreign molecules that provoke an immunological response, usually in the form of the production of defense proteins known as antibodies. There are two main classes of epitope, known as continuous and discontinuous, respectively. The former are short, continuous amino acid sequences of the protein in question, whereas discontinuous epitopes consist of groups of amino acids that are distant in the linear sequence but that can be brought into close proximity by the folding of the protein molecule. In Geysen's original method, the goal was to locate and identify (known as mapping) continuous epitopes of the VP1 protein of the coat of a strain of the foot-and-mouth disease virus. This protein consists of 213 amino acid residues and its amino acid sequence was subdivided into hexapeptide units, as shown in Scheme 6.1.

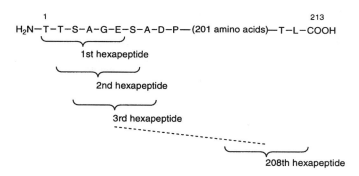

SCHEME 6.1   Epitope mapping using multiple peptide synthesis in parallel on polyethylene pins.

Epitope mapping required that each of these 208 different overlapping hexapep-tides be synthesized and screened. An apparatus was designed, consisting of an array of 96 polyethylene pins, each measuring 40 by 4 mm, mounted in a supporting block, so that each one fitted into one of the 96 wells of a standard microtiter plate. The heads of the pins were coated with a functionalized polymeric material, upon which peptide synthesis was to take place. Each well of the microtiter plate then serves as a separate reaction vessel for the amino acid–coupling steps, although washing and deprotection operations may be carried out in baths that can accom-modate the entire rack of pins.

In the original method, polyacrylic acid was radiation-grafted onto the heads of the pins and anchoring of all of the peptides was carried out by amide bond formation between the side chain N^{ε}-amino group of Lys derivative **1** and the carboxylic acid groups of the support. After N^{α}-Boc group removal, Ala was attached to each pin so that all peptides had the sequence Ala-Lys at the *C*-terminus, acting as a spacer between the desired peptide sequences to be synthesized and the solid support, as in **2**. SPPS was then carried out using normal Boc/Bzl protocols. When all peptides had seen synthesized (a different peptide on each of the 96 pins), acidolytic removal of all side chain–protecting groups was brought about by treat-ment of the whole rack with boron tris(trifluoroacetate) in trifluoroacetic acid. Although the procedure affords peptides of the type **3**, still attached to the solid support, their interaction with a solution of an antiserum against the foot-and-mouth disease virus particle could be determined by the enzyme-linked immunosorbent assay (ELISA) technique.[14] In this assay an enzyme such as alkaline phosphatase is attached to an acceptor molecule, which is often a monoclonal antibody. The enzyme can then catalyze a reaction with a dye, such that those pins having peptides that have interacted with the antibody become colored, allowing their easy visual iden-tification (see Section 6.2.6 below for a discussion of library screening). The pro-cedure for peptide synthesis on polyethylene pins is summarized in Scheme 6.2.

The results of the ELISA tests showed that hexapeptides 146 (GDLQVL) and 147 (DLQVLA) in Scheme 6.1 bound the antibody most strongly. This indicated either that the epitope was contained in the five amino acids common to the two peptides, that is to say, the pentapeptide DLQVL or, alternatively, that it was the sum of the two, the heptapeptide GDLQVLA. In order to distinguish more clearly

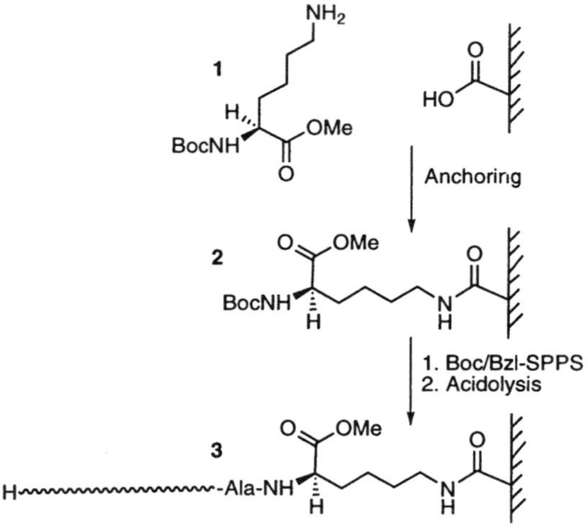

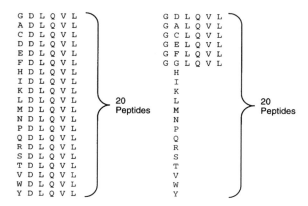

SCHEME 6.2 Peptide synthesis on the heads of polyethylene–polyacrylic acid-grafted pins.

between these two possibilities, all of the hexapeptides that differed from peptide 146 (GDLQVL) by only a single amino acid in each position were synthesized. Each of the other 19 proteinogenic amino acids was substituted in each of the six amino acid positions within the peptide. This gave a total of 120 peptides, 6 of which were exact copies of the original sequence. The different possibilities produced by substitution of the 20 proteinogenic amino acids at the first two positions of hexapeptide 146 are shown in Scheme 6.3.

All of these peptides were then screened by the ELISA technique. Positions at which all, or at least the majority of substitutions, resulted in the loss of antibody-binding activity indicate residues that are important for the specificity and binding to the antibody. To determine the contribution of the Ala residue at the *C*-terminus of peptide 147 (DLQVLA), a further 20 peptides were synthesized. These consisted

G D L Q V L			G D L Q V L	
A D L Q V L			G A L Q V L	
C D L Q V L			G C L Q V L	
D D L Q V L			G E L Q V L	
E D L Q V L			G F L Q V L	
F D L Q V L			G G L Q V L	
H D L Q V L			H	
I D L Q V L			I	
K D L Q V L			K	
L D L Q V L	20		L	20
M D L Q V L	Peptides		M	Peptides
N D L Q V L			N	
P D L Q V L			P	
Q D L Q V L			Q	
R D L Q V L			R	
S D L Q V L			S	
T D L Q V L			T	
V D L Q V L			V	
W D L Q V L			W	
Y D L Q V L			Y	

SCHEME 6.3 Single-residue substitution at the first two positions of the hexapeptide GDLQVL.

of the complete sequence of peptide 146 (GDLQVL) with one of the possible 19 remaining proteinogenic amino acids added at the C-terminus. The results of the ELISA assays indicated that the epitope is represented by the 146-152 heptapeptide sequence (GDLQVLA). Furthermore, it was possible to affirm that the Leu residues at positions 148 and 151 were essential for reaction with antisera raised against the intact virus while the Gln and Ala residues at positions 149 and 152, respectively, contributed to a lesser extent.

Although good results were obtained in the case discussed above, and are often obtained in others when the peptides remain bound to the solid support, the production of free peptides in solution is desirable in a number of situations. This can be achieved by modification of the method, in particular by introducing a handle between the solid support and the peptides to be synthesized, as in conventional linear SPPS (see Chapter 2 for a detailed discussion). It is advisable, however, that the removal of side chain protection be carried out as a separate step before cleavage from the solid support, so that the necessity of purifying each and every one of the 96 different peptides produced is circumvented. This is more easily done using Fmoc/tBu SPPS, since acidolytic removal of side chain protection, without concomitant cleavage of the peptide from the resins can be brought about under milder conditions. Appropriate choice of handle is, of course, necessary in order to achieve this.

The benzyl-based handles and 4, 5, and 6 have all been applied in multiple-peptide synthesis on pins.[15,16] Free peptides can be cleaved from supports incorporating 4, after protecting group removal, either by ammonolysis with ammonia vapor or saponification on treatment with sodium hydroxide solutions. In the former case, peptide amides are produced and, in the latter, peptide acids. The use of handles 5 and 6 allows free peptide acids and amides, respectively, to be obtained directly on trifluoroacetic acid treatment of the peptide–resin. With these handles, side chain group removal and cleavage occur concomitantly, however, so that a purification step of some kind is usually required after cleavage.

An alternative method, which allows the production of free-peptide C-terminal analogues, was developed by Geysen and is based upon the detachment of peptides from the solid support by diketopiperazine formation.[17-21] Peptides are synthesized using the Fmoc/tBu approach on a solid support incorporating a benzyl-based handle like 4, in such a way that all have the C-terminal sequence Lys-Pro, as in 7. Chain elongation takes place at the amino group of the Lys side chain, giving peptide–resins such as 8. After acidolytic removal of protecting groups, treatment of the deprotected

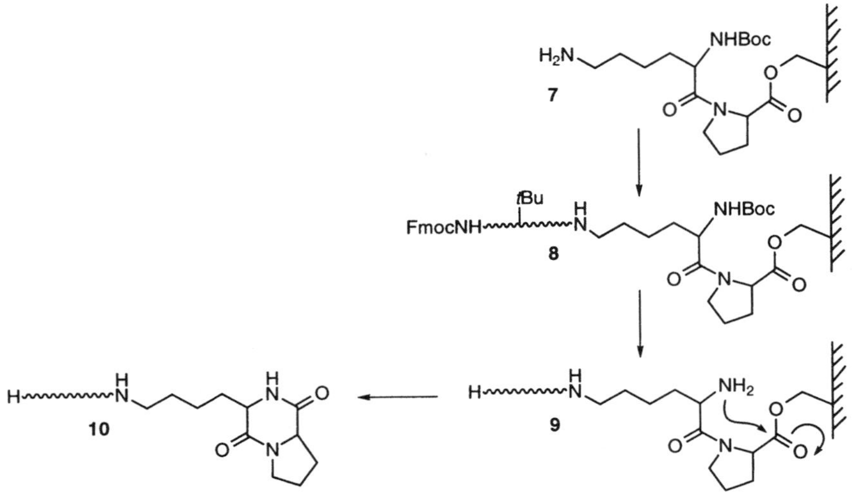

SCHEME 6.4 Cleavage of peptides from polyethylene pins by diketopiperazine formation.

peptide–resin **9** with mild base brings about diketopiperazine formation (see Section 2.4.2.1.1) and detachment of the peptide derivative **10**, as outlined in Scheme 6.4. The presence of the *C*-terminus diketopiperazine apparently does not affect biological activity in many cases.

Free peptides have also been produced using the pin method by introducing special enzymatic or acidolytic cleavage points.[22,23] If all peptides are synthesized with the *C*-terminal sequence Asp-Pro-Gly, then acidolytic cleavage of the Asp–Pro bond (see Section 2.5.2.3) can be brought about on treatment with formic acid. If the Lys-Gly sequence is incorporated at the *C*-terminus, then enzymatic cleavage of free peptides with trypsin is possible.

6.1.2 Peptide Synthesis in Tea Bags

In the "tea bag" method for simultaneous multiple-peptide synthesis, developed by Houghten,[24-26] batches of resin (usually about 100 mg) are sealed inside labeled, porous polypropylene packets, measuring 3 by 4 cm. In principle, all of the solid supports, handles, and protection strategies commonly applied in SPPS can be used. Amino acids are coupled to the resins by placing the bags in solutions of the appropriately protected and activated individual amino acid monomers. All common steps, such as resin washing and N^α-amino group deprotection, on the other hand, are usually performed simultaneously in a single reaction vessel. At the end of the synthesis each tea bag contains a single peptide sequence and the peptides may be cleaved from the resin using a multiple cleavage apparatus.[27] Partial automation has allowed the simultaneous synthesis of as many as 150 peptides.

The advantages of the method are that it offers considerable synthetic flexibility and that peptides can be produced in sufficient quantities (500 μmol) for purification and complete characterization, if desired. As a consequence, the tea bag method has been used quite extensively in methodological investigations. The effects of different bases for neutralization, different coupling reagents, and different methods for capping unreactive terminal N^α-amino groups have been systematically assessed.[28,29]

Multiple-peptide synthesis in parallel using tea bags was initially used[24] to prepare a series of replacement analogues of a 13-residue peptide corresponding to the 75-110 region of the hemagglutinin (HA1) protein. This peptide has the sequence YPYDVPDYASLRS and each of the 20 proteinogenic amino acids was systematically substituted at each position in the sequence, giving a total of 247 different peptides and 13 copies of the original sequence (260 peptides in all). The procedure is similar to that represented in Scheme 6.3. Each of the peptides was then assayed for antibody binding compared with the native sequence, using the ELISA technique. The results indicated that the Asp residue at position 101 is of unique importance for binding, while two other residues, Asp 104 and Ala 106, were found to play a lesser but significant role in the process.

6.1.3 Peptide Synthesis on Special Solid Supports

A number of attempts have been made to facilitate multiple-peptide synthesis in parallel by using nontraditional solid supports. One of the first, reported by Frank[30–32], was carried out on functionalized cellulose paper disks. These may be cut out from standard, commercially available sheets of filter paper using a punch. Each disk has a diameter of 1.55 cm and can be labeled simply with a pencil. The benzyl-based handle **11** or the allyl handle **12** (see Section 4.1.1.4) can then be attached to the hydroxyl groups of the cellulose, providing a point of attachment for the first amino acid. Chain elongation must be carried out using the Fmoc/tBu approach, since the disks are not stable to the repetitive acidolysis required in Boc/Bzl synthesis. For each coupling, the paper disks must be sorted and stacked in columns, and as many as 100 disks can be reacted simultaneously with the same amino acid derivative. Cleavage of the completed peptide from supports incorporating handle **11** can be brought about using a 50% solution of trifluoroacetic acid in dichloromethane. Allyl-based handle **12** is stable to trifluoroacetic acid treatment and allows side chain protection to be removed prior to cleavage from the solid support by palladium-catalyzed allyl transfer.

Other, similar procedures relying on the use of segmented solid supports have also been reported. Cotton (another form of cellulose) is mechanically more stable than paper and the substantial amount of solvent retained by the fabric makes a reaction vessel unnecessary — solvent and soluble by-products may be removed by centrifuging.[33-35] Polystyrene-grafted polyethylene film[36] segments (measuring 1.5 by 3 cm) have been used in a manner similar to the tea bag method. Yields per gram are reported to be similar for the two procedures.

In the spot synthesis technique, developed by Frank,[37,38] different peptides are synthesized at different locations on a single sheet of cellulose paper. This sheet is first derivatized by esterifying Fmoc-β-alanine onto the free hydroxyl groups of the cellulose. After washing and removal of the Fmoc group, this amino-functionalized sheet is dried. The different locations at which peptide synthesis is to take place are then marked at regular intervals with a pencil. Peptides can either be synthesized directly on the sheet, in which case they are bound irreversibly or, alternatively, handles can be incorporated to facilitate their cleavage. The different reagents (handles, amino acid derivatives, and solvents) are then pipetted onto the pencil points either manually or using a robot. After amino acid incorporations, the entire sheet is immersed in appropriate solutions for washing, acetylation, and deprotection. The sheet is then washed with ethanol and dried before coupling of the next amino acid, again by distribution of aliquots of the appropriate activated derivative. The advantages of the method are that it is simple and cheap and provides sufficient quantities of peptides for various applications.

6.1.4 Light-Directed, Spatially Addressable Parallel Peptide Synthesis

A considerably more sophisticated approach to the simultaneous synthesis of peptides in parallel is that based on a combination of SPPS and photolithography, a technique developed for computer microchip construction.[40-43] Synthesis is carried out on amino-functionalized glass microscope slides and the nitroveratryloxycarbonyl (Nvoc) group **13** is used as a photolabile N^α-amino protecting group.[44,45]

Structure is drawn to include the N^α-amino group of the amino acid.

Initial amino-functionalization of the microscope slide is brought about by its treatment with aminopropyltriethoxysilane. Nvoc N^α-protected γ-amino butyric acid or ε-aminocaproic acid is then coupled to the surface, forming a linker for peptide synthesis. Any residual unreacted amines or hydroxyl groups are capped by treatment with acetic anhydride. Photodeprotection of a specific region of the slide is effected by illumination of the substrate with 365 nm ultraviolet light, through a lithographic mask. Photolysis is carried out in the presence of a dilute (5 mM) solution of sulfuric acid in dioxane, which renders the free amino groups unreactive toward the photolysis by-products and greatly increases the photodeprotection yields. The amino groups of the illuminated zone are now available for coupling to another Nvoc N^α-protected amino acid derivative. The surface of the slide is neutralized with a solution of diisopropylethylamine, and amino acid incorporation can be brought about using the HBTU coupling reagent (see Section 2.4.1.2). Those regions of the

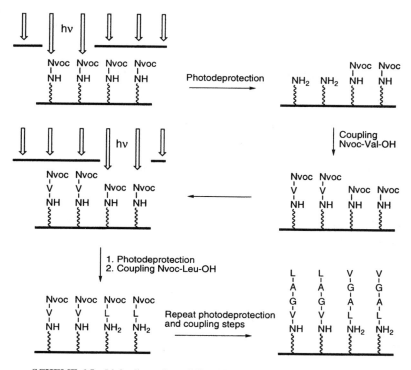

SCHEME 6.5 Light-directed, spatially addressable parallel peptide synthesis.

plate that were not illuminated do not undergo photodeprotection and consequently do not incorporate the amino acid. After coupling, the plate is illuminated through a second lithographic mask, which photodeprotects a different region, in turn preparing it for the incorporation of another Nvoc N^α-protected amino acid. Further illumination and coupling operations are then carried out as necessary. The overall result is to produce spatially addressable arrays of peptides on the microscope slide. The procedure is represented in Scheme 6.5.

The products formed depend upon the pattern and order of the photolithographic masks used and on the order of addition of the monomeric building blocks. By varying the lithographic patterns, many different sets of compounds can be synthesized in the same number of steps. In a binary masking strategy, represented in Scheme 6.6, the masks allow one half of the surface area to be illuminated at any one time and a different zone is illuminated after each amino acid–coupling step. Such a strategy ensures that the maximum number of compounds are generated in the fewest steps, so that for n chemical steps, 2^n compounds are produced. It does, however, generate peptides of different length so that, if chains having a fixed number of amino acids are required, another masking pattern must be used.

A high degree of miniaturization is possible and thousands of different peptides can be synthesized in an area only slightly greater than 1 cm^2. This capability for the simultaneous synthesis of such large numbers of peptides has led to the technique also being called very large scale immobilized polymeric synthesis (VLSIPS). The

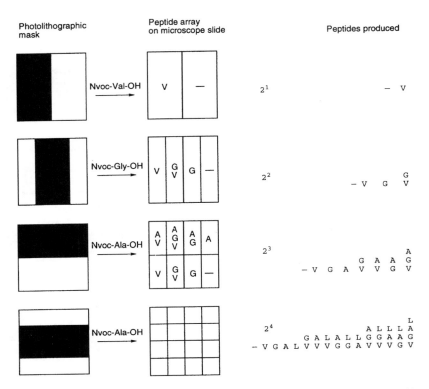

SCHEME 6.6 Binary masking strategy for light-directed, spatially addressable multiple-peptide synthesis. 2^n separate regions are produced in each step, one of which is always the unsubstituted solid support, denoted —.

binary masking strategy has been used to synthesize 1024 (2^{10}) different peptides, each one localized on a 50-μm square. The exact number of different compounds that can be synthesized is limited only by the number of synthesis sites that can be addressed with appropriate resolution.

After synthesis has been completed the side chain–protecting groups, which are chosen so as to be labile to trifluoroacetic acid, are removed by acidolysis, leaving the free peptides bound to the glass support. One of the ways in which these can be screened is to incubate the slide with fluorescein-labeled antibodies, which can be produced by treating them with fluorescein isothiocyanate. The efficiency of antibody binding at the different zones of the surface is reflected in the intensity of the fluorescence and can be measured by epifluorescence microscopy.

6.2 PEPTIDE LIBRARY TECHNIQUES

The iterative synthetic methods discussed above are useful for the production of relatively large numbers of peptides, up to the hundreds of thousands. However, a drawback of most of these methods is that each and every one of the various peptides produced must be screened, which can be laborious and time-consuming. Such a bottleneck can be avoided, however, if mixtures of peptides, rather than individual ones, are tested for activity. Fortunately, this can be done because the majority of serological tests rely upon the specificity of antibodies to detect a given antigen in the presence of large amounts of irrelevant protein. This means that receptors are usually highly selective and have a high affinity for active substances even in an environment consisting of vast amounts of other inactive molecules. Mixtures containing thousands or even millions of peptides can therefore be used to screen for active molecules. The problem then, of course, becomes one of identifying the active peptide or peptides in the mixture. In order for such screening to give meaningful results it is important that each of the peptides of the mixture or library be present in as near to equimolar quantities as possible.

Peptide mixtures can be produced using the synthetic methods described in Section 6.1 above, if the various peptides produced are mixed together after synthesis and cleavage from the solid support. However, such an approach is not really adequate for the creation of large libraries of compounds consisting of millions or even billions of different peptide structures. The production of such large numbers of compounds requires a more random generation of molecular diversity where, within each chemical coupling step, multiple compounds are generated simultaneously such that each synthesis cycle results in an exponential increase in library size. There are currently two general strategies for peptide library synthesis.

6.2.1 Synthesis of Peptide Libraries by Coupling Mixtures of Amino Acids

If, instead of a single activated amino acid derivative, mixtures of activated derivatives are added to one or more resins, then mixtures of peptides will be produced on the solid support. This approach to peptide combinatorial synthesis was first investigated by Geysen in an attempt to find peptides that bind to antibodies that recognize discontinuous epitopes.[12,46] The problem is to ensure an equimolar representation of each member of the library, owing to the difference in amino acid–coupling rates. If equal amounts of all the proteinogenic amino acids are applied in a standard SPPS coupling step (one using an excess of the acylating component), some will be incorporated preferentially over others.[47] The β-branched amino acids, Ile, Thr, and Val, for example, couple more slowly than, say, Ala, Asp, and Glu, which have no such branching.[48,49] This can be corrected by adjusting the relative proportions of the activated amino acid derivatives of the mixture so that greater amounts of amino acids that couple more slowly are present.

Alternatively, the problem can be overcome by applying less than one equivalent of an equimolar mixture of amino acids, relative to the resin-bound amino groups.[50-52] As long as the coupling reaction is left long enough, since the resin-bound amino groups are in excess, the different coupling rates of the amino acids are not reflected in the final composition. Equimolar mixtures of peptides have been obtained by double-coupling 0.8 equivalents of a mixture of 19 amino acids[50] that did not contain Cys. This has been termed *multiple substoichiometric addition.*

The spatial location of any given peptide sequence now cannot, of course, be determined and the identification of the active peptides produced in the synthesis must be done by a process of screening and resynthesis of ever smaller pools of compounds. This has been called *deconvolution* and is discussed below in Section 6.2.3.

6.2.2 Synthesis of Peptide Libraries by the Split-and-Combine Method

A second approach to peptide combinatorial synthesis circumvents the problem of competitive coupling reactions by physically segregating the solid support into multiple aliquots or spatially discrete fractions, to which single activated monomers are then added. The first report was that of Furka,[53] who called it "portioning-mixing," but it has also been referred to as "divide, couple, and recombine" by Houghten[54] or "split synthesis" by Lam[55] The procedure is simple and involves dividing the resin into a number (n) of equal batches. A single, different monomer is then coupled to each of them, and, after the couplings, they are combined and mixed thoroughly. This mixed resin is then separated again into n batches and another series of coupling reactions carried out. Repeating the protocol for a total of x cycles can, in principle, produce a collection of n^x peptides, as governed by the Poisson distribution. A schematic illustration of the split and combine method, for the synthesis of all the possible tripeptides of the amino acids Ala, Gly, and Val, is given in Scheme 6.7.

After division of the initial quantity of resin into three equal batches, Ala is coupled to one batch, Gly to another, and Val to the third. These three separate amino acid–resin portions are then recombined and mixed thoroughly. The mixed resin is again separated into three equal portions and each of these is again coupled with one of the amino acids Ala, Gly, or Val. A further mixing and separation operation followed by a final coupling then gives a resin in which all 27 possible tripeptides are present.

Of course, a given tripeptide cannot be located at this point, but, in principle, all 27 are present in equal quantities, since competitive coupling reactions have been avoided. Furthermore, any one resin bead contains one single structure, that is to say, mixtures of peptides on a single bead are not produced. This has come to be known as the "one-bead/one-peptide" concept, which is discussed in more detail in Section 6.3. When larger libraries are to be synthesized by the split-and-combine method, care must be taken to use a sufficiently large initial amount of resin. All possible library members can only be produced if the number of resin beads is around an order of magnitude greater than the total number of possible structures.[56]

Both the coupling of amino acid mixtures and the split-and-combine approach have been used extensively in the preparation of peptide libraries and each method has its own advantages and disadvantages. Split-and-combine synthesis is more involved, but it does produce only one structure per resin bead. While it is very useful for the generation of libraries of relatively small peptides, say, up to the hexapeptide level, the amount of resin required becomes impractical as the peptides

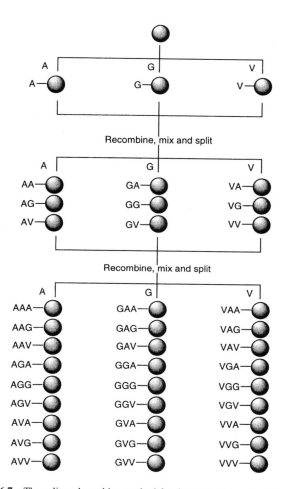

SCHEME 6.7 The split-and-combine method for the production of peptide mixtures.

get bigger. The synthesis of each of the 64 million possible hexapeptides of the 20 proteinogenic amino acids, requires 64 g of resin. That of all the possible heptapeptides, requires 1.28 kg, and after the decapeptide level literally tons of resin would be required.[57] The coupling of amino acid mixtures, on the other hand, is much less laborious and can easily be automated. It requires less resin and libraries of larger peptides can, in principle, be generated. It does, however, give rise to resin beads carrying a variety of different structures. With regard to their capacities for the production of peptide libraries containing equimolar quantities of each component, both methods can, in principle, give similarly good results.

6.2.3 Peptide Library Deconvolution

Once a peptide library has been synthesized, the problem is to determine which of the possibly millions of components is responsible for the observed activity. A number of methods have been described for doing this, depending upon the type and size of the library and the screening process used. The term *deconvolution* has been used to describe the process whereby the active molecule or molecules in a library are identified, usually by the iterative testing and resynthesizing of ever smaller collections of compounds, present in the initial library.

6.2.3.1 The Dual-Defined Method

One of the most popular procedures involves the resynthesis of pools of peptides that have defined amino acids at one or, more often, two positions.[58] At the remaining positions all possible combinations of amino acids are present. The optimal amino acids at the defined positions are then selected by determining which of the pools has the greatest biological activity. A second round of synthesis is then carried out with the selected amino acids in place at their initial positions, in order to prepare pools where the next defined position is introduced. Each of the pools is again evaluated for biological activity in order to select the optimal amino acid at the additional defined position. This process of iterative resynthesis and evaluation is then repeated until all of the positions of the peptide have been defined.

One of the first examples of this deconvolution strategy was reported by Geysen[46] who required a library of octapeptides for a study directed toward the identification of discontinuous epitopes. Synthesis of all the 25,600,000,000 (20^8) different octapeptides was judged to be impractical. As an alternative, mixtures of octapeptides, having the general sequence $XXXO_1O_2XXX$ were synthesized. Positions O_1 and O_2 were occupied by single, defined amino acids and the X positions represented randomly incorporated amino acid residues, resulting from coupling mixtures of activated amino acids. These octapeptides were produced by competitive coupling of amino acid mixtures on polythene pins (see Section 6.1.1), such that each pin had its own mixture of peptides.

Since all 20 proteinogenic amino acids were used, 400 such mixtures were possible — the number (20^2) of possible permutations of the dipeptide sequence O_1O_2 — and each of these mixtures would ideally contain 64 million (20^6) different sequences. Different mixtures of peptides were synthesized on different pins, and the peptide mixture on each pin was screened for antibody binding by the ELISA technique, in order to identify an optimum dipeptide sequence, represented as BJ. This then provided the basis for the synthesis of two further sets of peptide mixtures (40 in total) having the general form XXOBJXXX, in one case, and XXBJOXXX, in the other, in which the number of degenerate positions was reduced by one. After screening by ELISA the sequence exhibiting the most favorable binding provided the basis for a further round of synthesis and so on. In this way the progressive resolution of all residues in the sequence was achieved.

In Geysen's approach, the peptide mixtures remained bound to the polythene pins throughout the deconvolution process but the "dual defined" method can just as easily be applied to peptide libraries in solution. Houghten[54] synthesized a hexapeptide library with a view both to mapping the epitope of an antigen and to discovering new antimicrobial peptides. The library, consisting of more than 34 million members, each having an acetylated *N*-terminus and an amide *C*-terminus, was synthesized on a methylbenzhydrylamine resin (see Section 2.3.2) using the tea bag method described above in Section 6.1.2. Only 18 of the 20 proteinogenic amino acids were used, both Cys and Trp being excluded, in order to avoid the problems caused by disulfide bridge formation in the case of the former residue and side reactions in the latter. The tetrapeptide portion of the library was synthesized using the split-and-combine approach, where 18 polypropylene tea bags of resin were coupled to each of the 18 amino acids used. The packets were then opened and their contents combined and mixed thoroughly. This resin mixture was then divided into 18 portions, which were again sealed in tea bags, subjected to N^{α}-protecting group removal and amino acid coupling, again using only 18 of the proteinogenic amino acids. The above process was repeated twice more, yielding a mixture of 104,976 (18^4) protected tetrapeptide resins. This XXXX-resin was then divided into 324 (18^2) batches and placed in numbered tea bags so that the incorporation of the next two specifically defined amino acids could be carried out. The result of the library synthesis was, therefore, the generation of 324 different hexapeptide mixtures, each of which may be represented as Ac-O_1O_2XXXX-NH$_2$, where O_1 and O_2 are defined positions and X represents an equimolar mixture of the 18 possible amino acids. Each of the 324 peptide mixtures were then deprotected and cleaved from their solid supports by treatment with liquid hydrogen fluoride, giving peptide mixtures in solution.

Of the 324 hexapeptide mixtures examined, the mixture Ac-DVXXXX-NH$_2$ was found to be the most active. Twenty new peptide mixtures were then synthesized in which the third position was defined, as Ac-DVOXXX-NH$_2$. Of these new mixtures, Ac-DVPXXX-NH$_2$ was found to be the most active. The above iterative process, which reduced the number of peptide sequences by 18-fold each time it was repeated, was then carried out for the remaining three positions, leading to the hexapeptide Ac-DVPDYA-NH$_2$ as the most active member of the library. The same procedure but using a different assay, designed to measure the bacteriocidal activity of the peptide mixtures, led to the sequence Ac-RRWWCR-NH$_2$ that showed significant antibacterial activity against *Staphylococcus aureus*.

6.2.3.2   The Positional Scanning Method

In an alternative deconvolution strategy, known as *positional scanning*, Houghten[59,60] used a similar hexapeptide library, which was synthesized by the incorporation of amino acid mixtures rather than by the split-and-combine method. Here six different sublibraries of peptides were prepared, all having a free *N*-terminus and an amide *C*-terminus. In each of these, one position was individually and specifically defined, whereas all other positions consisted of mixtures of 18 of the proteinogenic amino acids. Cys and Trp were again omitted to simplify matters. Each of the six sublibraries, therefore, consisted of 18 different peptide mixtures, and the defined position was moved progressively from the *N*-terminus to the *C*-terminus, as follows: O_1XXXXX-NH$_2$, XO_1XXXX-NH$_2$, XXO_1XXX-NH$_2$, XXXO_1XX-NH$_2$, XXXXO_1X-NH$_2$, and XXXXXO_1-NH$_2$. These six sublibraries are referred to as positional synthetic combinatorial libraries.

The 18 different peptide mixtures making up the first positional library were screened for activity and the mixture having Tyr at the first position, YXXXXX-NH_2, was found to be the most active. In the second positional library, the peptide mixture having Gly at the defined position, XGXXXX-NH_2, was the most active. For the third positional library, the most active mixture was XXFXXX-NH_2 with XXGXXX-NH_2 also showing significant activity, and for the fourth, mixtures having Phe at the defined position XXXFXX-NH_2 were the most active. Although peptide mixtures of the type XXXXFX-NH_2, again with Phe at the defined position, were the most active in the fifth library, here other amino acids, such as Tyr, XXXXYX-NH_2, Met, XXXXMX-NH_2, and Leu XXXXLX-NH_2, also showed very similar activity. An analogous situation was also observed for the last position where Phe was again the most active amino acid, XXXXXF-NH_2, but with Arg XXXXXR-NH_2 and Tyr XXXXYX-NH_2 showing similar activity.

The sequence YGFFFF, derived from the amino acids of the defined positions of the most active mixtures from each of the six positional libraries, showed no biological activity. This being so, hexapeptides consisting of all possible combinations of the amino acids showing significant activity in each of the six positional libraries were synthesized. These amino acids were Tyr in the first position; Gly in the second; Phe and Gly in the third; Phe again in the fourth; Phe, Tyr, Met, and Leu in the fifth; and Phe and Arg in the sixth. This gave 24 ($1 \times 1 \times 2 \times 1 \times 4 \times 2$) new peptides in total. Of these, the most active were YGGFMY and YGGFMR, and it is interesting to note that both had the sequence of Met-enkephalin at the first five residues.

6.2.3.3 The Bogus Coin Method

A related approach to peptide library deconvolution, for libraries produced by the coupling of amino acid mixtures has been reported by Blake and Litzi-Davis.[61] The name is derived from the well known brainteaser in which one has to discover which of a group of coins weighs less than the others, by using a minimum number of weighings. The amino acids used in library synthesis were divided into three different groups, α, β, and γ. Those belonging to group α were Leu, Ala, Val, Thr, Phe, and Tyr in a molar ratio of 1:1:2:2:1:1.4. Group β was made up of Gly, Ser, Pro, Asp, and Glu in a molar ratio 1:1:1:1:1 and group γ of Lys, Arg, His, Asn, and Gln in a molar ratio of 1:2:1:1:1. Cys, Trp, and Met were excluded in order to avoid the chemical problem they present, and Ile was also excluded because the authors considered that it can be substituted in a general sense by Val. This left 16 amino acids in total to form the library of 16,777,216 (16^6) hexapeptides, having the form XXXXXX, where X represents an equimolar mixture of all 16 amino acids present in groups α, β, and γ.

The deconvolution method is exemplified for a tetrapeptide library, consisting of 50,625 (15^4) peptides, also synthesized by the same authors. The number of possible peptides in the tetrapeptide library was only 15^4 because, in a further simplification, Thr was omitted from amino acid mixture α. A series of peptide mixtures was first synthesized in which, at each position of the tetrapeptide in turn, those amino acids from group α were omitted, those from group β were doubled in quantity, and those from group γ were present in the initial quantities. This amino acid mixture was denoted $\beta_2\gamma$, and in Table 6.1 the effect of its incorporation at a given position on the biological activity is shown. A significant decrease in activity indicates that the presence of amino acids from group α is necessary at a given position. Unchanged activity in any mixture indicates that it is the amino acids from group γ that are the main contributors to activity, since this group is the only one whose proportions are unchanged. An increase in activity would indicate that amino acids from group β were the most important contributors. From Table 6.1 it can be seen that for the tetrapeptide in question the most important amino acid mixtures at positions 1 to 4 are α, α, γ, α, respectively.

By using this information, new peptide mixtures were synthesized, but this time only incorporating the mixtures of amino acids corresponding to groups α and γ at the positions in question. As can be seen from Table 6.1, the result was to produce a peptide mixture with greatly increased biological activity. In order to narrow down the possibilities further, groups α and γ were themselves subdivided into three sections. For group α, Phe and Tyr were omitted from the coupling mixture, Ala and Leu were unchanged and Val was increased. This mixture is represented as LAV_3 in Table 6.1. Now, a significant decrease in activity of the mixture is indicative of the importance of Phe or Tyr, while unchanged activity is indicative of the importance of Leu or Ala and an increase in activity of the importance of Val. Similarly for group γ, Asn and Gln were omitted, Lys and Arg were unchanged, and the amount of His was tripled. This mixture is represented as KRH_3 in Table 6.1. As before, a significant decrease in activity implies the requirement of Asn or Gln at a given position, while unchanged activity implies the importance of Lys or Arg; an increase would indicate that His was necessary at a given position. As can be seen from Table 6.1, this substitution allowed the positions of the tetrapeptide to be narrowed down to Phe or Tyr at position 1, Ala or Leu at position 2, Lys or Arg at position 3, and Phe or Tyr at position 4, respectively. The sequence of the most active tetrapeptide could then be deduced as Phe-Leu-Arg-Phe, by carrying out single amino acid substitutions.

An analogous procedure was then carried out for the considerably larger hexapeptide library, leading to the deduction of the sequence RQVGHD as the most active present in the library. Activity was a measure of the efficiency with which the binding of a 28-residue peptide to a monoclonal antibody was inhibited. Since the sequence PQAGID was present in this 28-residue peptide, it was postulated that these amino acids represented its epitope. A hexapeptide having the deduced sequence RQVGHD was twice as active as one having the sequence of the probable epitope PQAGID, and a single amino acid substitution gave a peptide PQVGHD that was 35 times more active.

6.2.4 One-Bead/One-Peptide Libraries

Although the library deconvolution methods discussed above allow the structure of the active components of large peptide libraries to be deduced, they are somewhat cumbersome, requiring as they do, several stages of iterative testing and resynthesis in order to reach a conclusion. Much would be gained if a single screening step could identify the most active component of a peptide library directly. This is possible in light-directed, spatially addressable peptide synthesis (see Section 6.1.4.), where

TABLE 6.1 Deconvolution of a Tetrapeptide Library by the Coupling of Amino Acid Mixtures

Peptide Sequence[a]				Activity[b]	Key Residues[c]
Aa[1]	Aa[2]	Aa[3]	Aa[4]		
X	X	X	X	1400	—
$\beta_2\gamma$	X	X	X	2500	α
X	$\beta_2\gamma$	X	X	2500	α
X	X	$\beta_2\gamma$	X	1450	γ
X	X	X	$\beta_2\gamma$	2500	α
α	α	γ	α	35	—
LAV$_3$	α	γ	α	500	F or Y
α	LAV$_3$	γ	α	44	A or L
α	α	KRH$_3$	α	41	K or R
α	α	γ	LAV$_3$	500	F or Y
FY	AL	KR	FY	3	—
F	AL	KR	FY	2	F
FY	**A**	KR	FY	20	L
FY	AL	**K**	FY	70	R
FY	AL	KR	**F**	1.5	F
F	L	R	F	0.5	—

[a] Aa[1] represents the amino acid at position one, Aa[2] that at position two, and so on.
[b] The lower the number, the higher the biological activity of the peptide mixture.
[c] Deduced from the biological screening results.

all products remain bound to the solid support and a single screening step is sufficient for identifying the most active of the various arrays produced. The screening of peptides still bound to solid supports can, in fact, also be done in large peptide libraries, if they have been generated on resin beads using the split-and-combine method. As has been seen in Section 6.2.2, this produces an equimolar distribution of peptide structures, with a given resin bead having one single structure. In other words, a library of beads is produced.

Acidolytic detachment of the peptides from these resin beads would generate a peptide library in solution, as has already been discussed in Sections 6.2.2 and 6.2.3. An alternative, however, is to remove only the peptide side chain–protecting groups and to screen the library of beads itself, with the peptides still attached, using suitable soluble receptor molecules such as monoclonal antibodies. These must be labeled or modified in some way so as to allow the visual identification of an interaction with a resin-bound peptide. As in light-directed, spatially addressable peptide synthesis, this can be done using acceptor molecules that have been attached to fluorescein. Alternatively, ELISA-based techniques (see Section 6.1.1) can be used, such that those beads that have interacted with the antibody become colored, allowing their easy visual identification since they stand out sharply against the majority, which remain colorless. Active beads can then be physically separated from the rest and analyzed. Unlike the VLSIPS technique, however, which is spatially addressable, here the structure of the active peptide bound to the bead is not known and must be determined. Since a given bead carries only a small quantity of peptide (typically, 50 to 100 pmol), there are, basically, only two methods that are sufficiently sensitive to be used for direct structure determination, namely, Edman degradation (see Section 2.4.3.4) or mass spectrometry.[62,63]

The first of these can be carried out directly on the peptide bead itself, while mass spectrometry is carried out on the solution produced by cleavage of all, or part, of the peptide from the resin (see Section 6.2.4.1). A useful mass spectrometric approach to the structure determination of peptides produced in combinatorial libraries is to add a capping reagent, such as acetic anhydride, at each coupling step. A small percentage of the growing peptide chains are therefore capped at each step. After cleavage from the solid support, the peptide sequence can be read directly from the molecular ions corresponding to the truncated and capped sequences that are present, in addition to the full sequence. An alternative to both of these methods of direct structure determination is to deduce the structure of the peptide indirectly, by encoding the library. This is discussed in Section 6.2.4.2.

Since both peptide synthesis and biological screening are carried out while the peptide is attached to the solid support, the nature of the bead material is important. As for any resin in peptide synthesis, the beads must be resistant to all of the conditions required for elaborating the peptide but, additionally, must be compatible with the use of aqueous media, in which the receptor-binding studies are performed. The polymeric support itself should not, however, bind to receptor molecules; otherwise detection of the desired receptor peptide interaction will be obscured. The choice of handle or spacer moiety on which peptides are synthesized can also influence the accessibility of the support-bound peptides to macromolecular receptors in solution. In general, the most successful solid supports in this type of application up to now have either been polyamide-based resins or polyethylene glycol-grafted polystyrene (see Section 2.1).

The generation of peptide libraries by split-and-mix synthesis followed by solid-phase screening is also known as the "Selectide process" (after the company that developed it)[57,64] and has several advantages, not the least of which is its rapidity. Very large numbers of beads (10^7 to 10^8) can be screened in a few hours. In addition, the color intensity or fluorescence is normally proportional to the binding affinity of the peptide in question. Furthermore, the library may be reused in other biological assays, after suitable washing steps to remove the receptor complex.

6.2.4.1 Screening of One-Bead/One-Peptide Libraries in Solution

Although primarily designed as a technique for the screening of solid support-bound peptides, the Selectide process for peptide library generation also allows screening to be carried out in solution. Of course, a peptide library in solution can be produced by cleavage of all the different peptides from the beads, but library deconvolution must then be carried out in order to identify the active components. An alternative is to cleave only a portion of the peptide on a given bead, maintaining the rest on the solid support for Edman sequencing in order to determine the structure.

Special handles have been developed in order to achieve such "staged release" of peptides from the solid support.[65-67] The application of one such handle–resin complex is shown in Scheme 6.8.

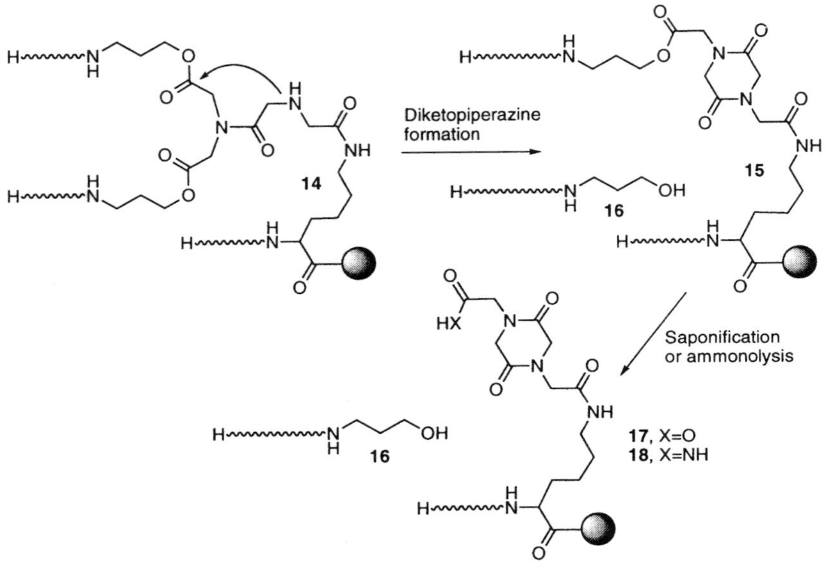

SCHEME 6.8 Staged release of peptide portions from a solid-phase bead in library synthesis.

Chain elongation can be brought about at three different sites, using the Fmoc/tBu approach. When the desired sequences have been elaborated, treatment with trifluoroacetic acid removes all side chain protection, while the peptides themselves remain bound to the resin, as in **14**. Release of the first portion of the peptide is brought about by treatment of **14** with organic base, which leads to the formation of a diketopiperazine **15**, with concomitant release of peptide as the hydroxypropylamide **16**. The second portion of the peptide can be released, again as hydroxypropylamide **16**, either by treatment with sodium hydroxide solution or alternatively with ammonia vapor. The third portion of peptide remains bound to the solid support for structure determination by Edman degradation, if necessary.[66]

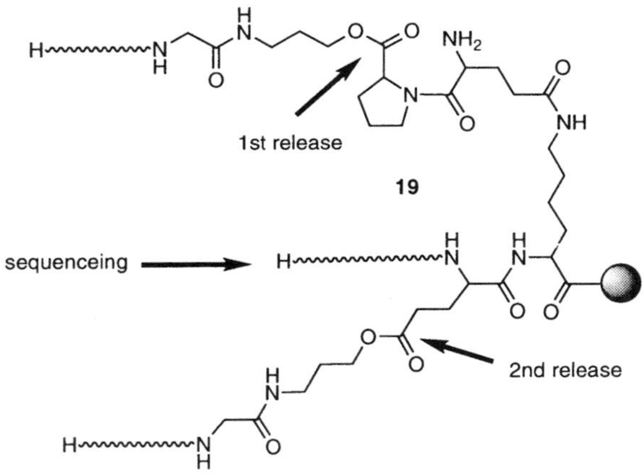

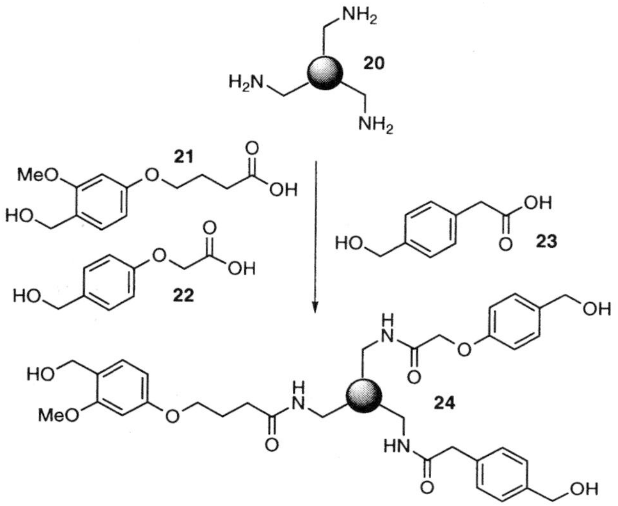

SCHEME 6.9   Resin for the staged release of peptides by graded acidolysis.

Similar staged release of peptides with a slightly different *C*-terminus, by similar mechanisms of diketopiperazine formation in the first instance and saponification or ammonolysis in the second, is also possible[67] from peptide–resin **19**. Another method for staged release relies upon graded acidolysis and is based on the attachment of three different benzyl-based handles to a bead of the solid support.[68] This is illustrated in Scheme 6.9.

The handles **21**, **22**, and **23** (see Section 4.1.1.1, and Section 2.3.1, respectively) are all attached to an aminomethyl resin **20**, such that each bead contains each handle in approximately equal proportions. After Fmoc/*t*Bu peptide synthesis, the first portion of peptide can be cleaved from handle **21** by acidolysis with 1% trifluoroacetic acid. The second portion is cleaved from handle **22** by treatment with 95% trifluoroacetic acid, and the third remains bound to the solid support for Edman degradation.

When staged cleavage of peptide is to be applied in library screening, the beads are distributed into batches of several hundred units. The first portion of peptide attached to the beads of a batch is then cleaved, resulting in an approximately equimolar mixture of peptide solution. Aliquots of these mixtures are then tested using some suitable assay. Active mixtures are identified and the beads from the batch responsible for activity are now redistributed singly. The second portion of peptide is then cleaved and the biological test repeated. The bead or beads that provide positive results are identified and structure determination of the peptides in question is carried out, either by mass spectrometry of the soluble peptide or by Edman degradation of the part remaining on the solid support.

6.2.4.2   Encoded Combinatorial Peptide Synthesis

The small amounts of individual peptides produced in the one-bead/one-peptide library synthesis approach mean that the only techniques currently available for direct structure determination are Edman sequencing or mass spectrometry. However, the former is only useful for peptides composed of the proteinogenic amino acids, and, even then, sequencing can be the slowest step in the screening process. A general solution to the problem of the structure determination of specific library members that may be present only in tiny quantities, especially those that are nonsequenceable, is library encoding.[69] This involves the attachment of readable "tags" to the resin bead at each coupling step in the synthesis. Each tag then identifies which amino acid (or other type of building block) is coupled at a particular step of the synthesis. The final structure can then, in principle, be deduced by reading the set of tags on the bead in question. In this way the reaction sequence that produces a given peptide is recorded and the library is said to be "encoded." As long as the tags are judiciously chosen, encoding can result in increased sensitivity and speed of analysis.

Ideally, tags should have a high information content, be detectable at very low concentrations, and be stable to the conditions of peptide synthesis. Library encoding using molecular tags can be done in one of two general ways. On the one hand, tags can be attached in such a way as to provide an "encoding sequence" in which the structure of a tag, or the sequence of a series of tags, encodes a given amino acid or building block. The position of the tag or tag sequence in the encoding sequence also then encodes the synthesis step at which this amino acid is incorporated. This situation is represented as **25**. On the other hand, each tag may be added individually to the resin, in which case each one must identify both the building block itself and the synthesis step at which it is incorporated, as represented in **26**. A generalized tagging strategy for peptide library synthesis is represented in Scheme 6.10.

Although tagging can be used in peptide library synthesis, it tends to be more important either in small-molecule organic combinatorial synthesis or in other situations where Edman sequencing, or related methods, are for one reason or another impractical.

6.2.4.2.1 Library Encoding with Oligonucleotides

The first method proposed for library encoding relied on oligonucleotide sequences as tagging molecules.[69] Oligonucleotide encoding has two main advantages. First, only a minute quantity of the tagging sequence is required, because it can be amplified using the polymerase chain reaction, and, second, the oligonucleotide sequence can be read rapidly using DNA sequencing methods. Each amino acid monomer used in peptide synthesis is assigned a distinct contiguous oligonucleotide sequence or "codon." The sequence of a peptide assembled on any bead is, consequently, reflected in the oligonucleotide sequence of the corresponding tag. Peptides

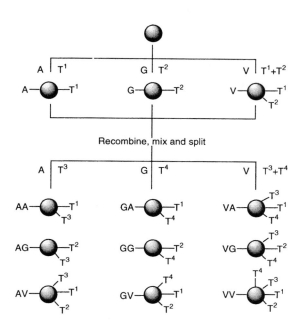

SCHEME 6.10   Encoded peptide library synthesis using molecular tags. Tagging molecules are represented as T^1, T^2, and so on.

and oligonucleotides are synthesized in parallel so that each bead carries copies of a single peptide sequence and a unique oligonucleotide identifier tag.[70,71] The average stoichiometry of peptide to oligonucleotide per bead is readily controlled by varying the ratio of differentially functionalized handles coupled to the beads. It can be heavily weighted in favor of the peptide, in this way reducing the possibility of oligonucleotide–receptor interactions confusing the results of the biological screening of the bead-bound peptides.

The main disadvantage of the use of oligonucleotides is that rather intricate orthogonal protecting strategies are required in order to allow the elaboration of both types of oligomer on the same bead. Furthermore, often quite long coding sequences must be synthesized, considerably increasing the time and cost of library synthesis. For the preparation of small-molecule combinatorial libraries, the stability of an oligonucleotide-encoding sequence to the synthesis conditions should be carefully evaluated.

6.2.4.2.2 Library Encoding with Peptides

Library encoding using peptides as the tagging entities is, of course, only logical in situations where the principal structure to be synthesized is either composed predominantly of nonproteinogenic residues or is nonpeptidic in nature.[72,73] Again, both structures are synthesized in parallel on the same resin bead, employing different handles as attachment points for each. One (or several) proteinogenic amino acids are attached to the resin bead at every step of the synthesis of the nonsequenceable molecule. When synthesis and screening have been completed, the structure present on an active bead can then be decoded by Edman degradation of the tagging sequence.

The problems are similar to those described for oligonucleotide encoding, in that orthogonal protecting group strategies are necessary to allow each of the different structures to be synthesized in parallel. Furthermore, the possibility that the peptide-encoding structure might interact with a given receptor molecule in binding assays must be borne in mind. A rather ingenious solution to this problem is to synthesize the encoding sequence in the interior of the bead and the main sequence on the exterior. Interaction of the encoding sequence with a receptor is then avoided.[74]

As is the case for encoding with oligonucleotide tags, encoding with peptides may not always be advisable for small-molecule organic combinatorial libraries, because of the instability of the encoding sequence to some of the conditions that might be used in synthesis.

6.2.4.2.3 *Library Encoding with Aryl Halide Tags*

One of the most satisfactory general approaches to library encoding is that developed by Still[75,76] based upon the use of aryl halide tagging molecules. These can be detected at the sub-picomolar level by electron capture gas chromatography, so that each tag can be attached to the solid support at less than 1% of the loading level of the synthesized compound. This minimizes the chance of complications in library evaluation resulting from tag–receptor interactions when solid-phase library screening is carried out. The aryl halide-based tag molecules have the general structures **27** and **28**, where y typically has values between 2 and 11 and X_n represents a combination of halogen atoms, (F, Cl, and Br), attached to the aromatic ring. A set of chromatographically resolvable tags can then be created by varying y and X_n. With 20 such tags, 1,048,576 (2^{20}) syntheses can be uniquely encoded.

Molecular tag **27** incorporates a photolabile carbonate linker that can be cleaved on irradiation at 365 nm. Prior to each amino acid–coupling step a substoichiometric amount of tag is added to the beads. This does effectively cap the amino group of, say, a peptide chain, but since subpicomolar quantities of tag can be detected by gas chromatography the amount of capping of the desired peptide on a bead is usually less than 5%, even in the most complex cases. After assembly and screening of the library, positive beads are subjected to photolysis, which releases the aryl halide tags. The resulting solution is then subjected to gas chromatographic analysis.

Molecular tag **28** has the advantage that it requires no specific functional group for its attachment to the resin bead. The aryl halide tag is again attached to the peptide–resin by a linker but, in this case, it is a diazoketone, which can be converted into the extremely reactive acyl carbene. The carbene reacts with either the polymer matrix or the attached peptide. However, because it is rather unselective it inserts predominantly into the solid support, on account of its greater contribution to the mass of the bead. The phenol tags can then be released by oxidative cleavage of the catechol diether unit, on treatment with ceric ammonium nitrate. Analysis is again carried out by gas chromatography.

This approach to tagging, whether using **27** or **28**, is that represented in **26** above. The tagging molecules are not sequentially connected so that intricate orthogonal protection schemes are not required for library synthesis. The tags, however, must encode for both the amino acid incorporated and the step at which it is incorporated. A binary synthesis code has been devised for accomplishing this, where each amino acid is represented by a three-digit binary code (001, 010, 011, etc.) with the presence or absence of a particular tag corresponding to 1 or 0, respectively.

6.2.4.2.4 Radio Frequency Encoding

Advances in microelectronics and chemistry have made it possible to produce resin constructs in which electronic sensors are embedded.[77,78] These are completely inert to the solvents and conditions required for peptide synthesis and permit the electronic labeling of each synthetic operation in a library synthesis. They can then be scanned and the stored information retrieved in order to allow the identification of a given peptide on a given batch of resin. The advantages they provide are that chemical steps for the decoding of the library are avoided and that decoding is instantaneous. Their use has been demonstrated in real-case library synthesis but they have not, as yet, been widely adopted.

6.2.5 Quality Control

The question of quality control in large peptide libraries is one that has not been extensively addressed up to now. The reasons are not difficult to fathom. The huge numbers of different structures that might be involved places stringent demands upon analytical techniques and makes the detection of a given sequence, or of subproducts resulting from deletion, insertion, or incomplete protecting group removal, a truly daunting proposition. Nevertheless, quality control in the synthesis of peptide libraries is important, and the issue should be borne in mind whenever they are generated.

The more complex a library is, the more difficult it is to get useful information on its individual components. The attainment of an equimolar representation of each library member is made difficult by, among other things, differences in amino acid–coupling rates, incomplete deprotection and coupling reactions, and any other side reactions that might occur. Although the ideal situation would be a library consisting of equimolar amounts of each of the possible members, it has been argued that the reproducibility of the synthesis, rather than library fidelity is, in fact, more important.[51] In any event, it is essential to have analytical methods that can provide some idea of library quality, and, in spite of the difficulties involved, much useful data can be obtained using standard analytical techniques.

6.2.5.1 Amino Acid Analysis and Edman Sequencing

Amino acid analysis is the most commonly applied method for the analysis of the quality of peptide libraries.[6] While it does give a preliminary and useful indication of fidelity, analyses of complex mixtures of peptides are, of course, only capable of highlighting gross variations in composition and are not capable of indicating the presence or absence of a particular sequence. Neither can any information be provided on incomplete deprotection or on the many side reactions that might have occurred during synthesis. Despite this, and taking into account factors such as the partial destruction of certain residues during hydrolysis, the ratio of all amino acids present in the library should be globally that predicted theoretically, given the characteristics of the particular library synthesized.

A good idea of the quality of a peptide library can also be obtained by solid-phase Edman degradation of a representative sample of the library beads.[79]

6.2.5.2 Mass Spectrometry

The analysis of the molecular diversity and integrity of peptide libraries can be done by mass spectrometry. Various modes of instrument operation have been evaluated,[80] but the best results are usually obtained with soft ionization techniques, such as electrospray,[79,81] or matrix-assisted laser desorption.[82] The mass spectrum of a complex peptide mixture does not have the form of a broad continuum, but rather that of a bell curve, whose maximum confirms the overall size of the peptides in the library (whether tetra-, penta-, or hexapeptides, etc.). The peak heights usually also reflect the number of isobaric peptides. As a first step, the comparison of the theoretical and experimental distribution of the protonated molecular ions gives a good indication of the composition of the library. If the library gives peaks containing masses that are too high compared with theoretical values, this is indicative of incomplete removal of protecting groups. The presence of significant quantities of values less than theoretical is indicative of deletion sequences. The presence and structure of side products derived from incomplete protecting group removal or other side reactions such as amino capping can be done using tandem mass spectrometry.[83] Mass spectrometry is especially useful for the analysis of libraries containing chemical modifications, such as acetylation or phosphorylation, since these are not detectable by amino acid analysis.

6.2.5.3 Other Techniques

Other techniques that have been applied to the analysis of peptide libraries include high-performance capillary electrophoresis and NMR.[81] The former separates the peptides of the libraries into families, based upon the number of charges they carry, and gives a measure of the number of peptides in each of these families. Such information is useful for gauging the extent of side chain deprotection, since the number of charges that a peptide can carry is influenced by the amino acid residues that it contains and whether or not the side chains are protected. Comparison of the expected theoretical numbers for each family with that obtained experimentally then allows the extent of deprotection of the side chains to be estimated.

Two dimensional NMR spectroscopy of peptide libraries, in spite of the complexity of the spectra obtained, can be used to confirm the presence of all of the amino acids or, alternatively, to provide a measure of their incorporation. Such techniques might also be useful if nonproteinogenic amino acids were to be incorporated into the library. New advances in gel-phase NMR spectroscopy can also be expected to provide alternative methods for the characterization of polymer-bound libraries.[84,85]

6.2.6   Screening

The screening of a peptide library is the key step upon which all else depends, and the necessity of choosing the correct assay is paramount. The purpose for which the library has been generated is particularly relevant to the choice of screening method. If it has been created to search for novel lead compounds, then the detection of rare sequences having moderate or, if possible, high affinities for the target receptor is probably the most desirable objective. Any promising peptide structures can then form the basis for new libraries in order to discover more active compounds. On the other hand, if the library has been generated for screening a set of analogues, to establish structure–activity relationships, and to increase the potency of a lead compound, a different assay may be used.

Whatever the application, however, it is obvious that the various methods that have been considered for peptide combinatorial library construction are highly dependent upon the sensitivity and the specificity of the screening processes used to identify the target peptides.[86]

6.2.6.1   Screening Libraries of Immobilized Peptides

In library techniques in which peptides are screened while still attached to the solid support, be it a glass microscope slide, polyethylene pin, or resin bead, the ability of the soluble receptor to interact with a tethered peptide must be considered. This may depend upon the position and nature of the attachment of the peptide to the support. Tethering at the C-terminus is by far the most common mode of attachment, so that screening with antibodies that require a free carboxylic acid at this position would, in all likelihood, not give a positive result. Many analogues that would bind when in free solution would be missed in a screening process in which they were bound at their C-termini. Methods for "reversing" peptides on solid supports so that they become tethered at their N-termini have been reported.[57]

Immobilized peptide screening always requires that the receptor be labeled in some way so that binding can be detected. Colorimetric enzymes, radioisotopes, or fluorophores have been used most commonly. This labeling must be done in such a way that the receptor retains its ability to bind to the peptide in question and should be checked by binding a known natural ligand to the labeled receptor complex prior to library screening. The successful screening of immobilized libraries also depends upon the ability to distinguish peptides that bind with high affinity and those that bind with low affinity. High-affinity ligands are, in principle, the most desirable, and the initial conditions for screening random libraries can be adjusted so that only the binding of these is detected. This can usually be done by using low receptor concentrations, competing ligand-mediated dissociation, or stringent washing conditions. However, it may well be that a more reliable approach to library utilization in the long run is to select screening conditions that allow the isolation and detection of lower-affinity ligands. These initial peptides can then serve as starting points for the generation of new peptide libraries or as lead compounds for refinement by analogue construction.

However, caution should always be exercised in the interpretation of binding data. The distribution of peptide on the solid support may not necessarily be uniform, and with different compound loadings one must be wary of using the absolute quantity of bound receptor as an index of the affinity of any given peptide. The screening of immobilized peptides by direct receptor binding can lead not only to the identification of peptides that bind to the receptor specifically, that is to say, that compete with the natural ligand for the active site, but also to the identification of peptides that bind nonspecifically to other portions of the receptor or even to secondary detection reagents. The ability to distinguish nonspecific binding from specific binding is a major factor in the success of the screening process. Such phenomena can be detected by performing sequential assays, first to test for receptor binding of any kind, followed then by an assessment of nonspecific binding, in order to identify compounds that interact with the receptor in the desired manner.

6.2.6.2 Screening of Peptide Libraries in Solution

When peptide mixtures are screened in solution in deconvolution strategies[58] (see Section 6.2.3), the results of a set of assays often do not indicate a clear preference for a unique amino acid residue at a given position in the peptide sequence. Rather, similar results may be obtained for several different substitutions and a decision must be made as to which of these partial solutions should then be more fully deconvoluted. The number of peptide mixtures to be synthesized and tested increases dramatically as the number of alternative sequences selected for complete resolution at each cycle is increased. Moreover, the deconvolution of different partial solutions often produces quite different fully resolved sequences, because the contribution of a given amino acid to the peptide–receptor interaction is quite often dependent on other nonneighboring residues within the peptide. Perhaps the greatest problem with the screening of peptide mixtures in solution is that the activity of a given pool of peptides is based on the cumulative activity of all the peptides in the pool. That is to say, pools with comparable activities may either contain many low-affinity peptides or a few high-affinity ones. The identification of the most effective ligand in a complex mixture of peptides is then complicated by the relative abundance of lower-affinity ligands.

The best results are obtained in iterative deconvolution strategies if the minimal fragment having activity has the same number of amino acids as that used for the preparation of each library member; an active tetrapeptide is more easily located in a tetrapeptide library than in a hexapeptide library. The identification of active peptides is also facilitated if the receptor has specific requirements for a fixed position within the peptide, such as the N- or C-terminus. Nevertheless, many if not all of the possible initial peptide mixtures having two adjacent or nonadjacent fixed residues may have to be screened. Although this increases the number of initial mixtures that must be synthesized, it also increases the probability that a critical residue is fixed in at least one pool so that pool can then be deconvoluted. However, the possibility always exists that the pool with the highest activity is composed of many moderately active compounds while the most active compounds are, in fact, to be found in other pools.

The testing of mixtures of soluble peptide is also limited by the concentration that can be achieved of the individual members. Pools containing as many as 160,000 different peptides have been tested, with each member being present at an approximately 10 nmol concentration. Because of limitations on the solubility of the total pool, the concentration of individual compounds present in larger libraries may be diminished, which can limit the identification of moderately active peptides. Nevertheless, although the screening of peptide libraries in solution does have certain drawbacks, it does avoid several of the problems associated with the screening of tethered libraries. More conventional biological assays can be used, and it is, consequently, often the best method for screening peptide mixtures in library techniques.

6.3 EXAMPLES OF PEPTIDE LIBRARY SYNTHESIS

6.3.1 An All-D-Amino Acid Opioid Peptide

One of the main problems with the development of peptides as therapeutic agents is that they are rapidly degraded by protease enzymes under physiological conditions. Peptides composed of D-amino acids, on the other hand, are not so degraded and remain intact for much longer periods of time. One of the options for developing peptide drugs is, therefore, to search for biologically active peptides, composed wholly or partially of D-amino acids. Houghten[87] have used a hexapeptide combinatorial library to identify an all-D-amino acid opioid peptide with analgesic activity.

The strategy for library synthesis adopted in this case was the dual-defined method discussed in Section 6.2.3.1. All peptides had an acetylated N-terminus and an amide C-terminus and all were composed exclusively of D-amino acids. Each D-amino acid residue is represented by a lowercase letter, corresponding to the single letter code for the proteinogenic amino acids. Although glycine has no enantiomer, for the sake of simplicity, it too is considered to be a D-amino acid and is represented by the letter g. The first two positions of each peptide were individually and specifically defined, while the last four positions represented an equimolar mixture of 19 amino acids, Cys being excluded. The peptide mixtures therefore had the general formula Ac-$o_1o_2$xxxx-NH$_2$, where o_1 and o_2 represent defined positions and x represents an equimolar mixture of the 19 amino acids used. Initially, 400 peptide mixtures were synthesized (Cys was included at this stage), and each contained 130,321 (19^4) peptides. These mixtures were screened for their ability to inhibit the binding of [^{3}H][D-Ala^2, MePhe^4, Gly^5-ol]enkephalin (DAMGO) to rat brain homogenates. The mixture Ac-rfxxxx-NH$_2$ was the most active of those tested. Each further position of the peptide was then defined iteratively, each cycle of resynthesis and testing reducing the number of peptides within a mixture by a factor of 19. The most active hexapeptide in a library consisting of 52,128,400 peptides in total was found to be Ac-rfwink-NH$_2$. This peptide exhibited moderate analgesic activity in mice and could be used as a basis for further refinement, again using library techniques.

6.3.2 Novel Ligands for the SH3 Domain of Phosphatidylinositol 3-Kinase

In attempting to find peptides that bind to SH3 domains, relatively short (50- to 70-amino acid residues) regions of proteins that are found in many intracellular signaling proteins, Schreiber[88] synthesized two separate peptide libraries, one of 2 million linear hexapeptides having the general form XXXXXX, where X is any proteinogenic acid other than Cys, and the other of 2 million cyclic heptapeptides having the formula CXXXXXC, where cyclization was brought about by intramolecular disulfide formation. Both of these libraries failed to yield peptides that bound strongly to the SH3 domain.

In order to increase the possibility of finding useful ligands, a new library was prepared that was intrinsically biased toward binding to SH3 domains. These are known to bind to proline-rich peptide motifs, and several peptides that bind specifically to SH3 domains are known to possess the amino acid sequence PPXP. On the basis of this observation a library having the basic formula XXXPPXPXX was constructed, where X represents any amino acid other than Cys. Approximately 2 million peptides with this structural bias were generated by the split-and-combine approach and were screened on the solid support, using bacterially expressed recombinant SH3 domain, having a fluorescein label attached to it. Analysis by fluorescence microscopy led to the isolation of 17 beads, which exhibited strong binding. Edman sequencing of these peptides yielded a general sequence RXLPPRPXX, where X has a tendency to be a basic amino acid such as Arg, Lys, His, etc. The peptide with the sequence RKLPPRPRR was found to exhibit the strongest binding.

In a second study, Schreiber[89] synthesized a new library based upon the biasing sequence PLPPLP, but incorporating a range (33 in total) of nonproteinogenic amino acids, together with a range (again, 33 in total) of capping reagents, that would terminate the peptide chain. After the synthesis of the common sequence PLPPLP, three cycles of split-and-combine synthesis were carried out, encoding the library with the haloaromatic molecular tags developed by Still (see Section 6.2.4.2.3). An encoded "blank step" was carried out during monomer incorporation, in order to increase library diversity. In this way, all representatives of the sublibraries Cap-ZZZPLPPLP, Cap-ZZPLPPLP, Cap-ZPLPPLP, and Cap-PLPPLP were produced, where Z represents an equimolar distribution of the nonproteinogenic amino acid residues used and Cap an equimolar distribution of the capping reagents.

29   **30**   **31**

Decoding of the library after screening with SH3 dominain led to a series of peptides that showed high-affinity binding and which incorporated combinations of the nonproteinogenic amino acids **29**, **30**, and **31** at the *N*-terminus of the PLPPLP sequence. These studies illustrate the utility of biased combinatorial libraries for ligand discovery in systems where something is known of the ligand-binding characteristics of the receptor.

6.3.3 Sequence-Selective Peptide Binding with a Peptido-Steroidal Receptor

In the search for synthetic receptor molecules that bind peptides specifically, Still[90] designed a peptido-steroid of the general form **32**, which has two short peptide arms extending from the steroid core. Previous work from the same group[91] had shown that similar receptors, based on steroids having a *cis* junction between the A and B rings, do in fact bind peptides, such as Leu-enkephalin, with a good degree of selectivity. In an attempt to improve matters even further, receptor **32**, which in addition to having a *trans* junction between the A and B rings has shorter peptide arms, was designed.

32

In order to find the optimal amino acid sequences for binding, a library was generated, using ten amino acids, Ala, Phe, Pro, Ser, and Asn, together with their D-enantiomers, 10^4 members in total. Synthesis was carried out by the split-and-combine method, using the encoding strategy developed by Still (see Section 6.2.4.2.3). Screening was carried out with an enkephalin derivative labeled with a dye molecule, and after decoding of the positive beads obtained it was discovered that the peptido-steroidal receptors of the type **32** that bound most strongly all had the same amino acid residues at positions 1 to 3, namely, L-Asn, D-Asn, L-Phe, with the amino acid at position 4 being less critical. The peptido-steroidal receptors having either Ser or Ala at this position were found to bind enkephalin most strongly.

REFERENCES

1. Cwirla, S. E., Peters, E. A., Barrett, R. W., and Dower, W. J., *Proc. Natl. Acad. Sci. U.S.A.,* 87, 6378, 1990.
2. Gorman, J. J., *Anal. Biochem.,* 136, 397, 1984.
3. Krchnák, V. and Vagner, J., *Pept. Res.,* 3, 182, 1990.
4. Gausepohl, H., Boulin, C., Kraft, M., and Frank, R. W., *Pept. Res.,* 5, 315, 1992.
5. Beck-Sickinger, A. G., Dürr, H., Hoffmann, E., Gaida, W., and Jung, G., *Biochem. Soc. Trans.,* 20, 847, 1992.
6. Zuckermann, R. N., Kerr, J. M., Siani, M. A., and Banville, S. C., *Int. J. Pept. Protein Res.,* 40, 497, 1992.
7. Luu, T., Pham, S., and Deshpande, S., *Int. J. Pept. Protein Res.,* 47, 91, 1996.
8. Krchnák, V., Cabel, D., and Lebl, M., *Pept. Res.,* 9, 45, 1996.
9. Albericio, F., Ruiz-Gayo, M., Pedroso, E., and Giralt, E., *React. Polym.,* 10, 259, 1989.
10. Geysen, H. M., Meloen, R. H., and Barteling, S. J., *Proc. Natl. Acad. Sci. U.S.A.,* 81, 3998, 1984.
11. Geysen, H. M., Barteling, S. J., and Meloen, R. H., *Proc. Natl. Acad. Sci. U.S.A.,* 82, 178, 1985.
12. Geysen, H. M., Rodda, S. J., Mason, T. J., Tribbick, G., and Schoofs, P. G., *J. Immunol. Methods,* 102, 259, 1987.
13. Geysen, H. M. and Mason, T. J., *Bioorg. Med. Chem. Lett.,* 3, 397, 1993.
14. Harlow, E. and Lane, D., *Antibodies: A Laboratory Manual,* Cold Spring Harbor Laboratory Press, Plainview, NY, 1988.
15. Bray, A. M., Maeji, N. J., Jhingran, A. G., and Valerio, R. M., *Tetrahedron Lett.,* 32, 6163, 1991.
16. Valerio, R. M., Bray, A. M., and Maeji, N. J., *Int. J. Pept. Protein Res.,* 44, 158, 1994.
17. Bray, A. M., Maeji, N. J., and Geysen, H. M., *Tetrahedron Lett.,* 31, 5811, 1990.
18. Bray, A. M., Maeji, N. J., Valerio, R. M., Campbell, R. A., and Geysen, H. M., *J. Org. Chem.,* 56, 6659, 1991.
19. Valerio, R. M., Benstead, M., Bray, A. M., Campbell, R. A., and Maeji, N. J., *Anal. Biochem.,* 197, 168, 1991.
20. Valerio, R. M., Bray, A. M., Campbell, R. A., DiPasquale, A., Margellis, C., Rodda, S. J., Geysen, H.M., and Maeji, N. J., *Int. J. Pept. Protein Res.,* 42, 1, 1993.
21. Maeji, N. J., Tribbick, G., Bray, A. M., and Geysen, H. M., *J. Immunol. Methods,* 146, 83, 1992.
22. Van der Zee, R., Van Eden, W., Meloen, R. H., Noordzij, A., and Van Embden, J. D. A., *Eur. J. Immunol.,* 19, 43, 1989.
23. Jung, G. and Beck-Sickinger, A. G., *Angew. Chem. Int. Ed. Engl.,* 31, 367, 1992.
24. Houghten, R. A., *Proc. Natl. Acad. Sci. U.S.A.,* 82, 5131, 1985.
25. Houghten, R. A., *Protides of the Biological Fluids,* 34, 95, 1986.
26. Houghten, R. A., DeGraw, S. T., Bray, M. K., Hoffmann, S. R., and Frizzell, N. D., *BioTechniques,* 4, 522, 1986.
27. Houghten, R. A., Bray, M. K., DeGraw, S. T., and Kirby, C. J., *Int. J. Pept. Protein Res.,* 27, 673, 1986.
28. Houghten, R. A. and Lynam, N., in *Peptides 1988. Proceedings of the 20th European Peptides Symposium,* Jung, G. and Bayer, E., Eds., Walter de Gruyter, Berlin, 1989, 214.
29. Jezek, J. and Houghten, R. A., in *Peptides 1990. Proceedings of the 21st European Peptides Symposium,* Giralt, E. and Andreu, D., Eds., ESCOM, Leiden, 1991, 74.
30. Blankemeyer-Menge, B. and Frank, R., *Tetrahedron Lett.,* 29, 5871, 1988.
31. Frank, R. and Döring, R., *Tetrahedron,* 44, 6031, 1988.
32. Blankemeyer-Menge, B. and Frank, R., in *Innovation and Perspectives in Solid Phase Synthesis and Related Technologies. Peptides, Polypeptides and Oligonucleotides. Macro-Organic Reagents and Catalysts,* Epton, R., Ed., SPCC (U.K.) Ltd, Birmingham, 1990, 465.
33. Eichler, J., Beyermann, M., and Bienert, M., *Coll. Czech. Chem. Commun.,* 54, 1746, 1989.
34. Eichler, J., Bienert, M., Sepetov, N. F., Stolba, P., Krchnák, V., Smékal, O., Gut, V., and Lebl, M., in *Innovation and Perspectives in Solid Phase Synthesis and Related Technologies. Peptides, Polypeptides and Oligonucleotides. Macro-Organic Reagents and Catalysts,* Epton, R., Ed., SPCC (U.K.) Ltd, Birmingham, 1990, 337.
35. Eichler, J., Bienert, M., Stierandova, A., and Lebl, M., *Pept. Res.,* 4, 296, 1991.

36. Wang, Z. and Laursen, R. A., *Pept. Res.,* 5, 275, 1992.
37. Frank, R., *Tetrahedron,* 48, 9217, 1992.
38. Frank, R., *Bioorg. Med. Chem. Lett.*, 3, 425, 1993.
39. Malin, R., Steinbrecher, R., Jannsen, J., Semmler, W., Noll, B., Johannsen, B., Frömmel, C., Höhne, W., and Schneider-Mergener, J., *J. Am. Chem. Soc.*, 117, 11821, 1995.
40. Fodor, S. P. A., Read, J. L., Pirrung, M. C., Stryer, L., Lu, A. T., and Solas, D., *Science*, 251, 767, 1991.
41. Gruber, S. M., Yu-Yang, P., and Fodor, S. P. A., in *Peptides. Chemistry and Biology. Proceedings of the 12th American Peptides Symposium*, Smith, J. A. and Rivier, J. E., Eds., ESCOM, Leiden, 1992, 489.
42. Fodor, S. P. A., Rava, R. P., Huang, X. C., Pease, A. C., Holmes, C. P., and Adams, C. L., *Nature (London),* 364, 555, 1993.
43. Holmes, C. P., Adams, C. L., Kochersperger, L. M., Mortensen, R. B., and Aldwin, L. A., *Biopolymers (Pept. Sci.)*, 37, 199, 1995.
44. Patchornik, A., Amit, B., and Woodward, R. B., *J. Am. Chem. Soc.*, 92, 6333, 1970.
45. Amit, B., Zehavi, U., and Patchornik, A., *J. Org. Chem.,* 39, 192, 1974.
46. Geysen, H. M., Rodda, S. J., and Mason, T. J., *Mol. Immunol.*, 23, 709, 1986.
47. Birkett, A. J., Soler, D. F., Wolz, R. L., Bond, J. S., Wiseman, J., Berman, J., and Harris, R. B., *Anal. Biochem.*, 196, 137, 1991.
48. Ragnarsson, U., Karlsson, S., and Sandberg, B., *Acta Chem. Scand.*, 25, 1487, 1971.
49. Ragnarsson, U., Karlsson, S. M., and Sandberg, B. E. B., *J. Org. Chem.,* 39, 3837, 1974.
50. Kramer, A., Volkmer-Engert, R., Malin, R., Reineke, U., and Schneider-Mergener, J., *Pept. Res.,* 6, 314, 1993.
51. Andrews, P. C., Boyd, J., Loo, R. O., Zhao, R., Zhu, C.-Q., Grant, K., and Williams, S., in *Techniques in Protein Chemistry*, Vol. 5, Crabb, J. W., Ed., Academic Press, San Diego, 1994, 485.
52. Quesnel, A., Delmas, A., and Trudelle, Y., *Anal. Biochem.*, 231, 182, 1995.
53. Furka, A., Sebestyén, F., Asgedom, M., and Dibó, G., *Int. J. Pept. Protein Res.*, 37, 487, 1991.
54. Houghten, R. A., Pinilla, C., Blondelle, S. E., Appel, J. R., Dooley, C. T., and Cuervo, J. H., *Nature (London),* 354, 84, 1991.
55. Lam, K. S., Salmon, S. E., Hersh, E. M., Hruby, V. J., Kazmierski, W. M., and Knapp, R. J., *Nature (London),* 354, 82, 1991.
56. Burgess, K., Liaw, A. I., and Wang, N., *J. Med. Chem.*, 37, 2985, 1994.
57. Lebl, M., Krchnák, V., Sepetov, N. F., Seligmann, B., Strop, P., Felder, S., and Lam, K. S., *Biopolymers (Pept. Sci.)*, 37, 177, 1995.
58. Houghten, R. A., in *Peptides. Synthesis, Structures and Applications*, Gutte, B., Ed., Academic Press, New York, 1995, 396.
59. Pinilla, C., Appel, J. R., Blanc, J. R., and Houghten, R. A., *BioTechniques*, 13, 901, 1992.
60. Dooley, C. T. and Houghten, R. A., *Life Sci.*, 52, 1509, 1993.
61. Blake, J. and Litzi-Davis, L., *Bioconjugate Chem.*, 3, 510, 1992.
62. Brummel, C. L., Lee, I. N. W., Zhou, Y., Benkovic, S. J., and Winograd, N., *Science*, 264, 399, 1994.
63. Youngquist, R. S., Fuentes, G. R., Lacey, M. P., and Keough, T., *J. Am. Chem. Soc.*, 117, 3900, 1995.
64. Lam, K. S., Salmon, S. E., Hersh, E. M., Hruby, V. J., Al-Obeidi, F., Kazmierski, W. M., and Knapp, R. J., in *Peptides. Chemistry and Biology. Proceeding of the 12th American Peptide Symposium*, Smith, J. A. and Rivier, J. E., Eds., ESCOM, Leiden, 1992, 492.
65. Lebl, M., Patek, M., Kocis, P., Krchnák, V., Hruby, V. J., Salmon, S. E., and Lam, K. S., *Int. J. Pept. Protein Res.*, 41, 201, 1993.
66. Kocis, P., Krchnák, V., and Lebl, M., *Tetrahedron Lett.*, 34, 7251, 1993.
67. Salmon, S. E., Lam, K. S., Lebl, M., Kandola, A., Khattri, P. S., Wade, S., Patek, M., Kocis, P., Krchnák, V., Thorpe, D., and Felder, S., *Proc. Natl. Acad. Sci. U.S.A.,* 90, 11708, 1993.
68. Cardno, M. and Bradley, M., *Tetrahedron Lett.*, 37, 135, 1996.
69. Brenner, S. and Lerner, R. A., *Proc. Natl. Acad. Sci. U.S.A.,* 89, 5381, 1992.
70. Nielsen, J., Brenner, S., and Janda, K. D., *J. Am. Chem. Soc.*, 115, 9812, 1993.
71. Needels, M. C., Jones, D. G., Tate, E. H., Heinkel, G. L., Kochersperger, L. M., Dower, W. J., Barrett, R. W., and Gallop, M. A., *Proc. Natl. Acad. Sci. U.S.A.,* 90, 10700, 1993.

72. Kerr, J. M., Banville, S. C., and Zuckermann, R. N., *Bioorg. Med. Chem. Lett.*, 3, 463, 1993.
73. Nikolaiev, V., Stierandová, A., Krchnák, V., Seligmann, B., Lam, K. S., Salmon, S. E., and Lebl, M., *Pept. Res.*, 6, 161, 1993.
74. Vágner, J., Krchnák, V., Sepetov, N. F., Strop, P., Lam, K. S., Barany, G., and Lebl, M., in *Innovation and Perspectives in Solid Phase Synthesis. Peptides, Proteins and Nucleic Acids. Biological and Biomedical Applications*, Epton, R., Ed., Mayflower Worldwide, Birmingham, UK, 1994, 347.
75. Ohlmeyer, M. J., Swanson, R. N., Dillard, L. W., Reader, J. C., Asouline, G., Kobayashi, R., Wigler, M., and Still, W. C., *Proc. Natl. Acad. Sci. U.S.A.*, 90, 10922, 1993.
76. Nestler, H. P., Bartlett, P. A., Still, W. C., *J. Org. Chem.*, 59, 4723, 1994.
77. Moran, E. J., Sarshar, S., Cargill, J. F., Shahbaz, M. M., Lio, A., Mjalli, A. M. M., and Armstrong, R. W., *J. Am. Chem. Soc.*, 117, 10787, 1995.
78. Nicolaou, K. C., Xiao, X.-Y., Parandoosh, Z., Senyei, A., and Nova, M. P., *Angew. Chem. Int. Ed. Engl.*, 34, 2289, 1995.
79. Stevanovic, S., Wiesmüller, K.-H., Metzger, J., Beck-Sickinger, A. G., and Jung, G., *Bioorg. Med. Chem. Lett.*, 3, 431, 1993.
80. Thompson, L. A. and Ellman, J. A., *Chem. Rev.*, 96, 555, 1996.
81. Boutin, J. A., Hennig, P., Lambert, P.-H., Bertin, S., Petit, L., Mahieu, J.-P., Serkiz, B., Volland, J.-P., and Fauchère, J.-L., *Anal. Biochem.*, 234, 126, 1996.
82. Egner, B. J., Cardno, M., and Bradley, M., *J. Chem. Soc. Chem. Commun.*, 2163, 1995.
83. Metzger, J. W., Wiesmüller, K.-H., Gnau, V., Brünjes, J., and Jung, G., *Angew. Chem. Int. Ed. Engl.*, 32, 894, 1993.
84. Giralt, E., Rizo, J., and Pedroso, E., *Tetrahedron*, 40, 4141, 1984.
85. Sarkar, S. K., Garrigipati, R. S., Adams, J. L., and Keifer, P. A., *J. Am. Chem. Soc.*, 118, 2305, 1996.
86. Gordon, E. M., Barrett, R. W., Dower, W. J., Fodor, S. P. A., and Gallop, M. A., *J. Med. Chem.*, 37, 1385, 1994.
87. Dooley, C. T., Chung, N. N., Wilkes, B. C., Schiller, P. W., Bidlack, J. M., Pasternak, G. W., and Houghten, R. A., *Science*, 266, 2019, 1994.
88. Chen, J. K., Lane, W. S., Brauer, A. W., Tanaka, A., and Schreiber, S. L., *J. Am. Chem. Soc.*, 115, 12591, 1993.
89. Combs, A. P., Kapoor, T. M., Feng, S., Chen, J. K., Daudé-Snow, L. F., and Schreiber, S. L., *J. Am. Chem. Soc.*, 118, 287, 1996.
90. Cheng, Y., Suenaga, T., and Still, W. C., *J. Am. Chem. Soc.*, 118, 1813, 1996.
91. Boyce, R., Li, G., Nestler, P., Suenaga, T., and Still, W. C., *J. Am. Chem. Soc.*, 116, 7955, 1994.

Index

MICHIGAN MOLECULAR INSTITUTE
1910 WEST ST. ANDREWS ROAD
MIDLAND, MICHIGAN 48640